Smart Trajectories

This book highlights the developments, discoveries, and practical and advanced experiences related to responsive distributed computing and how it can support the deployment of trajectory-based applications in smart systems.

Smart Trajectories: Metamodeling, Reactive Architecture for Analytics, and Smart Applications deals with the representation and manipulation of smart trajectories in various applications and scenarios. Presented in three parts, the book first discusses the foundation and principles for spatial information systems, complex event processing, and building a reactive architecture. Next, the book discusses modeling and architecture in relation to smart trajectory metamodeling, mining and big trajectory data, and clustering trajectories. The final section discusses advanced applications and trends in the field, including congestion trajectory analytics and real-time Big Data analytics in cloud ecosystems. Metamodeling, distributed architectures, reactive programming, Big Data analytics, NoSQL databases, connected objects, and edge-fog-cloud computing form the basis of the concepts and applications discussed. The book also presents a number of case studies to demonstrate smart trajectories related to spatiotemporal events such as traffic congestion and pedestrian accidents.

This book is intended for graduate students in computer engineering, spatial databases, complex event processing, distributed systems, and geographical information systems (GIS). The book will also be useful for practicing traffic engineers, city managers, and environmental engineers interested in monitoring and security analysis.

Smart Trajectories

This book highlights the developments, discoveries, and practical and advanced experiences related to responsive distributed computing and how it can support the deployment of trajectory-based applications in smart systems.

Smart Trajectories: Metamodeling, Reactive Architecture for Analytics, and Smart Applications deals with the representation and manipulation of smart trajectories in various applications and scenarios. Presented in three parts, the book first discusses the foundation and principles for spatial information systems, complex event processing, and building a reactive architecture. Next, the book discusses modeling and architecture in relation to smart trajectory metamodeling, mining and big trajectory data, and clustering trajectories. The final section discusses advanced applications and trends in the field, including congestion trajectory analytics and real-time Big Data analytics in cloud ecosystems. Metamodeling, distributed architectures, reactive programming, Big Data analytics, NoSQL databases, connected objects, and edge-fog-cloud computing form the basis of the concepts and applications discussed. The book also presents a number of case studies to demonstrate smart trajectories related to spatiotemporal events such as traffic congestion and pedestrian accidents.

This book is intended for graduate students in computer engineering, spatial databases, complex event processing, distributed systems, and geographical information systems (GIS). The book will also be useful for practicing traffic engineers, city managers, and environmental engineers interested in monitoring and security analysis.

Smart Trajectories
Metamodeling, Reactive Architecture for Analytics, and Smart Applications

Edited by
Azedine Boulmakoul, Lamia Karim and Bharat Bhushan

CRC Press
Taylor & Francis Group
Boca Raton London New York

CRC Press is an imprint of the
Taylor & Francis Group, an **informa** business

Cover image: © Shutterstock
First edition published 2023
by CRC Press
6000 Broken Sound Parkway NW, Suite 300, Boca Raton, FL 33487-2742

and by CRC Press
4 Park Square, Milton Park, Abingdon, Oxon, OX14 4RN

CRC Press is an imprint of Taylor & Francis Group, LLC

© 2023 selection and editorial matter, Azedine Boulmakoul, Lamia Karim and Bharat Bhushan, individual chapters, the contributors

ISBN: 978-1-032-18281-0 (hbk)
ISBN: 978-1-032-18666-5 (pbk)
ISBN: 978-1-003-25563-5 (ebk)

DOI: 10.1201/9781003255635

Typeset in Times
by KnowledgeWorks Global Ltd.

Contents

Preface

Classical trajectory modeling provided in the literature captures the movement of an object in space-time over a given period. Currently, the movement of people and goods in a given geographic area can be observed from digital traces deposited by personal or vehicular mobile devices and collected by wireless network infrastructures. For example, mobile phones leave positioning logs, which specify their location, and at all times they are connected to the mobile network; similarly, equipped with global positioning system (GPS)–type geolocation equipment, portable devices can record their latitude-longitude geolocation at each instant when they are exposed to a location device and can transmit their trajectories to a collection server. Automatic teller machines (ATMs), electronic payment services online or in the field, speed cameras, etc., keep track of the spatiotemporal geolocation of events.

The central issue dealt with in this book concerns the representation and manipulation of smart trajectories in various fields. Metamodeling, distributed architectures, reactive programming, Big Data analytics, NoSQL databases, connected objects, and edge-fog-cloud computing form the basis of the axes developed in this book. The omnipresence of pervasive technologies guarantees an increasing availability of large volumes of spatiotemporal data. Therefore, there is an opportunity to discover online, from these trajectories, spatiotemporal patterns that convey useful knowledge. For example, consider the traffic control system in an urban network. The sensors report congestion events on a continuous time basis. With such data, it is possible to support city traffic control decisions in real time. Such a discovery can help in the search for an efficient reengineering of traffic and develop innovative solutions in terms of informatics of urban systems: urban computing in the context of smart cities. The main benefit of this book is to have a consistent set of trajectory modeling patterns and advanced reactive manifesto–based software practices with real-world applications. This book is an essential concept and an engineering guide for the analysis of trajectories of complex systems.

The objective of this book is to combine the latest advances in terms of the model and applications of trajectories in various fields. It presents a synthesis of the authors' current and past research work and opens perspectives on future developments to be deepened in the context of smart trajectories. It is the convergence of several research works on trajectories, which gave rise to several defended PhDs. All of our research activities are presented in a factual manner.

This book can serve as a useful reference for a larger audience in different domain areas:

- Graduate students: Computer engineering, spatial databases, complex event processing, distributed systems, GIS
- Industry professionals
- City management engineers, urban traffic monitoring operators, and traffic engineers
- Security analysis and monitoring engineers

- Stakeholders in the field of road safety
- Stakeholders in the surveillance of pandemics
- Stakeholders in the monitoring of hazmat transportation
- Multidomain social media analysts, data scientists from various fields, and geomarketing
- Environmental monitoring engineers
- The state departments: interior, territorial surveillance, and crisis management
- Researchers, academics in sciences and engineering, etc.

FEATURES

The book highlights developments, discoveries, and practical and advanced experiences related to responsive intelligent distributed computing. Further, it describes how to support the deployment of trajectory-based applications in intelligent systems. The following points summarize the contributions:

1. The ecosystems and distributed intelligent computing paradigm deploying and instantiating a smart trajectories meta-model to take advantage of innovative trends in generic reactive architecture for analytics and smart city applications.
2. Concept of a smart trajectory concerning the transport of hazardous materials, the management of urban traffic congestion, the Internet of Vehicles (IoV), or more generally connected objects (IoV and IoT) with their integration into edge-fog-cloud computing, Big Data analytics of trajectories in general, calculation of fuzzy intuitionist risk, monotonic reasoning and modeling of pedestrian risks, fuzzy pheromone, and intelligent transport systems.
3. Meta-modeling with new trajectory patterns which are very useful for intelligent systems developers (intelligent transportation systems).
4. The case studies presented in the book will be of interest to researchers and professionals. In fact, the proposed case studies correspond to real industrial projects supported by institutional funding.
5. The balance between practice and theory is respected in the book in order to also interest young computer science students.

Editor Biographies

Prof. Azedine Boulmakoul: https://www.researchgate.net/profile/Azedine-Boulmakoul

Azedine Boulmakoul is a professor in the Department of Computer Science at FST Mohammedia-Hassan II University of Casablanca, where he has been since 1994. He received his PhD and hold a Habilitation à Diriger des Recherches from the Claude Bernard University of Lyon in 1990 and 1994, respectively. In 1987, he was a lecturer assistant at the ENTPE School (Ecole Nationale des Travaux Public de l'Etat Lyon France). For the period 1987-1988, he was appointed assistant professor at INSA de Lyon. Between 1987 and 1994, he joined the National Institute for Research on Transportation and Safety (INRETS-Paris, www.ifsttar.fr) for real-time intelligent systems research development. He is president of the Innovative Open Spatial Information Systems Association: www.iosis.ma. He has received many awards for his contributions to research and business development, including the French Academy of Sciences, R&D Maroc Awards and Distinction Medal from Moroccan Ministry of Higher Education and Research, and Distinction Medal from Istanbul International Inventions. He also contributes to intellectual property of industrial patents. He spent several years in R&D and in studies and business consulting. He has chaired several conferences and program committees. He is the author of over 300 papers (in books, journals and conferences, patents, etc.) on computer science. He holds 10 patents deriving from his research. His research interests include intelligent real-time systems, logic for artificial intelligence, fuzzy systems theory, distributed frameworks, data analytics, complex systems, information systems engineering, transportation, and computer engineering. In his spare time, he enjoys karate (third-degree black belt), running, cooking, and reading. He is amazed by beauty in the universe and likes to walk in the mountains and listen to silence to write a poem.

Prof. Lamia Karim: https://www.researchgate.net/profile/Lamia-Karim

Lamia Karim is a professor of computer science at Hassan First University of Settat, Ecole Nationale des Sciences Appliquées, Berrechid, Morocco. She is a Huawei-certified instructor in Big Data starting in May 2022. In 2020, she obtained the University Habilitation diploma at ENSAB Hassan First University. In 2015, she obtained the PhD degree in computer science at the Faculty of Science and Technology of Mohammedia, Hassan II University of Casablanca, Morocco. She obtained a Masters of Science and Technical in computer sciences in 2006 from the Faculty of Sciences and Technologies (FSTM), Morocco, and a computer engineer degree in software engineering from the National School of Mineral Industry (ENIM), Rabat, Morocco, in 2008. Her research interest includes spatial data engineering, mobile and web geocomputing, real-time intelligent transportation systems, complex systems computing, soft computing, and reactive intelligent systems. She is general secretary of the Innovative Open Spatial Information Systems Association: www.iosis.ma. She has been the session chair and a panelist in several conferences.

She earned numerous international certifications such as Microsoft Certified Technology Specialist (MCTS), Cisco Certified Network Associate (CCNA), and International Business Machines Corporation (IBM). She has published more than 80 research papers in various renowned international conferences and SCI indexed journals. She has served as a reviewer for several reputed international journals. She holds several national and international patents (OMPIC and WIPO). She has contributed several book chapters in various books. She has co-edited the book *Internet of Things: Frameworks for Enabling and Emerging Technologies* (CRC Press, Taylor & Francis Group) and is currently in the process of editing books from well-known publishers like Elsevier and CRC Press.

Dr. Bharat Bhushan: https://www.researchgate.net/profile/Bharat-Bhushan-22

Dr. Bharat Bhushan is an assistant professor of the Department of Computer Science and Engineering (CSE) at the School of Engineering and Technology, Sharda University, Greater Noida, India. He is an alumnus of the Birla Institute of Technology, Mesra, Ranchi, India. He received his undergraduate degree (B-Tech in computer science and engineering) with distinction in 2012, received his post-graduate degree (M-Tech in information security) with distinction in 2015, and his doctorate degree (PhD in computer science and engineering) in 2021 from the Birla Institute of Technology, Mesra, India. He has earned numerous international certifications such as CCNA, MCTS, MCITP, RHCE, and CCNP. In the last three years, he has published more than 80 research papers in various renowned international conferences and SCI indexed journals including *Wireless Networks* (Springer), *Wireless Personal Communications* (Springer), *Sustainable Cities and Society* (Elsevier), and *Emerging Transactions on Telecommunications* (Wiley). He has contributed more than 25 book chapters in various books and has edited 11 books from well-known publishers like Elsevier, IGI Global, and CRC Press. He has served as a reviewer/editorial board member for several reputed international journals. In the past, he worked as an assistant professor at HMR Institute of Technology and Management, New Delhi, and network engineer in HCL Infosystems Ltd., Noida. He has qualified GATE exams for successive years and gained the highest percentile of 98.48 in GATE 2013. He is also a member of numerous renowned bodies, including IEEE, IAENG, CSTA, SCIEI, IAE, and UACEE.

Contributors

Fatima-Ezzahra Badaoui
LIM/Innovative Open Systems
    FSTM, Hassan II University of
    Casablanca
Casablanca, Morocco

Hassan Badir
IDS Data Engineering and System
National School of Applied Sciences
Abdelmalek Essaâdi University
Tangier, Morocco

Wadii Basmi
LIM/Innovative Open Systems
FSTM, Hassan II University of
    Casablanca
Casablanca, Morocco

Kaoutar Bella
LIM/Innovative Open Systems,
    FSTM, Hassan II University of
    Casablanca
Casablanca, Morocco

Zineb Besri
National School of Applied Sciences of
    Tetouan
Tetouan, Morocco

Bharat Bhushan
School of Engineering and Technology
    (SET)
Sharda University
Greater Noida, India

Omar Bouattane
Informatics, Artificial Intelligence
    and Cyber Security (2IACS) Lab
    ENSET, Hassan II University of
    Casablanca
Casablanca, Morocco

Azedine Boulmakoul
LIM/Innovative Open Systems
    FSTM, Hassan II University of
    Casablanca
Casablanca, Morocco

Ghyzlane Cherradi
LIM/Innovative Open Systems,
    FSTM, Hassan II University of
    Casablanca
Casablanca, Morocco

Adil El Bouziri
LIM/Innovative Open Systems
    FSTM, Hassan II University of
    Casablanca
Casablanca, Morocco

Hafsa El Hafyani
DAVID Lab, UVSQ
Université Paris-Saclay
Versailles, France

Fatima-Ezzahra Ezzrhari
Informatics, Artificial Intelligence and
    Cyber Security (2IACS) Lab ENSET,
    Hassan II University of Casablanca
Casablanca, Morocco

Zoltan Fazekas
Institute for Computer Science and
    Control (SZTAKI)
Eötvös Loránd Research Network
    (ELKH)
Budapest, Hungary

Péter Gáspár
Institute for Computer Science and
    Control (SZTAKI)
Eötvös Loránd Research Network
Budapest, Hungary

Younes Hajoui
Informatics, Artificial Intelligence and
    Cyber Security (2IACS) Lab
ENSET, Hassan II University of
    Casablanca
Casablanca, Morocco

Abdelfettah Idri
ENCG, Hassan II University of
    Casablanca
Casablanca, Morocco

Vassilis Kaburlasos
HUman-MAchines INteraction
    (HUMAIN) Lab
Department of Computer Science
International Hellenic University (IHU)
Kavala, Greece

Saptadeepa Kalita
School of Engineering and Technology
    (SET)
Sharda University
Greater Noida, India

Lamia Karim
LISA Lab.Ecole Nationale des Sciences
    Appliquées, Berrechid
Hassan First University of Settat
Berrechid, Morocco

Snigdha Kashyap
School of Engineering and Technology
    (SET)
Sharda University
Greater Noida, India

Ahmad Ktaish
DAVID Lab, UVSQ
Université Paris-Saclay
Versailles, France

Anuj Kumar
School of Engineering and Technology
    (SET)
Sharda University
Greater Noida, India

Avinash Kumar
SITAICS
Rashtriya Raksha University
India

Aziz Mabrouk
I.S Engineering Research Team
Abdelmalek Essaadi University
    FSJEST
Tetouan, Morocco

Soufiane Maguerra
LIM/Innovative Open Systems, FSTM,
    Hassan II University of Casablanca
Casablanca, Morocco

Meriem Mandar
LMAI Lab. ENS Casablanca
Hassan 2nd University
Casablanca, Morocco

Mouna Amrou Mhand
IDS Data Engineering and System
National School of Applied Sciences
Abdelmalek Essaâdi University
Tangier, Morocco

Mohamed Nahri
LIM/Innovative Open Systems
FSTM, Hassan II University of
    Casablanca
Casablanca, Morocco

Parma Nand
School of Engineering and Technology
    (SET)
Sharda University
Greater Noida, India

Mohammed Obaid
Department of Automotive Technology
Faculty of Transportation Engineering
    and Vehicle Engineering
Budapest University of Technology and
    Economics
Budapest, Hungary

Nandita Pokhriyal
School of Engineering and Technology
   (SET)
Sharda University
Greater Noida, India

Rachid Oulad Haj Thami
ADMIR Lab ENSIAS
Mohamed V University
Rabat, Morocco

Maroua Razzouqi
LIM/Innovative Open Systems
FSTM, Hassan II University of
   Casablanca
Casablanca, Morocco

Khushi Samridhi
School of Engineering and Technology
   (SET)
Sharda University
Greater Noida, India

Sonal Shriti
School of Engineering and Technology
   (SET)
Sharda University
Greater Noida, India

Yehia Taher
DAVID Lab, UVSQ
Université Paris-Saclay
Versailles, France

Tanmayee Prakash Tilekar
School of Engineering and Technology
   (SET)
Sharda University
Greater Noida, India

Laurent Yeh
DAVID Lab, UVSQ
Université Paris-Saclay
Versailles, France

Mohamed Youssfi
Informatics, Artificial Intelligence and
   Cyber Security (2IACS) Lab
ENSET, Hassan II University of
   Casablanca
Casablanca, Morocco

Karine Zeitouni
DAVID Lab, UVSQ
Université Paris-Saclay
Versailles, France

Acknowledgments

We would like to acknowledge the effort and time invested by all the contributing authors for their patience and excellent work. At least two reviewers have reviewed each chapter. We would like to thank the reviewers for their time and valuable contribution to the quality of the book.

This book is partially funded by the Ministry of Equipment, Transport, Logistics and Water, Kingdom of Morocco, the National Road Safety Agency (NARSA), and National Center for Scientific and Technical Research (CNRST). Road Safety Research Program# An intelligent reactive abductive system and intuitionist fuzzy logical reasoning for dangerousness of driver-pedestrians interactions analysis: Development of new pedestrians' exposure to risk of road accident measures.

This book is also partially funded by the Digital Development Agency (ADD) and the National Center for Scientific and Technical Research (CNRST) in partnership with the Ministry of Industry, Commerce and Green and Digital Economy (MICEVN) and the Ministry of National Education, Professional Training, Higher Education and Scientific Research (MENFPESRC). # LKHAWARIZMI Program #Intelligent & Resilient Urban Network Defender: A distributed real-time reactive intelligent transportation system for urban traffic congestion symbolic control and monitoring.

Finally, thanks to our publishers, their editors, and anonymous reviewers whose comments have helped us immensely in improving and completing the text.

1 Intelligent Distributed Computing Paradigm

Abdelfettah Idri

CONTENTS

1.1 INTRODUCTION

The rising discipline of intelligent distributed computing (IDC) is more focused on the development of advanced intelligent distributed systems allowing the combination and the adaptation of both research areas, namely intelligent computing and distributed systems. The field of distributed systems deals with the development of architectures and technologies composed of collaborating, reactive, and interacting components, while intelligent computing is concerned with the invention and innovation of intelligence-oriented solutions ranging from computational intelligence to machine learning and multiagent systems (MASs).

Distributed computing is based on the performance of a set of interconnected nodes, representing the individual computing units and operating independently. Seen from the end-user point of view, the whole infrastructure (logical or physical) is always considered as a single coherent and transparent system. Whether it is about data storage or service execution, the user would not be able to identify the node providing the service or the one storing the data.

In general, the problem complexity is reduced by means of transforming the global problem into smaller and well-managed subproblems that can be distributed on the different interconnected nodes. These nodes need to collaborate and

DOI: 10.1201/9781003255635-1

communicate by passing messages in order to resolve the problem and deliver the requested services to end users. As the nodes can act autonomously, there is in fact no global time reference, and consequently, the synchronization aspect remains vital in distributed architectures.

Owing to the fast growth of data collection today, according to the technological advances in different areas, distributed computing becomes a good alternative to respond to these issues, and therefore, distributed systems are largely involved in diverse application areas like machine learning, transport, social media, banking, and e-learning, in addition to others. The adoption of distributed computing highlights several aspects like dependability, concurrent access, availability, and replication.

To hide the complexity of the heterogeneousness of the different physical network hardware units composing a distributed system infrastructure, which is very common, an abstraction is adopted: a logic layer is inserted on top of the different operating systems in order to provide a middleware exposing a transparent service interface to the users.

In addition to these aspects and in order to handle the large datasets efficiently in a distributed environment, sophisticated algorithms have to be invented and rigorously implemented to enable parallel computing and data distribution.

This chapter discusses the major concepts of IDC. To achieve this, several architectures will be exposed, as well as many facets of distributed computing and intelligent techniques with a focus on MASs.

At last, two previous developed models of IDC architectures and their components are presented to bring into practice the theoretical components design and the logic residing behind it: data mining and multimodal transport.

1.2   DISTRIBUTED COMPUTING

1.2.1   TRENDS IN DISTRIBUTED COMPUTING

In contrast to centralized controllers that introduce several shortcomings like a single point of failure and performance issues, distributed computing presents the advantages of scalability and is more resilient.

While it is conceivable to improve the computation power of a single node following different methods, one prefers the scale-out strategy that means extending the network with more nodes. This choice is in general justified by the lower cost of individual nodes (Verbraeken et al. 2020), resilience against a single point of failure, and the increase in input/output (I/O) traffic (aggregated). This is known as distributed problem solving (DPS) and is concerned with dividing a global task into smaller subtasks that are further dispatched to the nodes. The partial results are then consolidated to construct the global solution. Despite this fact, the eventual communication issues related to the shared resources can sometimes limit the efficiency and the flexibility of the whole system (Dorri et al. 2018). Globally, there are two strategies to accelerate workloads:

- Acting vertically: improving the computation capacity of a single machine
- Acting horizontally: adding more nodes to the distributed system

Distributed systems (providing distributed computing) are more complex to design and control. The aim of innovative distributed systems is to benefit from the available tools and experience to manage unanticipated events. When distributed systems provide advanced features like adaptability, reactivity, scalability, autonomy, and sophisticated algorithms that can qualify the intelligent system, then advanced concepts and abstract layers become possible to build upon this infrastructure to resolve more complex problems.

Similarly to the distributed nodes, MASs consist of standalone entities possessing the ability of adapting and making decisions by exploiting the experience gained during learning (Jain and Srinivasan 2010). Each agent executes the assigned subtask to reduce the complexity of the global task according to its capacity expressed in terms of its knowledge, processing performance, and perspectives. The agent can then choose the most appropriate approach to solve its problem. In case of a failure, another agent takes over the current task transparently (Rezaee and Abdollahi 2015). The flexibility of MASs resides in their ability to interact with the environment and with other agents, and therefore they are suited to make decisions autonomously to solve problems in different domains (Dorri et al. 2018), (Kshemkalyani 2008). Moreover, the development of MASs encompasses several challenges like security, coordination between agents (Dayong et al. 2017), learning (Lea et al. 2006), and security (Dorri et al. 2018).

Agents should expose specific behavior fundamentally characterized by (see Dayong et al. (2017) and Lea et al. (2006)):

- Dynamism: consideration of environmental changes and learning ability
- Determinism: prediction of results
- Proactivity: adaptation of actions based on predictions and autonomous decision making
- Reactivity: action in response to their environment
- Sociability: cooperation with other agents and humans

A simplified model of an agent can be seen in Figure 1.1.

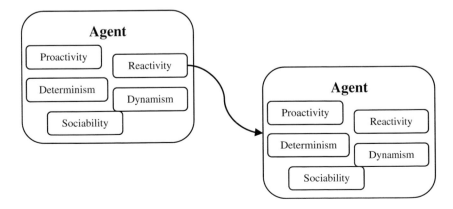

FIGURE 1.1 Model of an agent.

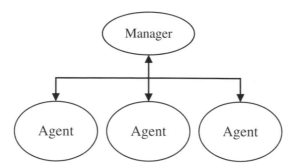

FIGURE 1.2 Manager-agent model.

Among the agents concerned by the resolution of the same problem and for which they are assigned tasks to perform, at least one agent should take into account the management aspects of the whole infrastructure, and that is the definition of tasks and goals, as well as the dispatching of tasks, the collection of results, and the communication of knowledge. Such an agent is called a manager or leader (Dorri et al. 2018; see Figure 1.2).

Several applications of MASs can be found in monitoring, meteorology, social networks, cloud computing, and telecommunication (Jiang and Jiang 2014, Gatti et al. 2013). However, a lot of challenges remain surrounding MASs like synchronization, learning, and coordination (Su et al. 2016).

The communication aspect within such an infrastructure is necessary to interconnect the different heterogeneous units and permit a transparent and coherent exchange of messages between them. Transactional mechanisms are used. Examples of well-formalized communication tools are Remote Method Invocation (RMI) and Remote Procedure Call (RPC), whereas natively distributed systems like mobile networks rely on sensors and actuators (Van Steen and Tanenbaum 2016, Veríssimo and Rodrigues 2001), Attiya and Welch2004). To optimize the communication process, several aspects have to be taken in account like the load distribution and parallelism (Xing et al. 2016).

Distributed artificial intelligence (DAI) falls within this category and has received a lot of attention from researchers in different areas. DAI deals with modeling knowledge based on MASs to address complex problems in a distributed way adopting the divide-and-conquer strategy (Verbraeken et al. 2020, Dorri et al. 2018): the global problem is divided into partial cooperative problems, sharing the knowledge of their partial solutions. MASs have the advantage of being involved in loosely coupled problem solving with a cooperative perspective (Zheng and Koenig 2008).

We will focus on MASs as a distributed approach because they exhibit the appropriate characteristics to mimic an intelligent behavior and serve the purposes of our theme (Dorri et al. 2018).

1.2.2 DISTRIBUTED COMPUTING PRINCIPLES

The role of computing systems is mainly executing applications, and when the targeted single-node performance is not affordable nor an option, then distributed computing becomes an alternative to surpass this obstacle. Examples of these situations are the huge amount of training data in the case of complex applications, where the order of terabytes is very common, or in the case of natively distributed data on different locations.

We speak about distribution when there is at least one case verified, namely, distribution of data, process, or infrastructure.

Distributed computing relies on distributed systems that are composed of independent interconnected, autonomous computing units working collaboratively and having the following characteristics (Kshemkalyani 2008):

- Heterogeneity: applies to hardware and software (e.g., communication networks and operating systems). A consistent solution for the heterogeneity problem is middleware that enables communication across heterogeneous networks. Middleware offers an abstract software layer to hide the heterogeneity of software and hardware components constituting the distributed system.
- Scalability: allows adding more nodes to the existing pool (horizontal scale) to enhance transparently the performance of the distributed system without impacting the existing nodes.
- Fault tolerance: the system must remain available and continue delivering the services even if a failure of single nodes occurs by delegating tasks to other available nodes and applying recovery and redundancy techniques.
- Parallelism: distributed systems are suitable for parallelism by dividing up a task into parallelizable subtasks and assigning these to different nodes.
- Lack of global clock: there should be no synchronous global clock. Instead, a logical clock is adopted (protocol) to maintain event ordering.
- No shared physical memory: nodes communicate by exchanging messages.
- Autonomy: each node belonging to the distributed system acts individually and independently and doesn't have information over the state of other nodes. Nodes are loosely coupled, and they may have different hardware and run different operating systems.
- Transparency: the entire distributed system should be perceived by end users as a single transparent system hiding the physical location of the provided services.
- Reliability: guarantees that the whole distributed system is up and running at any time. The probability that a set of nodes falls down is far less than that of a single node.
- Replication: maintains copies on different nodes to guarantee high availability and fault tolerance by exploiting the replicas.

Besides the benefits of distributed systems, it is relevant to highlight the design issues that may occur:

- Concurrency: simultaneous access to shared resources needs to be managed to keep resources operating correctly and coherently.

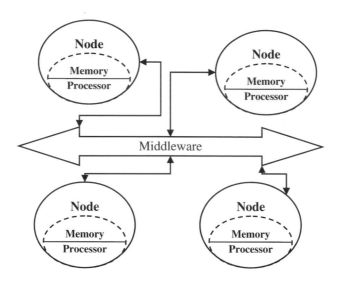

FIGURE 1.3   A distributed architecture model.

- Failure handling: this is complex and difficult to implement because it is partial and should take into account the distributed execution context.
- Security: encryption enables the protection of shared resources, and it is applied to an end-to-end exchange between nodes to keep the information within messages safe.

1.2.3 DISTRIBUTED ARCHITECTURE

The nodes of a distributed architecture may be heterogeneous and belong to different platforms. They cooperate by communicating through a communication network. From an architectural point of view and in order to break down the complexity of the architecture design, a layered approach is adopted using middleware that appropriately interfaces the applications and the network (Lea et al. 2006) by exchanging messages transparently across heterogeneous platforms (Zheng and Koenig 2008). Assuming that nodes operate independently, as described in the characteristics of a distributed system, Figure 1.3 shows the logical relation between these architecture components, and we see that each node has its own memory and processor. There exist several standards for implementing the middleware, like Common Object Request Broker Architecture (CORBA), RPC, and RMI.

1.3 INTELLIGENT DISTRIBUTED COMPUTING CONCEPT

For the description of this complex process, we will proceed by introducing progressive models of the fundamental components and their behavior, allowing them to be qualified by intelligent entities when involved in distributed computing. The purpose of these steps is to build up a generic model of a distributed architecture providing IDC. A functional view is first presented to serve as a milestone for the abstract architecture. An agent-based approach is adopted, as agents are the most suitable for modeling adaptable, flexible, and intelligent behaviors.

1.3.1 FUNCTIONAL VIEW

An agent is considered a computational unit that is capable of performing autonomous actions to meet its objectives. It can be a physical or a virtual component and has the capacity of reasoning about its state, environment, own actions, and interactions with other agents. This entity is considered the fundamental individual component on which the functional view will be projected. An agent should adapt itself to the current situation and therefore can deliver varieties of services depending on its context. From this perspective, the following main features are considered, in addition to the mentioned conventional features of distributed systems. These features are intended to categorize rigorously IDC based on MAS and to identify the functional content to be respected by agents, as well as the abstract levels to deal with (Van Steen and Tanenbaum 2016, Osrael et al. 2006, Xie and Liu 2017, Rodríguez 2008, Omicini and Mariani 2013):

- Reactivity: two categories can be distinguished relative to distributed systems: native reactive agents that map directly environmental situations to actions (Xie and Liu 2017) using only the current information and logic-based agents that may use historical information to make decisions based on logical reasoning and deduction. The objective is to deliver a responsive distributed system that can handle client requests and respond transparently in an accurate time despite failures of workloads. Reactivity requires resilience, scalability, and asynchronous communication between agents (Salvaneschi et al. 2013). Usually, reactive programming paradigms are adopted to develop reactive systems in an efficient and elegant way.
- Task allocation: reflects the ability of an agent to allocate its own tasks to other agents in case of, for example, incapacity or performance necessity. There are mechanisms of task allocation in MASs: centralized, decentralized, and self-organizing task allocation (Zheng and Koenig 2008). Decentralized mechanisms overcome single-point-of-failure issues related to the centralized approach. Self-organizing task allocation mechanisms are instead better suited for scalability, as they are decentralized and they don't need global information.
- Learning: agents may possess the ability of making decisions accordingly to the state of the environment to achieve their goals. To decide appropriately on the suitable actions, agents can take advantage of learning methods to predict the future state changes of the environment and adapt the actions on it (Bowling and Veloso 2002). To overcome some challenges, like the communication overheads raised by learning algorithms, collaborative learning can be considered by sharing knowledge between agents. Reinforcement learning is a good alternative based on the principle of trial-and-error interactions with the environment to learn behaviors. Herein, the agent changes its state and perceives the state changes in the environment and other agents

(Bianchi et al. 2014). Based on the environment feedback, the agent adopts the actions with a positive effect and ignores the other ones.

- Decision-making mechanisms: enables agents to select the appropriate actions to execute in order to achieve their goals. Decision-making algorithms are used to generate actions and allow agents to choose the optimal one (Rizk et al. 2018). Several decision-making models are available: Markov decision processes and its extensions, the graph theoretic model, and swarm intelligence, among others. Collective decision making begins to find its way to MASs. In this context, an agent decision is influenced by other agents and a global decision is made by consensus (Montes et al. 2011).

Based on the functional description of the fundamental features related to agents involved in an IDC context, a generic agent model is proposed that covers the main components suited to support such a paradigm. Figure 1.4 shows schematically these features and their functional dependencies.

According to the context of the problem to be solved, agent profiles can be created for the modeling of the abstract architecture and then for its instantiation later. A profile is obtained by the combination of the mentioned features. Not all the features need to be integrated within a profile.

Complex problems require collaborative approaches and can be better solved using intelligent distributed agents. To solve this category of problems coherently, a global problem is divided into subproblems with reduced complexity whenever it is

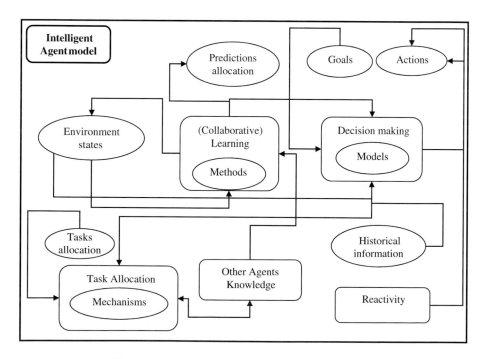

FIGURE 1.4   Intelligent agent model.

feasible. Besides, an effective collaboration of the distributed agents involved in this context is required. Consequently, the generated partial solutions issued from agents need to be consolidated and integrated into one complete solution. Therefore agent profiles like coordinators, managers, or mediators could be adopted. In such a situation, for the modeling of a coordinator agent that will synchronize the information and knowledge exchanges between the agents, the features "task allocation", "other agents' knowledge", and "historical information" can be activated. These features will allow the coordinator to interact with the other agents in order to assign the available agents new tasks and collect the generated results. The next tasks could be identified in this case:

- Complex task transformation: applies techniques to subdivide a consistent task into smaller subtasks like the concepts of partial/complete solutions and divide and conquer.
- Task allocation and dispatching: by sensing the environment states and exploiting the knowledge using features like "other agents' knowledge" and "environment states", the available useful agents are identified and assigned the tasks using suitable techniques as the alternative task decompositions.
- Solution integration and consolidation: includes the constitution of the global solution by gathering the agent partial solutions. This can be achieved again through the "task allocation", "other agents' knowledge", and eventually "collaborative learning" features in the case of machine learning.

1.3.2 INTELLIGENT DISTRIBUTED COMPUTING ARCHITECTURES

From the introduced functional specification of the intelligent agent, this later is considered as the main component that will compose the generic distributed architecture. As the architecture is intended to comprise several agents depending on the problem context and the required performance, the interconnection between agents is defined in terms of a middleware that will provide a transparent communication between agents on top of this logical layer. By means of the various profiles of intelligent agents that can be modeled, several architectures can be built according to the problem specificity.

There are alternatives for middleware infrastructure. It may consist of a service-oriented architecture (SOA) or brokers such as CORBA based on a software bus. Other technology frameworks can also support distributed architectures. Web services, RMI, and RPC fall into this category.

Figure 1.5 presents a reference generic architecture based on the introduced concepts from the perspective of an intelligent distributed multiagent system.

Each agent may have a different profile related to the role it should fulfill and the features it will require to achieve its goals. The communication between agents takes place through the middleware. Different configurations of this abstract architecture can be adopted to derive instances from it. To model a manager/agent architecture, we can make use of two profiles, namely a manager that orchestrates all the processes and an operator that mainly executes the assigned subtasks. Schematically, the architecture in Figure 1.6 reflects this model.

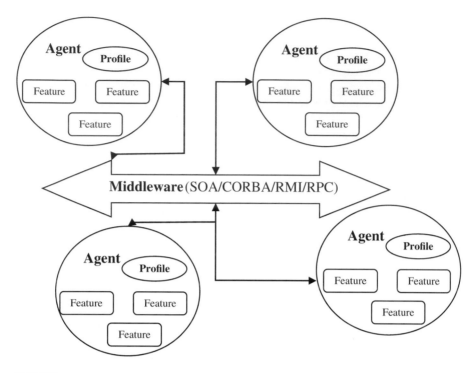

FIGURE 1.5 Generic intelligent agent-based architecture.

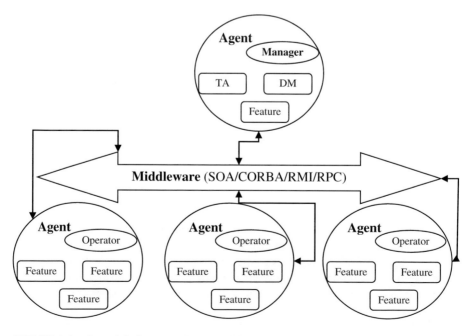

FIGURE 1.6 A model of manager/agent architecture.

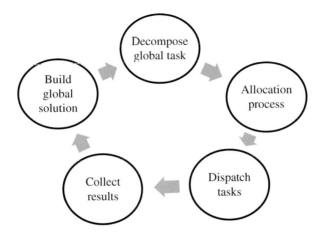

FIGURE 1.7   Allocation task workflow.

Among the underlying features of the manager, the "task allocation" (TA) should be activated, and then the corresponding workflow comprising the necessary steps will be executed. Typically, using this feature allows, in the first place, the slicing and dicing of the global task into smaller subtasks. In addition, an allocation approach is adopted to select the agent a subtask will be assigned to. In general, the appropriate agent is selected by rendering the goals/actions mechanism of the "decision making" (DM) feature. The subtasks are then dispatched to the available agents. Thereafter, the partial solutions are collected and consolidated to build the global solution. These steps are summarized in Figure 1.7.

On the other hand, the features that the operator may contain are "learning", "decision making", and "other agents' knowledge". The "learning" feature will help the agent perform some trainings on specific datasets. To select the appropriate learning method and algorithms, the "decision making" feature looks for a match regarding the predefined goals and the former experience cumulated in "historical information" to come out with the necessary actions. In addition, the "learning" feature makes use of "other agents' knowledge" to synchronize the learning method or the datasets with the other agents if needed. Schematically, a possible scenario of the operator workflow is presented in Figure 1.8.

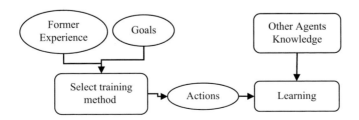

FIGURE 1.8   Learning task scenario.

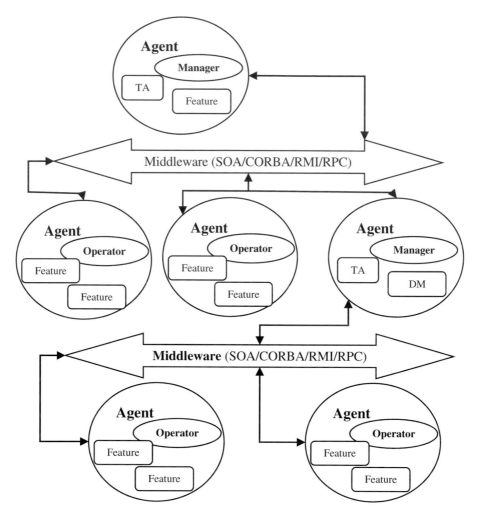

FIGURE 1.9   Hierarchical multiagent architecture.

Another model where a hierarchical architecture can be built is the case of an agent that decides to switch its role from operator to manager to deal with its assigned task from a prior manager. This scenario occurs when a complex task should be divided recursively to achieve the desired level of granularity. Figure 1.9 shows a hierarchical architecture of three levels. In this scenario, the manager at the first level assigns tasks to agents in the second level using the "task allocation" feature. In the second level, one of the agents decides to switch its role to a manager to deal with its complex tasks using the "decision making" feature. Consequently, its own "task allocation" feature becomes activated.

Whenever an agent decides to delegate its consistent tasks to other agents, an extra level in the architecture hierarchy will be added. In this way, a complex hierarchical architecture can be built.

1.3.3 Multiagent Systems Simulations

Several frameworks and platforms are available to model and simulate MAS in order to evaluate and analyze new methods and performance statistics (Bellifemine et al. 2007, Gama 2022):

- FLAME is used to simulate and model complex systems in various domains such as social sciences and economics and generate complete solutions (Flame 2020).
- Gama can be used in many domains and adopts a high-level agent-based language to build spatially explicit multiagent simulations (Gama 2022).
- JADE (Java Agent Development Framework) is based on a middleware and possesses a graphical user interface to support designing MAS (Jade 2022).

1.3.4 Case Studies (IDC Projects)

This section serves as an application of the architectures and concepts presented earlier to model a multiagent system in two different areas, namely data mining (Idri and Boulmakoul 2019) and transport (Idri et al. 2017a, 2017b, 2017c). As the interest is to highlight the benefit of using this architectural model, the focus will be on the necessary steps to build this architecture, rather than the business and technical details of the solution. Nevertheless, the main concepts and knowledge required to build up the multiagent system architecture will be presented.

1.3.4.1 Data Mining

Data mining explores large datasets to extract knowledge and detect trends in the data. The mining process is often intended to deal with huge amounts of data that cannot be gathered and processed in a single repository. Therefore, an advanced and efficient distributed architecture and algorithms are required in such a context. The method used in this use case adopts Galois lattices as an approach to extract useful information from datasets and generate a concept hierarchy that can be considered a well-structured container for frequent item sets (Idri and Boulmakoul 2019). Although the construction of Galois lattices is a complex process in formal concept analysis (FCA), it is a suitable candidate for performance improvement through distribution, especially when using a multiagent system. A consequent process that will benefit from this resolution is the generation of association rules also based on Galois lattices. The purpose of this use case is to build an intelligent scalable architecture derived from the multiagent system architecture model described earlier in order to build the Galois lattice and generate the related association rules.

1.3.4.1.1 Scope and Functional Requirements

The main process in this business context is the generation of association rules. This step is itself based on the generation of closed frequent item sets (CFISs), which is the most consistent step in the whole process. A suitable solution should provide the possibility to interact with the final outcome and allow the end user to modify the search parameters and adapt these according to his needs. This means the user

---

TABLE 1.1

Context Example

	1	2	3	4	5
a	x				
b	x		x	x	x
c	x			x	
d			x		x

---

should be able to explore individually selected association rules and accordingly adapt the support and confidence of the CFISs in order to avoid their regeneration. Therefore, to maintain permanent CFISs and make these available for repetitive investigations, the Galois lattice approach is chosen, as it provide the full generation of CFISs in a persistent way.

1.3.4.1.2   Terminology

Definition 1 Context: A context consists of a set of n objects O, a set of m attributes M, and a binary relation $I \subseteq O \times M$, denoted as (O, M, I). It can be represented by a table with objects in columns and attributes in rows, as shown in Table 1.1.

Definition 2 mapping functions: The function $attr : 2^O \rightarrow 2^M$ maps an object set $X \subseteq O$ to $attr(X) = \bigcap_{g \in X} nbr(g)$ and the function $obj : 2^M \rightarrow 2^O$ maps an attribute set $J \subseteq M$ to $obj(J) = \bigcap_{j \in J} nbr(j)$ where $nbr(i) = \{g \in O : (g,i) \in I\}$ and $nbr(g) = \{i \in M : (g,i) \in I\}$.

Definition 3 closure: An object set X is closed if $X = obj(attr(X))$. An attribute set J is closed if $J = attr(obj(J))$.

Definition 4 concepts: A concept is a couple $C = (X, J)$ where: $X = obj(J)$ and $J = attr(X)$. X is the extent of C. J is the intent of C (X and J are closed). $B(O, M, I)$ or B is the set of all concepts. The relation defined on B is:

$$(A_1, B_1) \prec (A_2, B_2) \Leftrightarrow A_1 \subseteq A_2 (B_2 \subseteq B_1)$$

where (A_1, B_1) and (A_2, B_2) are concepts of B and are a partial order relation on B.

Definition 5 Galois lattice: $L = \langle B, \prec \rangle$ is a complete Galois lattice.

1.3.4.1.3   Galois Lattice Generation Process

The fundamental idea of building the Galois lattice is to generate each concept successors, beginning with the root or parent concept. The depth first search or the breath first search algorithms are applied to process the concepts recursively. The Galois lattice building process passes through a few steps. A restriction will be made to only those steps related to the Galois lattice generation:

- Concept expansion: generation of its children (successors: concept candidates).
- Closure test: to decide whether a successor is a concept or not.

- Existence test: to guarantee the uniqueness of a concept in the global solution.
- Galois lattice building: progressively constructing the Galois lattice by capitalizing on the other steps.
- Memory management: use of Trie as an abstract data structure to enhance set management.
- Visualization: the graphical interface allows the visualization of the Galois lattice and provides the possibility to explore individually the association rules.

To improve the performance of these tasks, a specific abstract data structure is used to support a lexicographic codification of concepts and concept candidates: Trie.

A Trie is a multitree that adopts a lexicographic codification for storing concepts. As described earlier, a concept is identified by a couple of unique sets verifying the closure condition. This means that both components of a concept, intention (attributes) or extension (objects), can identify the concept uniquely. A Trie can then be based interchangeably on one component as a key and refer to the other component as a data container, assuming that all sets are sorted. Figure 1.10 shows an example.

The next step consists of identifying the agent profiles needed to accomplish all these tasks. Therefore, the generic and specific features will be specified and grouped into agent profiles. The nature of the tasks to be executed and the goals to be achieved has led to a manager/agent architecture (Idri and Boulmakoul 2019). The modeling of the manager and agent profiles are hereafter depicted using the specification of the Galois lattice generation process.

1.3.4.1.4 Agent Profile "Operator"

The "operator" profile consists of the following features:

- Concept expansion (CE): it is a pretty consistent task. From the approach adopted in (Idri and Boulmakoul 2019) and the FCA background, it is stated that the generation of children based on a given concept can be assigned to agents without the need for a global knowledge.

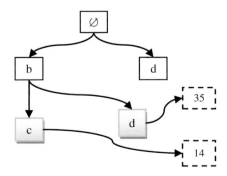

FIGURE 1.10 Trie containing concepts (bc,14) and (bd,35).

- Closure test (CT): the same applies to this test because it relies only on the intention and the extension of the parent concept, assuming the context is shared with all agents.
- Local Trie management (LTM): deals with concept-related functionalities (CE subtask).
- Distributed event handling (DEH): to ensure that the agents remain responsive, reactive, and adapt their configuration to the context requirements. In this case, the historical information on agents can be used to select the appropriate data mining algorithm according to the data sensitivity and the dataset's density.

1.3.4.1.5   Agent Profile "Manager"

- Concept expansion task allocation (CETA): the algorithm for generating the Galois lattice requests recursively through the expansion of concepts to determine the successors level by level until all concepts are generated. Each time a concept requires an expansion, the task allocation feature looks for an available agent (operator) and assigns the task to it.
- Existence test (ET): all generated concepts are maintained in one shared place to ensure the integrity of the final result and the uniqueness of each new inserted concept. That is the reason for adopting a global Trie for the codification of concepts. Collection and consolidation tasks are required to populate the global Trie and construct the final concept hierarchy. This task is to be fulfilled by the "manager" profile as it is centralized.
- Concepts collection (CC): the concepts generated by the active agents by applying the corresponding tasks CE and CT are collected and eventually inserted in global Trie if this hasn't been done yet.
- Galois Lattice building (GLB): This depends on the CC and the ET tasks. The selected new concepts by means of the ET are integrated one by one into the global Trie, with the necessary connections between parents and children. The Galois lattice is complete by the generation of the last concept.
- Global Trie Management (GTM): a subtask of the GLB playing the role of an interface between the Galois lattice and the Trie.
- Visualization (V): it is related to the concept container, that is the Galois lattice. The graphical interface relies on third-party software such as Galicia (Galicia 2003) in addition to specific components to support the exploration of a single association rule.
- Distributed event handling (DEH): similar to the operator profile. This is obvious, as the communication is full duplex between both profiles.

From the specifications of the main tasks with respect to the manager and operator profiles, a repository of the necessary features by profile can be drawn up, as shown in Table 1.2. The generic features are directly derived from the proposed intelligent agent model, while the specific features are domain related.

TABLE 1.2
Data Mining Feature Distribution by Profile

	Manager Profile	Operator Profile
Generic features	Task allocation: Concept expansion task allocation (CETA)	Decision making: Data mining algorithm selection, recursive CETA
	Reactivity: Distributed event handling (DEH)	Reactivity: Distributed event handling (DEH)
Specific features	Concepts collection (CC)	Concept expansion (CE)
	Existence test (ET)	Closure test (CT)
	Global Trie management (GTM)	Local Trie management (LTM)
	Galois lattice building (GLB)	
	Visualization (V)	

1.3.4.1.6 Architecture

According to this feature distribution, a multiagent system architecture proposal is presented in Figure 1.11. This architecture can be layered on several levels if required. The manager and operator profiles may differ from level to level in the hierarchy, as well as the nature of the middleware. The number of agents is determined with respect to the scalability needs.

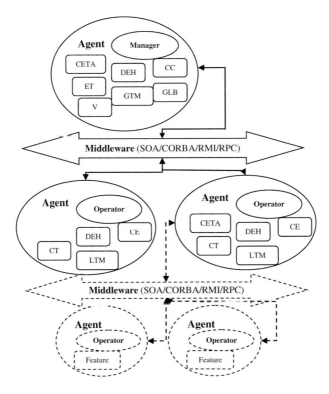

FIGURE 1.11 Data mining an intelligent multiagent system architecture.

1.3.4.2 Transport

Transport systems are still facing several challenges because of the increase in the complexity of transport networks due to the fast-growing development in all kinds of related infrastructures. When it comes to transport planning, this remains a fundamental issue, especially for sensitive domains and goods. The shortest-path algorithm is one of the crucial and central issues of transport systems. The variant of the approach considered in this case study is target oriented, time dependent, and multimodal (Idri et al. 2017a, 2017b, 2017c).

Similarly to the first case study, the generic multiagent system architecture model will be instantiated and projected on the transport domain to achieve the shortest-path generation goal with respect to multimodality and time-dependency constraints.

1.3.4.2.1 Scope and Functional Requirements

The aim of this process is to compute the shortest path from a source node to a destination node based on the information about the transport network, transport modes, and time tables.

1.3.4.2.2 Terminology

According to previous work (Idri et al. 2017a), a transport network is modeled as a direct graph with respect to the following definition:

Definition: A transport network is represented by a multimodal time-dependent graph $G = (V, E, M, T)$ where V is a set of vertices, E is a set of edges, M is a set of modes, and T is a set of travels. A couple (t_{jd}, t_{ja}) stands for a travel $t_j \in T$ indicating its departure and arrival time. A tuple (v_i, v_j, m_k) is an edge e_l connecting the vertices (nodes) v_i and v_j by mode m_k. The time dimension is expressed as a timetable comprising the possible set of travels from v_i to v_j using mode m_k. In addition, a cost function from v_i to v_j is defined on edge e_l at instant t and denoted by $C_{el}(t)$. A path is then defined as a set of connected edges. The cost of a path is the sum of the costs of all edges composing it. The shortest path is the path with the lowest cost.

The example in Table 1.3 consists of two modes and several edges with their corresponding timetables. The cost function refers to the travel time. Figure 1.12 presents the graphical view of the transport network.

1.3.4.2.3 Shortest-Path Search Process

The adopted approach as introduced in (Idri et al. 2017) considers a straightforward virtual path between the source and the destination and drives the search toward the closest nodes to this virtual path by updating continually a constraint d. The idea is to reduce the search space by considering first the nodes within the scope defined by the corridor having width 2d, length D, and the virtual path as its centerline. The search process converges to the target progressively by exploring the new candidate nodes with every iteration. The parameter d is increased with a step Δd whenever no connections are found without exceeding the maximum value d_{max}. The search process stops when the destination node is reached or no solution can be found.

TABLE 1.3

Transport Network Example

	Mode 1			Mode 2	
Edge	Timetable	Cost Function	Edge	Timetable	Cost Function
a → d	2 → 5	3	b → c	2 → 6	8
	4 → 8	2		8 → 11	5
d → e	3 → 7	2	a → d	1 → 4	6
	7 → 9	1		14 → 18	4
c → f	10 → 12	4	a → b	5 → 6	3
	5 → 8	2		8 → 10	2
c → f	6 → 9	3	d → c	9 → 13	1
	9 → 10	5			

The search process can be detailed in the following steps:

- Expand the node dimensions (END): this approach is time-dependent, multimodal, and constraint based (parameter d). The instances of each edge are generated and explored. The set of instances is obtained by combining the time tables, the available modes, and the current value of d.
- Compute the context components and parameters (CCCP): includes the virtual path, the start value of d, the initial search area (2d, virtual path length D), and $_{dmax}$.
- Constraints validation test (CVT): verifies the constraints satisfaction regarding the node candidates. Only nodes belonging to the restricted search area are considered for further processing.
- Select the search algorithm (SSA): based on the context specificities, like the network density and complexity, in addition to the historical data and machine learning outcomes, an adequate algorithm is selected automatically to deal with the current scenario.
- Compute the next-step candidates (CNSC): the selected algorithm generates the next candidates set. The parameter d is increased when no possible connections are available.

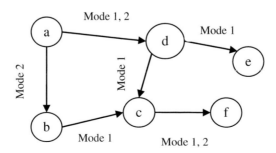

FIGURE 1.12   Transport network graphical presentation.

- Update the partial solution (APS): progressively builds up the shortest path by adopting a backtracking method that from a partial path looks to reach the shortest path (complete solution) by iterating recursively and moving back and forth to try different alternatives.
- Visualization (V): displays graphically the shortest path on a Geographical Information System (GIS) map.

By analogy with the data mining case study, the manager and operator profiles will be specified according to the generic and specific features. Both profiles are presented hereafter.

1.3.4.2.4 Agent Profile "Operator"
The "operator" profile consists of the following features:

- Select the search algorithm (SSA): Each agent decides on the search algorithm selection, making use of the "decision making" feature.
- Compute the next-step candidates (CNSC).
- Distributed event handling (DEH): to guarantee the reactivity of agents.

1.3.4.2.5 Agent Profile "Manager"
- Expand the node dimensions (END).
- Compute the context components and parameters (CCCP).
- Constraints validation test (CVT).
- Next-step task allocation (NSTA): delegates the CNSC task to operators, as it is a consistent task.
- Update the partial solution (APS).
- Visualization (V).
- Distributed event handling (DEH).

The following repository presented in Table 1.4 summarizes the feature distribution on both profiles.

TABLE 1.4
Transport Feature Distribution by Profile

	Manager Profile	Operator Profile
Generic features	Next-step task allocation (NSTA)	Decision making: recursive NSTA for
	Distributed event handling (DEH)	hierarchical delegations
		Distributed event handling (DEH)
Specific features	Expand the node dimensions (END)	Select the search algorithm (SSA)
	Compute the context components and	Compute the next-step candidates (CNSC)
	parameters (CCCP)	
	Constraints validation test (CVT)	
	Update the partial solution (APS)	
	Visualization (V)	

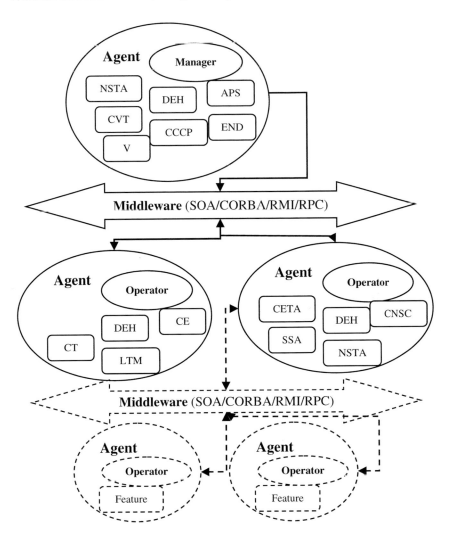

FIGURE 1.13   Transport intelligent MAS architecture.

1.3.4.2.6   Architecture

The resulting multiagent system architecture from the agent specification presented earlier is presented in Figure 1.13. In this same way, an hierarchical solution can be adopted if required.

1.4   CONCLUSION

This chapter categorizes IDC by investigating the functionalities of the components constituting an intelligent distributed system and the various architectures layered according to the problem context and requirements. First the classical concepts of distributed systems were introduced, as well as the characteristics and specificities of their components. The IDC paradigm is then presented by addressing the functional and the architectural models based on feature and profile notions depicted in a generic way. The focus was put on multiagent systems as a framework for the purposes of this chapter. Multiagent systems are very suitable to support IDC, assuming that an agent is capable of activating its intelligent features whenever it is required, like reactivity, learning, task allocation, and decision-making. To concretize these models, two case studies were presented showing the building process of all architectural components stepwise. The data mining case study was concerned with knowledge extraction from large datasets. This knowledge appears in the form of association rules derived from concepts holding the basic elements of the whole process, namely CFISs. The transport case study was concerned with the fundamental known thematic: shortest-path search problem. The search space is target oriented and is reduced using some constraints and a straightforward virtual path, considered as a reference path, is used to drive the search as close as possible from this ideal shortest path using different algorithms.

REFERENCES

Attiya H. and J. Welch (2004). Distributed Computing Fundamentals, Simulations and Advanced Topics. Hoboken, NJ: John Wiley & Sons, ISBN: 978-0-471-45324-6.

Bellifemine F. L., G. Caire, and D. Greenwood (2007). Developing Multi-Agent Systems with JADE. New York, NY: Wiley, vol. 7.

Bianchi R. A. C., M. F. Martins, C. H. C. Ribeiro, and A. H. R. Costa (2014). Heuristically-accelerated Multiagent Reinforcement Learning. IEEE, Trans. Cybern., vol. 44, no. 2, pp. 252–265.

Bowling M. and M. Veloso (2002). Multiagent Learning Using a Variable Learning Rate. Artif. Intell., vol. 136, no. 2, pp. 215–250.

Dayong Y., Z. Minjie, and V. Athanasios (2017). A Survey of Self-organization Mechanisms in Multi-agent Systems. IEEE Trans. Syst., Man, Cybern.: Syst., vol. 47, pp. 441–461.

Dorri A., S. S. Kanhere, and R. Jurdak (2018). Multi-Agent Systems: A Survey. IEEE Access, vol. 6, pp. 28573–28593. DOI:10.1109/ACCESS.2018.2831228.

Flame (2020). http://flame.ac.uk.

Galicia (2003)., http://www.iro.umontreal.ca/~galicia/publication.html.

Gama (2022) https://gama-platform.org.

Gatti M. et al. (2013). Large-scale multi-agent-based modeling and simulation of microblogging-based online social network. In Proc. Int. Workshop Multi-Agent Syst. Agent-Based Simulation, pp. 17–33.

Idri A. and A. Boulmakoul (2019). A parallel distributed Galois Lattice approach for data mining based on a CORBA infrastructure. Intelligent Computing, Proceedings of the 2019 Computing conference, London, vol. 998, pp. 99–117. DOI: https://doi.org/10.1007/978-3-030-22868-2_8.

Idri A. F., M. Oukarfi, A. Boulmakoul, and K. Zeitouni (2017c). A distributed approach for shortest path algorithm in dynamic multimodal transportation networks. EWGT 2017, Euro Group on Transportation Meeting, Transportation Research Procedia, volume 27, pp. 294–300.

Idri A., M. Oukarfi, A. Boulmakoul, and K. Zeitouni (2017b). Design and implementation issues of a time-dependent shortest path algorithm for multimodal transportation network. CEUR Workshop Proceedings, Vol 1929,TD-LSG@PKDD/ECML: 32–43, ISSN 1613-0073, urn:nbn:de:0074-1929-7.

Idri A., M. Oukarfi, A. Boulmakoul, K. Zeitouni, and A. Masri (2017a). A new time-dependent shortest path algorithm for multimodal transportation network. Procedia Computer Science,, vol. 109, pp. 692–697. DOI: 10.1016/j.procs.2017.05.379.

Jade (2022). https://jade.tilab.com.

Jain L. C. and D. Srinivasan (2010). Innovations in Multi-Agent Systems and Application. Berlin, Heidelberg: Springer.

Jiang Y. and J. C. Jiang (2014, 10). Understanding Social Networks from a Multiagent Perspective. IEEE Trans. Parallel Distrib. Syst., vol. 25, no. 10, pp. 2743–2759.

Kshemkalyani Ajay D. (2008). Distributed Computing: Principles, Algorithms, and Systems. New York, NY: Cambridge University Press, ISBN-13 978-0-521-87634-6.

Lea D., S. Vinoski, and W. Vogels (2006). Guest Editors' Introduction: Asynchronous Middleware and Services. IEEE Internet Computing, vol. 10, no. 1, pp. 14–17.

Montes M. A., E. Ferrante, A. Scheidler, C. Pinciroli, M. Birattari, and M. Dorigo (2011). Majority-rule Opinion Dynamics with Differential Latency: A Mechanism for Self-organized Collective Decision-making. Swarm Intelligence, vol. 5, pp. 305–327.

Omicini A. and S. Mariani (2013). Agents & Multi-agent Systems: En Route Towards Complex Intelligent Systems. Intelligenza Artificiale, vol. 7, no. 2, pp. 153–164.

Osrael, J., L. Froihofer, K. M. Goeschka, S. Beyer, P. Galdámez, F. D. Muñoz Escoi (2006). A system architecture for enhanced availability of tightly coupled distributed systems. In: Proceedings of the 1st International Conference on Availability, Reliability and Security, IEEE Computer Society, Los Alamitos.

Rezaee H. and F. Abdollahi (2015, 11). Average Consensus Over High-order Multiagent Systems. IEEE Trans. Autom. Control, vol. 60, no. 11, pp. 3047–3052.

Rizk Yara, M. Awad, and E. Tunstel (2018, 5). Decision Making in Multi-Agent Systems: A Survey. IEEE Transactions on Cognitive and Developmental Systems, vol. 10, no. 3, pp. 514–529. DOI: 10.1109/TCDS.2018.2840971.

Rodríguez E. J. P. (2008). Distributed Intelligent Navigation Architecture for Robots. AI Communications, vol. 21, no. 2–3, pp. 215–218.

Salvaneschi G., J. Drechsler, and M. Mezini (2013). Towards Distributed Reactive Programming. 15th International Conference on Coordination Models and Languages (COORDINATION), Jun, Florence, Italy. pp. 226–235.

Su S., Z. Lin, and A. Garcia (2016, 1). Distributed Synchronization Control of Multiagent Systems with Unknown Nonlinearities. IEEE Trans. Cybern., vol. 46, no. 1, pp. 325–338.

Van Steen, M. and A. S. Tanenbaum, (2016). A Brief Introduction to Distributed Systems. Computing, vol. 98, no. 10, pp. 967–1009. DOI: https://doi.org/10.1007/s00607-016-0508-7.

Verbraeken J., M. Wolting, J. Katzy, J. K.,T. Verbelen, J. S. Rellermeyer (2020). A Survey on Distributed Machine Learning. ACM Comput. Surveys. vol. 53, no. 2, pp. 1–33. DOI: 10.1145/3377454.

Veríssimo P. and Rodrigues L. (2001), Distributed Systems for System Architects: Advances in Computing and Middleware, Springer, New York USA.

Xie Jing and C.-C. Liu (2017). Multi-agent Systems and Their Applications, J. Int. Council Electr. Eng., vol. 7, no. 1, pp. 188–197, DOI: 10.1080/22348972.2017.1348890.

Xing E. P, Qirong Ho, Pengtao Xie, and Dai Wei (2016). Strategies and Principles of Distributed Machine Learning on Big Data. Engineering, vol. 2, no. 2, pp. 179–195.

Zheng X. and S. Koenig (2008). Reaction function for task allocation to cooperative agents. In Proc. of AAMAS'08.

2 Multi-Micro-Agent System Middleware Model Based on Event Sourcing and CQRS Patterns

Mohamed Youssfi, Fatima Ezzahra Ezzrahari, Younes Hajoui, Omar Bouattane, and Vassilis Kaburlasos

CONTENTS

2.1 INTRODUCTION

Distributed and multiagent systems are a natural solution for solving complex distributed intelligence problems that require high-performance computing (Basmi et al. 2021, Ali and Bagchi et al. 2019, Ranjan et al. 2015, Chen et al. 2017, Benchara and Youssfi 2021, Hajoui et al. 2019). However, setting up such solutions requires making wise choices to resolve emerging issues such as load balancing, high availability, fault tolerances, failure recovery, and distributed agent communication models (Rajani and Garg. 2015, Iranpour and Sharifian 2018). In recent years, micro-service architectures have reached a high level of maturity due to the development of containerization solutions such as Docker, container orchestration solutions such Kubernetes, cloud computing such as Amazon Web Services (AWS),

DOI: 10.1201/9781003255635-2

25

security standards such Oauth2 and Open Id Connect protocols, the Big Data tools ecosystem, and the DevOps tools ecosystem (Wan et al. 2018, O'Connor et al. 2017). The challenges of micro-services architectures have been accompanied by the search for suitable patterns and the development of a framework to find the simplest and the best way to implement and orchestrate such architectures. The Command Request Responsibility Segregation (CQRS) and Event Sourcing patterns, which are based on distributed asynchronous event-driven architectures, are a couple of patterns that have demonstrated their effectiveness for the implementation of massively distributed architectures based on micro-services (Lima et al. 2021). However, the middleware and framework for the development of multiagent systems have not evolved and have not followed the same trend and have remained in architectures hampered by old specifications such as the Foundation for Intelligent Physical Agents (FIPA). Therefore, frameworks for multiagent systems such Java Agent Development (JADE) that meet its specifications have been widely consumed by the scientific community to solve complex problems of distributed artificial intelligence (Bellifemine et al. 2007, Bellifemine et al.2001). However, the production environment of such solutions is handicapped by the performance of this kind of framework because of their building model architecture, which uses a non–high-scalable data distribution and processing model. These problems appear clearly when high-performance computing is required, with a very high level of load growth and for massively distributed architectures in which the number of agents is very large. The development of a new generation of brokers such KAFKA opens new opportunities to build a new generation of frameworks for multiagent systems with a very high level of performance compared to the existing models.

In this chapter, we propose a new framework model for developing massively distributed architectures based on micro-agents. This model will attempt to transpose the evolutions of micro-services architectures into a framework dedicated to distributed artificial intelligence using the multiagent systems approach. This middleware is mainly based on the two patterns: CQRS and Event Sourcing, using a polymorphic broker model, allowing events to be consumed with a polling model, and offering the back-pressure mechanism (Wang et al. 2020, Betts et al. 2013). The chapter will be limited mainly to the description of the agent communication model and will leave its cognitive model for further study.

2.2   BACKGROUND

Domain-driven design (DDD) is an approach for software development that centers the development on programming a domain model that has a rich understanding of the processes and rules for a target domain. This approach comes from (Cao et al. 2019) and describes the approach through a catalog of patterns. The approach is particularly suited to complex systems, where a lot of often-messy logic needs to be organized. The idea is that each subsystem must be based on a data model that is most appropriate for the logic of the target domain. Micro-services architecture, which consists of splitting the functional scope of a large project into several small independent micro-services, has reached a high level of maturity thanks to the development of the DevOps tools ecosystem. To delimit the contexts of each

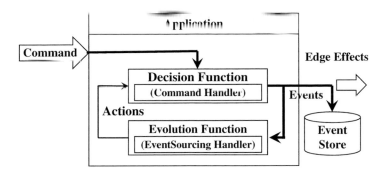

FIGURE 2.1   Event sourcing pattern architecture.

domain, a balance should be found between two constraints. First, we seek to initially create small micro-services with the finest level of granularity possible. On the other hand, chatty communications between micro-services should be avoided. For complex systems, including distributed artificial intelligence issues, multiagent systems are a very good solution. To resolve the technical performance of the classic middleware used to develop multiagent systems, we think it has become obvious to exploit the maturity of micro-services architectures. The idea is to transform a micro-service into an intelligent micro-agent equipped with artificial intelligence algorithms.

As shown in Figure 2.1, event sourcing is an architecture pattern that consists of capturing all changes in the state of an application as a sequence of events. The idea is not to focus on the current state of the application, but on the sequence of state changes represented by events that led to the current state. From this sequence of events, we can aggregate the current state of the application. Any change in the state of the application has a single cause, which is the event.

A micro-service based on event sourcing receives external requests called "commands", which can be user interface requests. To react to the command, the micro-service triggers a "decision function" that executes the business logic. Depending on this command and the current state of the application, the decision function fires a set of "events" throws an "event bus". These events are persisted in an "event store". To change the state of the application, an "evolution function" as an "event sourcing handler" reacts to published events in order to update the final state of the application. Once the state of the application is in a stable state, we transmit (Broadcast) the fired events to the neighborhood agents through an event bus, which leads events to other micro-services that depend on it.

The advantages of event sourcing are numerous, such as:

- The possibility of creating an audit database for the system
- We can easily find the origin of bugs in production
- Possibility of data recovery by replaying business events of the event store
- High level of performance by using asynchronous communications with message buses that scale very well
- Micro-services can create several projections with different data models using event handlers

The event sourcing pattern is very useful for event-driven architecture. This pattern considers the event store as the single database for writing and reading application data. This could constitute a major handicap for the performance of the application. To overcome this problem, another pattern that is often used in combination with this pattern is CQRS.

CQRS is a pattern that consists of separating the reading part from the writing part of the application. In this pattern, "commands" are used as external items to modify the state of an application like Insert, Update, and Delete requests. After the execution of a command, each change in the application state is translated by firing events, which are stored in the event store as the writing database of the application. The reading side of the application uses event handlers emitted by the writing side to project the event data into the reading database. This makes it possible to synchronize the reading with the writing databases and choose a data model that is most appropriate to the reading domain. On the reading side, "queries" are used as an intention to read the state of the application such as a Select request.

In addition to event sourcing advantages, the CQRS pattern, as shown in Figure 2.2, brings other important advantages such as:

- Each micro-agent in the application can use a different data model that is more appropriate to its domain. This should speed up data consultation operations.
- Scaling of the read side independently of the write side. Experience shows that 90% of the operations concern the reading of application data against 10% of the operations, which concern the writing data operations. To deal

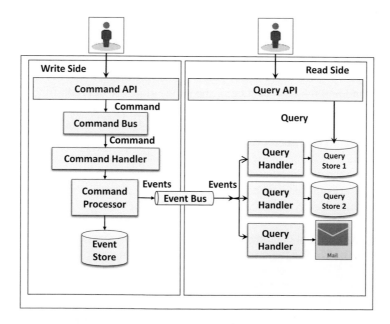

FIGURE 2.2   CQRS pattern architecture.

with the problem of scaling up, we can therefore apply the read part differ
ently from the write part, which saves on resources.

- Ability to apply two different security policies for the read part and the
write part of each micro-agent.

2.3   MICRO-AGENTS COMMUNICATION MODEL

To show the main components of the proposed architecture, we represent in Figure 2.3 a
communication diagram based on a classic blocking communication model that
shows different micro-agents:

- The UI agent, which represents the user interface, is used to interact with
application users. This can be a web, desktop, mobile, or external embedded system application.
- Two functional micro-agents: Agent A deployed in a single instance and
Agent B deployed in dual instances. Each micro-agent has its own database
representing the model of its business domain.
- To be able to easily contact micro-agents by their identifier names, the
architecture deploys two technical agents which must be provided by the
framework.
  - The first one is the Directory Facilitator agent (DFA), which is designed
  to register all micro-agents in the platform. At start-up, each micro-
  agent publishes to this DF agent its uniform resource identifier (URI),
  including the technical information for the agent location, namely the
  name of the agent, the Internet Protocol (IP) address, the port, and
  agent services description.
  - The second technical micro-agent in the architecture is the Gateway
  agent. All external interactions with the application are sent through
  this agent. The Gateway agent contacts the DF agent to retrieve the
  location of the micro-agent whose name is provided in the user request
  in order to dispatch user requests to the target agent.

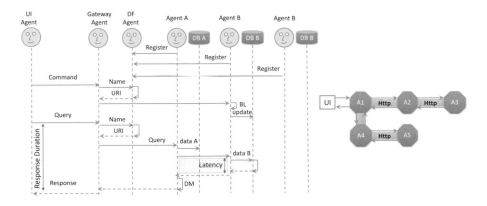

FIGURE 2.3   Blocking interaction model.

There are two types of external solicitations: commands and queries. Commands are intentions to change the state of the application. In this communication diagram, Agent B receives the command, then executes its business logic and makes the changes in its own database in its domain. The second solicitation of the UI agent is a query targeted to Agent A with the intent to request some data of the agent domain. To respond to this request, Agent A needs to consult its local database, which falls under its domain, but it also needs to communicate with Agent B to consult information from the domain B to aggregate the received information and then make the right decision. This leads the model into a blocking synchronous communication whose network latency negatively impacts the application response time. If Agent B is not available, the customer will not be served. Performance will be even more impacted if several sequential data consultation calls between agents are necessary. Network latencies will add to the overall response time, and the chances of failure will increase.

To overcome this problem and eliminate these rather expensive interactions, an event-driven communication model can be used (Michelson 2006, Taylor et al. 2009). Figure 2.4 shows a representative communication diagram based on event-driven architecture (Michelson. 2006, Tragatschnig et al. 2018). In this model, all agents use event handlers to listen for events published in the event bus by other distributed agents. In this scenario, when a command is dispatched to Agent B, it executes its business logic and fires an event reflecting changes of the state of the domain B. This event is listened for by Agent A and all instances of Agent B. Therefore, all those agents will handle this event to project event data into their own local database. This

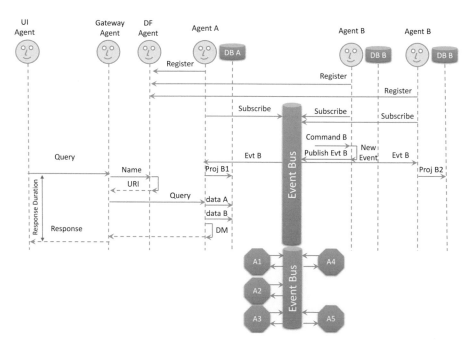

FIGURE 2.4   Event-driven interaction model.

guarantees that all the agents will be synchronized. Therefore, when the request comes to Agent A, it will no longer need to interact with Agent B to handle the requested query. This means we have a distributed model with duplicate data and with a data synchronization system. In the next section, we will detail how we can improve this event-driven architecture using the CQRS and Event Sourcing patterns (Betts et al. 2013, Debski et al. 2017).

2.4   MICRO-AGENT MODEL ARCHITECTURE

In Figure 2.5, we present the main components of the micro-agent model architecture. Each agent is split into two independent micro-agents: the command side and the query side. The command side is used to process all incoming commands representing intentions to modify the state of an agent. On this side a controller component is proposed using Google Remote Procedure Call (GRPC), Simple Object Access Protocol (SOAP), REpresentational State ansfer (REST) protocols to interact with the UI of the application. For each UI request, the controller dispatches a command in a command bus using an internal command dispatcher interface.

To handle this command, command handlers, listening to the command bus, are implemented in an aggregate component, representing the state of the micro-agent domain.

Each command handler performs the business logic representing the decision function. Then a set of events reflecting the change in the agent state is fired in the event bus and persisted in an event store. To mutate the state of the agent, event sourcing handlers, representing evolution functions, react to published events in order to update the final state of the aggregate. Once the state of the application is

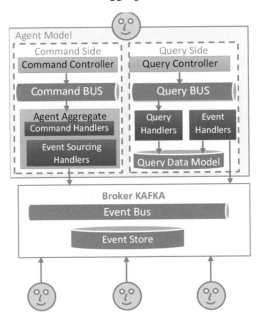

FIGURE 2.5   Agent architecture model.

stable, we we transmit (Broadcast) the fired events to the neighborhood agents through an event bus, which leads events to other micro-services that depend on it.

The query side of the agent model is a micro-agent, which is used to exploit the data of its environment to serve the rest of the application. To do so, events triggered by the command side are handled by event handler components. Data brought by those events are projected and stored in the query database model. The query model can be a relational, hierarchical structure or even a graph-oriented model in which the agent can easily reason. To query the data of the agent domain, we no longer need the command side of the agent. The micro-agent query side is made for this kind of task. To request the application state, the UI agent interacts with the query micro-service side via a GRPC, SOAP, or REST query controller component. Each UI request received by the query controller is translated by dispatching a query in a query bus component using a query dispatcher component. These requests are handled by a service component that implements query handlers listening to queries dispatched in the query bus component. To send a response to the UI agent, the query handler uses a data repository component to query the query database model without needing any interaction with other micro-agents of the multiagent system.

This separation of the micro-agent query side and the micro-agent command side brings several advantages:

- The micro-agents don't need any interaction with other agents of the multiagent system during reasoning tasks. This reduces the cost of communication between agents and therefore improves the performance of the multiagent system.
- At startup, the micro-agent can rebuild its own environment by replaying the events from the event store. So, if necessary, we can shut down any micro-agents without any negative consequences for other micro-agents.
- In the case of a load increase due to the number of consultation requests issued from the UI agent, only the query part needs to be scaled independently of the command part. This should optimize resources and increase multiagent system performance.
- To secure the micro-agents, we can use separate policies for command and query micro-agents. Indeed, in some cases, the command side needs to be more secured than the query side because the only way to make changes to system data is through the command side.

2.5   MICRO-AGENT STRUCTURE MODEL

Figures 2.6–2.8 show class and communication diagrams that describe the main components of a micro-agent. For each component of the framework, we define an abstraction layer which hides, for the developer, the technical complexity of the middleware with default implementations to reduce the amount of code we need to write and gives the possibility of making the framework work with the principle of auto-configuration. Then the implementation layer is open to the developer to customize their own implementations if necessary. As we apply the CQRS pattern, we need to have as a priority defining the commands, events, and queries of each micro-agent.

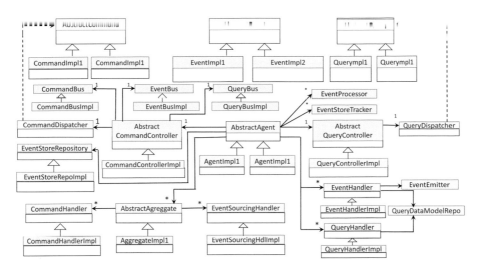

FIGURE 2.6 Micro-agent structure model.

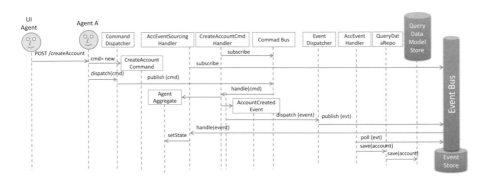

FIGURE 2.7 Micro-agent command-side structure model.

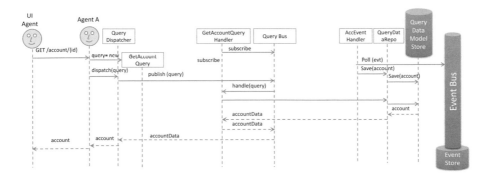

FIGURE 2.8 Micro-agent query-side structure model.

For each agent, we define three buses:

- A command bus for dispatching the commands triggered by incoming requests as the intention to change the state of the agent domain
- A query bus for dispatching the queries triggered by incoming consultation requests
- An event bus for dispatching the events fired by command handlers

For each type of bus, we define several possible implementations. To expose the services of the agent to the external application users, we associate each agent with a controller such as GRPC, SOAP, or REST services.

Upon receipt of an HTTP request of type POST, PUT, or DELETE, representing the intent to modify the agent state, the command controller creates a corresponding command and dispatches it to the command bus using the command dispatcher component. Each command sent on the command bus is handled by command handler components deployed inside the aggregate component, which represents the domain state of the micro-agent.

When performing business logic, command handlers publish events in the event bus. To store those events in the event store, an EventStoreRepository component is used. The framework provides several implementations of EventStoreRepository, which will record events in a database with a single table whose structure will be presented in the next section. To change the state of the aggregate from the published events, event sourcing handlers are defined inside the aggregate to react to replay the published events to change the state of the aggregate. A micro-agent can be deployed in two instances with two different profiles: the command profile representing the command side of the agent and the query profile representing the query side of the agent. For the query side of any micro-agent of the platform, we define event handlers to project the data of the captured events in the agent's query database using a QueryDataModelRepository component whose implementation is chosen according to the agent data structure model. To increase performance of the proposed model, all event handlers operate in an event loop driven by an independent single thread. Each UI request received by the query controller component is translated by dispatching a query throwing a "Query Bus" component using a "Query Dispatcher" component. These requests are handled by a service component where we implement query handlers. To send a response to the UI agent, the query handler uses a "Query Data Repository" component to query the query database model.

For real-time reactive UI applications, by using web sockets or a server sent event, we can use the Query Dispatcher component to subscribe to the query bus to listen to the backend results to push to the UI. Therefore, event handlers can emit data and push it to the controller's subscription using an Event Emitter component. To track the event store, the micro-agent query side needs to store an index that reminds it of the position of the last event consumed from the event store, thanks to an Event Tracker component. While restarting, the micro-agent can easily replay events from the event store that it has not yet consumed. To replay all the events on the query side, we just need to reset the Event Tracker index to zero. Therefore, the choice of broker for the event buses and the choice of the event store must be based on technologies ensuring high-availability and high-performance computing. In our implementation of this framework, we have chosen KAFKA (Birajdar et al. 2016) as a broker and as an event store. We think that it is the best choice right now. However, the framework integrates well with other brokers like RabbitMQ or ActiveMQ. From the multiagent system middleware, we propose to develop a new model of event store, which will be described in the next section.

2.6   EVENT STORE MODEL AND MESSAGE MODEL

The event store is a storage unit that allows all events published by agents to be persisted. In addition to the functional information of the message, technical and semantic information is added to the message to translate the fundamental characteristics of the Agent Communication Language (ACL) specification. A message event is published in the event store under the following structure:

- [event Offset]: A unique number of the events generated automatically in a sequence.
- [event Identifier]: A unique identifier of the Universal Unique Identifier (UUID)–type event.
- [timestamp]: The moment when the event was created.
- [speech Act]: Represents the act of communication conveyed by the message according to the FIPA specifications.
- [ontology]: The ontology of the message according to ACL.
- [language]: The language of the message according to ACL.
- [Agent Aggregate Identifier]: The identifier of the aggregate to which the event applies.
- [agent Aggregate Version]: The version number of the aggregate, which is very important in the case of concurrent accesses to the same aggregate by several commands.
- [Agent Group Id]: The identifier of the group of consumer agents.
- [Payload]: A JSON or XML representation of the event. All the properties of the event can be found there. The Payload is a serialization of the Event object.
- [Payload Type]: The type of the payload, which is a string indicating the package and the class representing the structure of the event. This information is very important for event handlers to listen to and deserialize targeted events.
- [Aggregate Type]: The type of the aggregate representing the name of the class of the Aggregate. This information is very important in reconstructing the state of an aggregate from the events produced by that aggregate.
- [Metadata]: A JSON or XML serialization of an object containing additional useful technical information such as tracked ID, correlation ID, etc.

All events produced by agents (Benjamin et al. 2018) are recorded in a single table in the topic of the event store. Using a broker like KAFKA makes it easier because

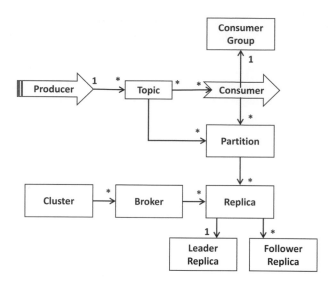

FIGURE 2.9　Topic event store structure.

the partitioning, high availability, and security of the event store are effectively managed internally at the broker level. As shown in Figure 2.9, to guarantee a high level of system performance, a cluster of several brokers is deployed with a coordination system based on a tool like Zookeeper. A producer agent can post messages on topics. The consumer agents can consume the messages from the topics, using the pulling mechanism. This ensures that each consumer can consume messages at the rate they choose for themselves to guarantee not to overflow the agents' consumers.

To create agent communities, each consumer agent must belong to a group. Each message published on a topic defines the identifier of the consumer group. By default, we define a group named EVENT_BUS to represent a single event bus for the entire application and EVENT_STORE_APP_ID for the name of the topic associated with each application representing the event store. To manage the large quantity of messages published in the event store and ensure high availability, each topic is divided into several partitions. Each partition has several replicas, one of which is a leader and is deployed in an instance of the cluster broker, and the other replicas are deployed in other instances of the cluster.

As shown in Figure 2.10, each published event in the event bus is given an offset that identifies it in a topic partition. Each event handler belongs to a group whose name must be unique. This name corresponds to the name of the event processor. This feature is very useful when you want to replay the events of a group of specific handlers. To know which offset of the topic begins to consume, in its data query model database, the event processor component keeps the value of the last message offset consumed.

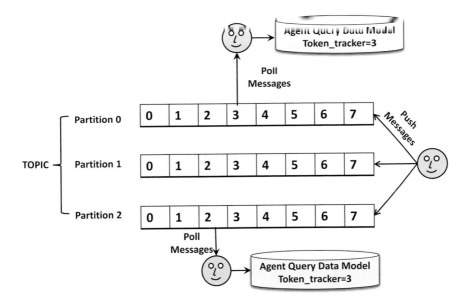

FIGURE 2.10   Topic event store structure.

2.7   DISTRIBUTED TRANSACTIONS MODEL ARCHITECTURE

A transaction is a single unit of logic or work, sometimes made up of multiple operations. Within a transaction, an event is a state change that occurs to a data entity, and a command encapsulates all information needed to perform an action or trigger a later event.

Transactions must be atomic, consistent, isolated, and durable (ACID). Transactions within a single service are ACID, but cross-service data consistency requires a cross-service transaction management strategy.

In multiagent architectures we can say that:

- Atomicity is an indivisible and irreducible set of operations that must all occur or none will occur.
- Consistency means that the transaction brings the data only from one valid state to another valid state.
- Isolation guarantees that concurrent transactions produce the same data state that sequentially executed transactions would have produced.
- Durability ensures that committed transactions remain committed even in the case of system failure or power outage.

Using a distributed service-oriented architecture (SOA) (Taylor et al. 2009, Yuan et al. 2009) based on micro-agents or micro-services with a database model for each micro-agent provides many benefits. However, ensuring data consistency across service-specific databases poses challenges.

In the proposed model, we use the Saga pattern as a way to manage data consistency across micro-agents in distributed transaction scenarios. A Saga is a sequence of transactions that updates each service and publishes a message or event to trigger the next transaction step. If one step fails, the Saga executes compensating transactions that counteract the preceding transactions.

There are two common Saga implementation approaches: choreography and orchestration. Each approach has its own set of challenges and technologies to coordinate the distributed workflows.

Choreography is a way to coordinate Sagas where participants exchange events without a centralized point of control. With choreography, each local transaction publishes domain events that trigger local transactions in other services. This model is good for simple workflows that require few participants and don't need any coordination logic component and don't require additional service implementation and maintenance. However, the distributed workflow can become confusing when adding new steps, as it's difficult to track which Saga participants listen to which commands. Figure 2.11 presents an example of choreographing Saga implementation for an e-commerce application based on a multiagent system. Before committing an order initiated by the order agent, are invited to participate in the workflow to validate the payment; update the inventory; perform the shipment; and send notification by the payment agent, inventory agent, the other agents of the multiagent system shipping agent, and notification agent, respectively. Choreography provides the ability to coordinate Sagas by applying publish-subscribe principles. With choreography, each micro-agent runs its own local transaction and publishes events to a message broker system and that triggers local transactions in other micro-agents.

The second type of Saga implementation is orchestration, which is a way to coordinate Sagas where a centralized controller tells the Saga participants what local transactions to execute. The Saga orchestrator handles all the transactions and tells the participants which operation to perform based on events. The orchestrator executes Saga requests and stores and interprets the states of each task and handles failure recovery with compensating transactions. This kind of Saga implementation

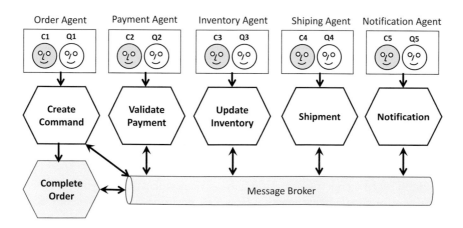

FIGURE 2.11 Choreography Saga example architecture.

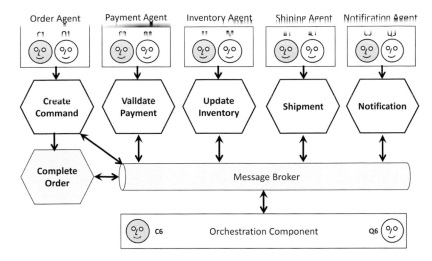

FIGURE 2.12   Orchestration Saga example architecture.

is the best choice for complex workflows involving many participants or when new participants are added over time, and Saga participants don't need to know anything about the commands for other participants. It brings a clear separation of concerns and simplifies business logic. However, additional design complexity requires implementation of a coordinator logic component, and this can be an additional point of failure, because the orchestrator manages the complete workflow. Figure 2.12 presents an example of orchestrating Saga implementation for the e-commerce application based of a multiagent system. Orchestration provides the ability to coordinate Sagas with a centralized controller micro-agent. This centralized controller micro-agent orchestrates the Saga workflow and coordinate the execution of local micro-agents' transactions sequentially. The orchestrator component executes the Saga transactions and manages them in a centralized way; if one of the steps fails, it executes rollback steps with compensating transactions.

2.8   MODEL PERFORMANCE

As shown in Figure 2.13, the model aims to eliminate micro-agents interactions at runtime. Using this model, each micro-agent interacts only with the event bus to publish or consume events. Each micro-agent can build its own environment representation in which it can perform tasks and make decisions using its own artificial intelligence algorithms without needing to interact with any other agent at runtime.

Several other advantages can be presented as solid arguments to prove the performance of the proposed model, such as:

- The separation of the micro-agent query side from the micro-agent command side brings advantages compared to the classic distributed architectures (Veiga et al. 2016, Umer et al. 2018):
  - The micro-agents don't need any interaction with other agents of the multiagent system during reasoning tasks. This reduces the cost of

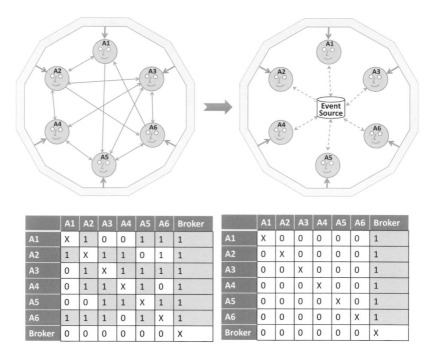

The following tables appear below the figure:

	A1	A2	A3	A4	A5	A6	Broker
A1	X	1	0	0	1	1	1
A2	1	X	1	1	0	1	1
A3	0	1	X	1	1	1	1
A4	0	1	1	X	1	0	1
A5	0	0	1	1	X	1	1
A6	1	1	1	0	1	X	1
Broker	0	0	0	0	0	0	X

	A1	A2	A3	A4	A5	A6	Broker
A1	X	0	0	0	0	0	1
A2	0	X	0	0	0	0	1
A3	0	0	X	0	0	0	1
A4	0	0	0	X	0	0	1
A5	0	0	0	0	X	0	1
A6	0	0	0	0	0	X	1
Broker	0	0	0	0	0	0	X

FIGURE 2.13   Communication model performances.

communication between agents and therefore improves the performance of the multiagent system.

- At startup, the micro-agent can rebuild its own environment by replaying the events from the event store. So, if necessary, we can shut
down any micro-agents without any negative consequences for other
micro-agents.
- In the case of a load increase due to the number of consultation requests
issued from the UI agent, only the query part needs to be scaled independently of the command part. This should optimize resources and
increase multiagent system performance.

- To secure the micro-agents, we can use separate policies for command and
query micro-agents. Indeed, in some cases, the command side needs to be
more secure than the query side because the only way to make changes to
system data is through the command side.
- The possibility of creating an audit database for the system.
- We can easily recover data and find the origin of bugs in production by
replaying events of the event store.
- High level of performance by using asynchronous communications using a
message bus based on brokers that scale very well.
- Micro-services can create several projections with different data models
using event handlers according to domain specificity.

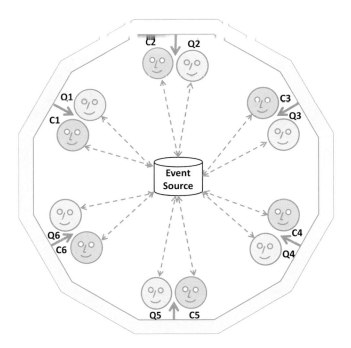

FIGURE 2.14   Micro-agent blockchain architecture.

To test the performance of the proposed model, we created an application implementing a blockchain application model. We have split the blockchain miner agents into two micro-agents representing the command side and query side of the agent.

The "Block Chain Miner Command" (C1) micro-agent implements features like creating a new blockchain, creating wallets, submitting a transaction, and mining a block.

The "Block Chain Miner Query" (Q1) micro-agent implements functionalities such as consulting blockchains, consulting blockchain wallets, consulting blockchain blocks, checking blockchain validation, and consulting the wallet balance.

The couple of the two micro-agents is deployed in several instances, and each couple (Ci, Qi) represents a blockchain miner. The architecture of the test application is shown in Figure 2.14.

To test the stability and performance of the proposed model, we performed two stress tests using the Apache JMeter tool and monitored the broker and the two micro-agents, C1 and Q1.

The first test concerns the consultation of the blocks of a blockchain. This test is targeting the query side of the agent. In this test, we simulated 2000 batches of HTTP requests. Each batch is for 20 concurrent users.

The second test aims to add a new wallet in a blockchain. This test is intended for the command side of the agent. In this test, we simulated 40 batches of HTTP requests. Each batch is for 10 concurrent users.

The results of the two tests are represented in Figures 2.15–2.17. These figures represent the state of consumption of memory, threads, and CPU, respectively, for the micro-agent Q1, the micro-agent C1, and the broker.

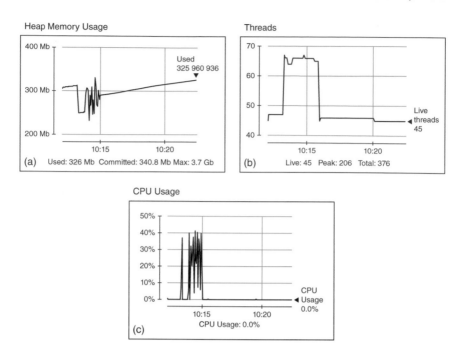

FIGURE 2.15 a) Micro-agent query heap memory, b) micro-agent query threads, and c) micro-agent query CPU usage.

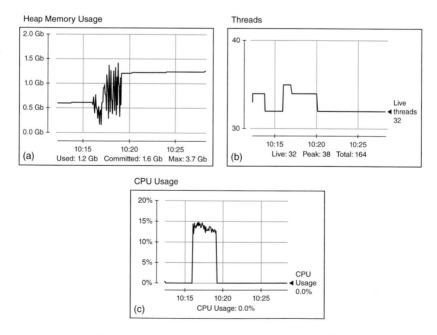

FIGURE 2.16 a) Micro-agent command heap memory, b) micro-agent command threads, and c) micro-agent command CPU usage.

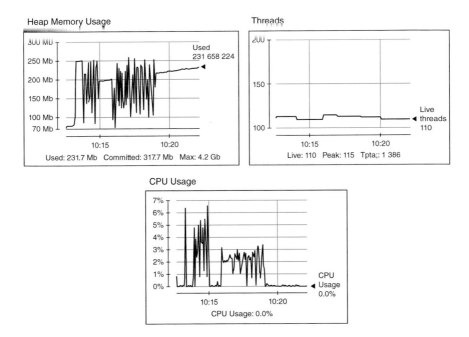

FIGURE 2.17 a) Broker heap memory usage, b) broker threads, and c) broker CPU usage.

The main results of these tests are summarized in Table 2.1.

From these tests, we can draw the following results:

- The micro-agents and the broker do not consume a lot of memory during their activities. The memory consumption status is stable.
- The CPU utilization rate is considered very satisfactory with a very stable state. The query part consumes more CPU than the command part. This is completely normal because the query part is busier than the command part. The proposed model is interesting because it makes it possible to scale the query part independently of the command part.
- The number of threads used by the different agents is stable with a very satisfactory level.

TABLE 2.1
Performance Test Results

	Max Heap Memory	Max Threads	Max CPU Usage	Stability
Query side	320 MB	67	40%	OK
Command side	1400 MB	32	15%	OK
Broker	255 MB	115	6%	OK

- The activity of the query part does not influence the state of the order part and the broker.
- The activity of the command part logically solicits the broker and the query part only slightly because the events emitted by the command part are processed by the query part.

2.9 CONCLUSION

In this chapter, we presented a new model of middleware for multi-micro-agent systems. The proposed model is based on event-driven architecture using event sourcing and CQRS patterns. The proposed model allows separating completely the read and write sides for each micro-agent. Therefore, the micro-agents don't need any interaction with other agents of the multiagent system at runtime during reasoning tasks. This reduces the cost of communication between agents and improves the performance of the distributed multiagent system. The separation of the micro-agent query side from the micro-agent command side brings several advantages, such as at startup, the micro-agent can rebuild its own environment by replaying the events from the event store. So, if necessary, we can shut down any micro-agents without any negative consequences for other micro-agents. In the case of a load increase due to the number of consultation requests issued from the UI agent, the query part can be scaled independently of the command part. This should optimize resources and increase multiagent system performance. The proposed event-driven framework allows agents to react to all or part of events from the event store to rebuild their environment representation and knowledge database. This feature provides agents a great deal of autonomy with possibilities of going offline, stopping their activity when necessary, or even recovering after failure or loss of data. The event bus model of this framework provides both the pub/sub and the polling patterns. So, we can use any broker model as the event bus and event store. In the current implementation, the KAFKA broker is used as the event bus and as the event store. In this platform micro-agents can easily and safely perform distributed transactions using the mechanism of the Saga pattern provided as a building block of the framework. To prove the performance of the proposed model, we have implemented a blockchain application base on micro-agents, and some stress tests have been performed. The results prove the high performance of the proposed model. However, some constraints need to be taken into consideration for the implementation of this model relating to the organizational level and the complexity of the development model, which require the developer to use other software engineering patterns to avoid code duplication and to minimize the code size needed to write. In future research works we plan to valid the proposed model to resolve some complex problems such as cyber-physical systems for special education relating to an H2020 CybSPEED project.

ACKNOWLEDGEMENT

This project has received funding from the European Union's Horizon 2020 research and innovation program under the Marie Skłodowska-Curie grant agreement No. 777720.

REFERENCES

Ali, M. and S. Bagchi, "Probabilistic Normed Load Monitoring in Large Scale Distributed Systems Using Mobile Agents", Future Gener Comput Syst, vol. 96, pp. 148–167, 2019. doi: 10.1016/j.future.2019.01.053

Basmi, W., A. Boulmakoul, L. Karim and A. Lbath, "Distributed and Scalable Platform Architecture for Smart Cities Complex Events Data Collection: Covid19 Pandemic Use Case", J Ambient Intell Humaniz Comput, vol. 12, no. 1, pp. 75–83, 2021.doi: 10.1007/s12652-020-02852-9. Epub 2021 Jan 2. PMID: 33425049; PMCID: PMC7778396.

Bellifemine, F., A. Poggi, and G. Rimassa, "Developing Multiagent Systems with a FIPA-compliant Agent Framework" Software - Practice & Experience, vol. 31(2), pp. 103–128, 2001.

Bellifemine, F., G. Caire and D. Greenwood, Developing Multi-agent Systems with JADE, Wiley Series in Agent Technology, Wiley, 2007. https://www.wiley.com/en-br/journal-editors

Benchara, F. Z., and M. Youssfi, "A New Scalable Distributed k-means Algorithm Based on Cloud Micro-services for High-Performance Computing", Parallel Computing, vol. 101, p. 102736, 2021. ISSN 0167-8191.

Betts, D., J. Dominguez, G. Melnik, F. Simonazzi, M. Subramanian, and G. Young, Exploring CQRS and Event Sourcing: A Journey into High Scalability Availability and Maintainability with Windows Azure, Microsoft, 2013.

Birajdar, P. M., K. Ujede, R. Yalawar, K. H. Biradar, Z. Khan, and S. Chaudhari, "Bidirectional Hadoop Kakfa Managing Messaging Bus", International Research Journal of Engineering and Technology (IRJET), vol. 03, no. 03, March 2016.

Cao, W., M. J. A. Patwary, P. Yang, X. Wang and Z. Ming, "An Initial Study on the Relationship between Meta Features of Dataset and the Initialization of NNRW", Proc. Int. Joint Conf. Neural Netw. (IJCNN), pp. 1–8, Jul. 2019

Chen, C., K. Li, A. Ouyang, and K. Li, "A Parallel Approximate SS-ELM Algorithm Based on MapReduce for Large-scale Datasets", Journal of Parallel and Distributed Computing, vol. 108, pp. 85–94, 2017.ISSN 0743-7315.

Debski, A., B. Szczepanik, M. Malawski, S. Spahr and D. Muthig, "A Scalable, Reactive Architecture for Cloud Applications," in IEEE Software, vol. 35, no. 2, pp. 62–71, March/April 2018, doi: 10.1109/MS.2017.265095722.

Erb, B., Meißner, D., Ogger, F. and Kargl, F. "Log Pruning in Distributed Event Sourced Systems", In Proceedings of the 12th ACM International Conference on Distributed and Event- based Systems. ACM. 2018, pp. 230–233.

Hajoui, Y., O. Bouattane, M. Youssfi and E. Illoussamen, "Q-learning and ACO Hybridisation for Real-time Scheduling on Heterogeneous Distributed Architectures", International Journal of Computational Science and Engineering, vol.20(2), pp.225–239, 2019.

Iranpour, E., and S. Sharifian, "A Distributed Load Balancing and Admission Control Algorithm Based on Fuzzy Type-2 and Game Theory for Large-scale Saas Cloud Architectures", in: Future Generation Computer Systems, vol. 86, Elsevier, pp. 81–98, 2018.

Lima, S., J. Correia, F. Araujo, and J. Cardoso, "Improving Observability in Event Sourcing Systems", Journal of Systems and Software, vol. 181, 111015, ISSN 0164-1212, 2021. https://doi.org/10.1016/j.jss.2021.111015.

Michelson, B. M., "Event-Driven Architecture Overview: Event-Driven SOA Is Just Part of the EDA Story," Patricia Seybold Group, 2006.

O'Connor, R.V., Peter Elger and Paul M. Clarke, "Continuous Software Engineering – A Microservices Architecture Perspective", Journal of Software: Evolution and Process, vol. 29, no. 11, 2017.

Rajani, S. and N. Garg, "A Clustered Approach for Load Balancing in Distributed Systems", Int. J. Mobile Comput. & Appl., vol. 2, pp. 2393–9141, 2015. SSRG-IJMCA, ISSN: 2393-9141.

Ranjan, R., L. Wang, A. Y. Zomaya, D. Georgakopoulos, X.-H. Sun and G. Wang, "Recent Advances in Autonomic Provisioning of Big Data Applications on Clouds", IEEE Trans. Cloud Comput., vol. 3, no. 2, pp. 101–104, Apr. 2015.

Taylor, H., A. Yochem, L. Phillips and F. Martinez, Event-driven Architecture: How SOA Enables the Real-time Enterprise, Boston: Pearson Education, 2009.

Tragatschnig, S., S. Stevanetic and U. Zdun, "Supporting the Evolution of Event-driven Service-oriented Architectures Using Change Patterns", Inf. Softw. Technol., vol. 100, pp. 133–146, 2018.

Umer, M., B. Mahesh, L. Hanson, M.R. Khabbazi and M. Onori, "Smart Power Tools: An Industrial Event-Driven Architecture Implementation", Procedia CIRP, vol. 72, pp. 1357–1361, 2018.

Veiga, J., R. R. Expósito, G. L. Taboada and J. Touriño, "Flame-MR: An Event-driven Architecture for MapReduce Applications", Futur. Gener. Comput. Syst., vol. 65, pp. 46–56, 2016.

Wan, X., Xinjie Guan, Tianjing Wang, Guangwei Bai and Baek-Yong Choi, "Application Deployment Using Microservice and Docker Containers: Framework and Optimization", Journal of Network and Computer Applications, vol. 119, pp. 97–109, 2018.

Wang, Z., L. Sui, J. Xin, L. Qu and Y. Yao, "A Survey of Distributed and Parallel Extreme Learning Machine for Big Data," IEEE Access, vol. 8, pp. 201247–201258, 2020, doi: 10.1109/ACCESS.2020.3035398.

Yuan, S.-T., and M-R Lu, "A Value-Centric Event Driven Model and Architecture: A Case Study of Adaptive Complement of SOA for Distributed Care Service Delivery". Expert Systems with Applications, vol. 36(2), Part 2, pp. 3671–3694, 2009, ISSN 0957-4174, https://doi.org/10.1016/j.eswa.2008.02.024.

3 Intelligent Distributed Computing Paradigm

Emergence, Challenges and Future Research Directions

Avinash Kumar, Snigdha Kashyap,
Anuj Kumar, Saptadeepa Kalita,
Sonal Shriti, and Khushi Samridhi

CONTENTS

3.1   INTRODUCTION

The current era of modernization is witnessing rapid developments in the field of machine intelligence, and intelligent systems are one of those crucial developments. Intelligent systems, as the name suggests, are the systems which are intelligently able to communicate with other machines and systems, with the use of the internet. It can be an embedded machine or a machine or computer connected to the internet. It can also be referred to as a collection of interconnected components (devices or collections of devices) which can communicate and interact with each other, to efficiently collect as well as analyze large sets of data (Torres et al. 2018). Intelligent systems also possess the flexibility and ability of connecting to networks, learning as well as adapting to heterogeneous data. Moreover, they instill flexibility, adaptability, and comprehensive power to comprehend information, providing robustness and improved efficiency to solve complexities. Due to their behavior, intelligent

DOI: 10.1201/9781003255635-3

systems are widely used, especially for remote monitoring and in security applications (Grzonka et al. 2018). One can relate intelligent systems to technologies such as artificial intelligence (AI) and machine learning (ML); expert systems, chatbots, facial recognition systems, and image processing systems are some of the most important applications of intelligent systems (Tzafestas 2020). AI can be categorized as humanistic and rationalistic. Humanistic AI focuses on the nature of machines to exhibit human-like behavior and enables machines to act and think similar to human beings. On the other hand, rational AI intends to replicate human behavior and form it as a basis for machines to behave intelligently. Domains which have benefited from intelligent systems and AI include education, healthcare, robotics, automation, surveillance, etc. Intelligent systems may or may not be self-adapting or self-learning in nature, but exhibit machine intelligence and reasoning to a large extent. These systems are efficient enough to construct methods and techniques to solve real-time, complex problems and are able to provide users with correct and reliable outputs. As per one of the studies, intelligent systems can be categorized based on the following processes: combination, integration, fusion, and association (Li et al. 2018). The given processes comprise hybrid architecture models. Intelligent systems arise as a product of hybrid systems, which combine the power of multiple individual technologies, including neural networks, fuzzy logic, some genetic algorithms, and similar ones (Posey et al. 2020). This hybridization improves the performance of machines along with throughput and reliability and thus provides a better vision of intelligence. Technologies such as cloud computing act as an important ingredient to the intelligent computing paradigm.

Also known as "on-demand computing", cloud computing is a technology which provides network, computations, and storage on the basis of a pay-as-you-go model, where users pay for the service only when they require it. Rather than using on-premises data centers, it involves the delivery of hostable services on the internet. Hence, it eliminates the need for a separate physical server room for offices and industries to store and manage data. The data can be conveniently stored on the cloud and made accessible to the desired users. Cloud computing aims at providing efficient and reliable services to the end users, and also ensuring the quality of service (QoS) parameters of the services (Potluri and Subba Rao 2017). It is a major leap in the field of computing and has transformed the way information and computational resources are managed. Cloud computing can be treated as a dynamic infrastructure for information technology (IT), and is considered a replacement for grid computing. It consists of three major types of service models: Infrastructure-as-a-Service (IaaS), Platform-as-a-Service (PaaS), and Software-as-a-Service (SaaS). In IaaS, the complete infrastructure of cloud services is on a pay-as-you-go basis, including the network, storage, and virtualization resources. PaaS provides a platform, i.e., software and hardware tools for creating and building software applications. SaaS enables end users to run their applications smoothly over the internet, without being concerned about installing and configuring them on their systems. It can be said that SaaS allows end users to run applications via a third party. Additionally, with more and more advancements and upgrades in the field of cloud computing, more models have been designed and practically implemented, one of them being Everything- or Anything-as-a-Service (XaaS) (Ganapathy 2020). XaaS encompasses a myriad of services, which can be provided to end users and enhance their convenience and ease of access, such as Desktop-as-a-Service (DaaS), Backup-as-a-Service (BaaS), and many more. Other than service models, cloud computing constitutes four deployment models: public cloud, private cloud, community cloud, and hybrid cloud. The hybrid cloud is a combination of both public and private clouds, and is considered the most widely adopted deployment model of cloud computing by the industries. Cloud computing provides seamless benefits to the end users, such as ease of access to information, consistent data backup and restore, low costs due to the pay-as-you-go model, and enhanced collaboration of people (Yu et al. 2016). However, it possesses some limitations, including vendor lock-in and dependency of users on the internet. Cloud computing can be enhanced with distributed computing, a newer paradigm shift.

Distributed computing deals with the computations and working with distributed systems, i.e., multiple devices and multiple systems which act and operate individually as software components but compose a single system. It allows different individual software components to share resources among multiple devices or systems. In other words, it can be stated that multiple computers and servers are linked together over a provided network as a cluster (Kuan Hon et al. 2021). The distributed systems can be located either locally or spread over distances in geographical areas. Each computer or device within a distributed system is referred to as a node. The nodes involved in a distributed system generally communicate using message-passing techniques (Lamport 2019). The aim as well as purpose of distributed computing is to allow functions including resource sharing and communication for end users. Some prominent examples of distributed computing or distributed systems are the World Wide Web (WWW), the internet and intranets, e-mail, cellular networks, etc. The Advanced Research Projects Agency Network (ARPANET) can be considered one of the earliest examples of the distributed computing paradigm, and it also opened the door for newer technologies and advancements in this field (Crocker 2019). A crucial advantage of distributed computing is that due to the involvement of multiple devices, failure of one device or computer does not impact the whole system. It is broadly used in order to solve complex computational problems; the computational tasks are divided into smaller subtasks, which are solved or executed by one or more computers separately (Alwakeel 2021). Distributed computing possesses some typical characteristics, out of which some are mentioned as follows: the system should be able to tolerate and withstand failures. The system must be flexible and adaptable to manage connections among different types of devices and communication links. Additionally, each computer or node in the system may or may not have a complete knowledge about the system. Distributed computing is different from parallel computing due to the fact that parallel computing involves a shared memory concept where different devices within the system share a common memory space (Jonas et al. 2017). On the contrary, in distributed computing, each node possesses a private memory space to work on, and information exchange takes place with the help of message passing. There are some major architectures constructed for distributed systems, some of which include client-server architecture, peer-to-peer (P2P) architecture, three-tier architecture and n-tier architecture (Sethi et al. 2020). Distributed computing, due to its versatile characteristics, has a major role in enabling the Internet of Things (IoT) and communication in an IoT network.

IoT has been a boon to modern-day technologies, and has emerged as a great player in technology. IoT is also referred to as the Internet of Everything, which is a modern technology paradigm and allows setting up of a global network of devices and machines (Goyal et al. 2020). "Things" in IoT may refer to living entities, including humans and plants, as well as nonliving entities such as food, clothing, gadgets, etc., which are connected within a network and are able to exchange information in real time. The IoT devices are able to interact and communicate with each other. Smart homes, smart cities, and similar applications are a boon of IoT. IoT involves specialized devices called sensors and actuators, which are responsible for gathering and analyzing information from the surroundings. Sensors are responsible for sensing, gathering, and collecting data from the surrounding environment. Actuators, as the name suggests, act on the data provided by the sensors and process as well as analyze it to further deliver it to the respective devices. Both sensors and actuators may be embedded in physical objects and are connected with the help of wired or wireless networks (Bhushan and Sahoo 2017). These physical objects can act as a tool for responding to real-time problems and situations. There exist some prerequisites which are required for setting up an IoT network and architecture. The prerequisites include hardware (sensors, actuators, embedded hardware for communication), middleware (computing and on-demand storage tools), and visualization, as well as interpretation tools for presenting the results (Aldowah et al. 2017). Radio frequency identification (RFID) is the inspiration for devising IoT, using which objects can be tagged to discover information (Puri and Bhushan 2019). Moreover, there is an emphasis on automating the tasks in the real world by reducing human intervention to operate IoT devices. According to statistics, around more than 50 million devices are connected to the IoT networks by 2020. IoT is given multiple aliases by different organizations and people, including the following; embedded intelligence, Internet of Objects, Web of Things, and many more. IoT is a revolution in the current era, having an objective to unify the devices and systems across the world with a common architecture (Malik et al. 2019). As of now, IoT does not have a unified, common architecture due to its vast and broad nature (Arora et al. 2019). However, many studies have represented some models to implement a practical, unified IoT architecture. Projects such as the FP7 project for research, International Telecommunication Union (ITU) Architecture, IoT Forum Architecture, and many more have contributed to the field of IoT and its frameworks.

This chapter addresses the approaches to state and explain distributed computing and intelligence paradigm as best it can. The major contributions can be summarized as follows:

- This chapter discusses the prerequisites of distributed computing.
- This chapter clearly articulates a distributed computing paradigm.
- This chapter details the applications of distributed computing with respect to intelligent systems.
- This chapter also tries to discuss the challenges faced in implementing distributed intelligent systems.

The remainder of this chapter is organized as per the following scheme: Section 3.2 provides a detailed overview of distributed computing and its relation with intelligent systems to form distributed intelligent systems. Section 3.3 addresses intelligent distributed paradigm. Section 3.4 details the constraints in intelligent distributed computing. Finally, the last section concludes, Section 3.5 the chapter, followed by stating the future research directions in the domain of distributed intelligent computing.

3.2 OVERVIEW OF DISTRIBUTED INTELLIGENT CONTROL TECHNOLOGY

Intelligence can be closely linked to technologies, including AI, ML, and IoT. Distributed intelligent control technology is a by-product of distributed artificial intelligence (DAI) and hence makes use of AI-based technologies including neural networks, fuzzy logic, reinforcement learning, and many more advanced and intelligent control techniques and approaches for designing control systems, especially in manufacturing and production industries (Yigitcanlar et al. 2020). Industrial control systems are critical in nature, as they have underlying physical components, devices, and processes in an environment which is susceptible to disruptions and disturbances of various kinds. An intelligent distributed control system is more in sync with the physical systems in an industry. In the model of distributed intelligent control, the components and agents of a distributed system are autonomous and cooperative in nature. This model is used for simulating real-time systems, for instance, real-time manufacturing systems and controls (Ding et al. 2018). Distributed intelligent control systems are highly helpful in assisting dynamic industries, as that of manufacturing. One of the challenges of distributed intelligent control systems is that they need to be compatible with the real-time conditions of associated hardware and devices in order to ensure operations are safe, secure, and convenient. Such challenges have given birth to control models such as closed-loop distributed control, which emphasize a flexible as well as adaptable approach for intelligent controls (Ding et al. 2019). Some prominent approaches to constructing distributed control systems include a global control methodology and an agent-based control methodology (Gómez-Cruz et al. 2017).

From the view point of manufacturing, the global control technique has cognitive systems as the basis. This model of intelligent control can be categorized into two types of planning: global and local (Dixit et al. 2018). Global planning includes construction of a rough plan and allocation of tasks. On the other hand, local planning is composed of creating a local resource graph based on the sequence of tasks to be processed according to the rough plan (Woods and Branlat 2017). The sequence for resource processing is decided according to the shortest path algorithms. Lastly, the global planning layer is updated with the help of production data. In case of any failure, the intercommunication among devices helps in overcoming the same through negotiation and task allocation (Zhang et al. 2018). Due to its characteristics, the method of global planning control is considered suitable and efficient mainly for maintaining stability of the order flow in manufacturing units.

Second, the agent-based distributed control method works in a different manner and makes use of different agents, who assist with the task allocation and execution. Agents are the key elements for maintaining the intelligence of these systems (Nguyen et al. 2017). In this control method, the distributed system is responsible for selecting the task and loading it into the manufacturing system with the help of a load agent. The load agent then assigns it to the agent for spare parts. The spare parts agent assists in the identification of work piece(s) and then detects and decides the operation required to be done (El Zaatari et al. 2019). With the help of the bidding process, the agent searches for necessary equipment for operation and uses it for transportation and processing actions. In the end, the products are dismounted from the production system using an unloading agent (Conti and Orcioni 2019). The use of agents makes this kind of control system an interactive one, thereby ensuring smooth and reliable operations. Additionally, agents also improve the system's flexibility and make it effective enough to deal with technical or physical failures. However, a system with fully distributed and intelligent control mechanisms is quite complex to be practically implemented in practice, since both communication and computing need to be looked at. In order to detail an agent-based distributed intelligent control mechanism, it can be divided into certain categories: production planning and control, predictive scheduling, and optimization (Fu et al. 2021). Planning and scheduling form a major part of an intelligent distributed control system. In production planning and control, the production data are categorized as product, knowledge, and resource. Physical entities collect and share the product information reliably. More importantly, RFID is used for data collection and product tagging due to having support for wireless transmission, dynamic reading-writing ability, enhanced capacity, and much more. Thus, production planning ensures closed-loop control and ensures system stability. Second, predictive scheduling involves prediction and estimation of processing time and carrying out transportation of materials in advance. This helps reduce waiting time for equipment and also ensures continuous and consistent operations. Moreover, predictive scheduling encompasses decision making of agents responsible for storage and upkeep of machines. There is an involvement of a buffer for prescheduling the tasks (Zhang et al. 2016). The tasks are updated or cancelled before their execution in order to avoid failure and create an impact on their execution. Finally, the optimization involves optimizing the processing and transportation time and costs. It is achieved by considering and studying and optimizing the real-time data, rather than relying on the static data. Hence, optimization can be understood as the method which allows processing time, transportation time, and the data of production to update timely as per the current status of manufacturing units, when the production system is active and running (Turker et al. 2019). This ensures the adaptable nature of distributed intelligent control systems.

In the current scenario, manufacturing and production systems require the following characteristics: agility, flexibility, traceability, reliability, and adaptability (Bueno et al. 2020). All of these characteristics need to be efficiently reflected in production systems for their smarter and intelligent operations and control. For ensuring the same, adaptive, reliable, and accurate methodologies for production planning as well as control are required. A distributed intelligent control system should be capable enough to implement distributed control, as well as perform the scheduling of tasks and processes dynamically in real time using agents. Overall, a distributed intelligent control system exhibits certain characteristics, including communication among the work pieces, devices, and other entities; decision making, learning, and ability; and techniques of handling disturbances. The disturbances can namely consist of equipment failures and emergency conditions. Despite having multiple advantages, since every coin has two faces, so do these intelligent systems. The decision processes involved in transportation make use of the shortest transportation time strategy of scheduling. Moreover, the discussed models are least validated as of now. Fully intelligent distributed systems are yet to be practically implemented (Steen and Tanenbaum 2016). Also, developing and constructing such systems for handling real-time situations is complex, and addressing the complexity of real-time systems poses a challenge to their implementation. Most importantly, being a field of research and due to its nature, the implementation of intelligent control systems requires specific skills and human resources. Distributed intelligent systems can be further improved to have enhanced decision-making power for real-time situations and data, on the basis of systems' operating status (Zheng et al. 2018). They can be enhanced with the capability to deal with heterogeneity in software and hardware and conduct dynamic configuration and reconfiguration in real time (Zhou 2020). As a whole, an intelligent distributed control system should not only emphasize and consider what actions are to be taken for a system but also consider and deploy the decision making for how and when the actions are required to be taken. With immense research and development in this domain, the industries and industrial systems are largely benefitting in terms of convenience, reliability, and throughput. Thus, it can be stated that distributed computing has made an appreciable contribution in designing, implementing, and revolutionizing distributed intelligent control systems.

3.3 PRACTICAL IMPLEMENTATIONS OF THE INTELLIGENT DISTRIBUTED PARADIGM

The paradigm of intelligent distributed computing and control has several applications which have transformed the industry of today and tomorrow. Some of the prominent applications can be explained in the following subsections.

3.3.1 DELIVERY MODELS FOR INTELLIGENT SYSTEMS USING CLOUD COMPUTING

Cloud services are accessible and purchased by users in any of the following forms: software, platform, or infrastructure. Hence, there are mainly three delivery models for cloud computing and services: SaaS, PaaS, and IaaS. SaaS is considered the highest level of abstraction of cloud services for end users, according to which the cloud service providers provide end users with applications. That is, SaaS acts as a portal or interface for application delivery. A software stack is provided, which includes the operating system (OS), a middleware along with databases and web servers, and a cloud application instance (Shahrestani 2017). All the stack elements are provided in a virtual machine. This enables convenience for end users, who then do not need to bother about managing the underlying complexities of the infrastructure, including network, storage, OS, and individual application functionalities, and the platform for using the applications. Overall, SaaS is efficient in provisioning user-specific applications as per the required configurations by end users. Moreover, SaaS eliminates the need for end users to maintain the software and also simplifies software development and testing. It is considered to be the simplest cloud delivery model. Second, the PaaS cloud delivery model is used for making the cloud more conveniently programmable. Along with the software stack, the cloud service providers also provide the platform, which is scalable in nature. PaaS is beneficial to developers as well as testers because they do not need to be concerned about managing the infrastructure, which makes application deployment easier for them. However, PaaS is less suitable for portable software and applications, which need specific software or hardware to execute. Service level agreements (SLAs) are used by PaaS providers for ensuring cloud services' accessibility, reliability, and accuracy (Gebreslassie and Prasad Sharma 2019). Lastly, IaaS offers complete infrastructure, including virtual resources (network, compute, storage, communication), on a pay-as-you-go basis. IaaS providers provision the software stack along with the virtual machines to the users for designing a custom server similar to a physical server, or applications as per the requirements. Moreover, in the case of multiple users, using IaaS, roles and privileges can be granted to the end users for controlling the system and performing specific operations and actions (Ngo et al. 2016). IaaS is generally used by start-ups as well as small and medium businesses (SMBs) in order to deal with a dynamic market and expansion of changes and advancements. However, the SLAs for IaaS are more complex compared to that in PaaS. It is so because most of the control of the users' systems lies with the IaaS providers. Some of the great players in cloud service delivery are Amazon Web Services (AWS), Google Cloud, Salesforce, IBM Cloud, and Alibaba (Paul et al. 2020). Cloud service delivery models have transformed the way businesses are established and industries operate. These models have highly contributed and assisted to the implementation and operations of distributed intelligent systems and have paved a way for more advancements in this domain.

3.3.2 Intelligent System Using Jungle Computing

Jungle computing is a new paradigm which can be referred to as the involvement of several distributed systems for computation and storage in order to achieve high performance and reduce the complexity of programming and implementation (Bazai et al. 2021). Jungle computing possesses a great deal of heterogeneity. It is so because jungle computing constitutes clouds, grids, devices and computers, clusters, supercomputers, and mobile devices (Tychalas and Karatza 2017). There may be involvement of graphic processing units (GPU) and other accelerators. In traditional distributed systems, there is a constellation of multiple core processors. Moreover, the platforms can have different interfaces, middleware, policies, and mechanisms for security and protection (Lin et al. 2017). For instance, a cloud architecture needs pay-as-you-go frameworks, whereas a machine that is standalone in nature has permanent availability. This heterogeneous nature of traditional distributed systems allows them to become more complex and vaster in terms of programming and usage. The modern-day intelligent systems based on jungle computing consist of a myriad of computation resources, including supercomputers, grids, mobile devices, the cloud, and many more. These resources are connected to each other via wireless or wired networks. Each resource consists of some nodes, which are mainly responsible for computation tasks, as they contain their own local processing units and memory (Portugal et al. 2017). Due to these characteristics, the nodes are able to communicate over a connection, either a local area network (LAN) or wide area network (WAN). This interconnection of networks of resources is referred to as a "jungle", and the computation is called jungle computing. However, in the worst scenario, jungle computing constitutes almost all of the platforms and interfaces for computation: grid interface, mobile device interface, standalone machine interface, desktop grid interface, cloud interface, supercomputer interface, and cluster interface.

Each of them is a part of the middleware, which allows arbitrary applications to run. The worst-case scenario can be divided into various possible subsets, each subset depicting a realistic situation. Hence, jungle computing needs to ensure that the solutions for real-time problems apply to every possible situation within the system. That is, there are some constraints and challenges associated with jungle computing and its implementation. The arbitrary application which is run onto the architecture may need to be recompiled or rewritten as per the requirements (Coppolino et al. 2019). Thus, the architecture and the underlying software and hardware should be efficient and flexible enough to handle the changes in configurations of software and hardware based on the demands. Different interfaces for middleware need to be available to maintain each resource within the system. Another considerable challenge is to connect and interconnect the resources present in the jungle. Since resources such as clusters and supercomputers are not meant for communicating externally, it poses a challenge to jungle computing implementation (Pal 2021). Jungle computing, hence, needs to possess certain important characteristics. There should be abstraction of resources, and the architecture should be completely resource-independent (Zarrin et al. 2017). A jungle should be interoperable, which is possible with the use of interfaces. A jungle must have robust connectivity, and the resources should be named uniquely in order to be identified for service provision (Qin et al. 2016). A jungle must be adaptable to withstand heterogeneity and a dynamic environment and resources. It should be fault tolerant in nature. A jungle also needs to be integrated well to the external entities (Hu et al. 2017).

3.3.3 Intelligent System Using fog Computing

Fog computing can be referred to as a newer paradigm shift in the domain of computing and the cloud. The fog is an additional layer to the cloud, which enables a decentralized computing architecture for faster service delivery and efficient computation. The term was coined by Cisco in 2012 (Mahmud et al. 2017). To elaborate, fog computing is a secure, virtualized, systematic, and efficient platform which is network-integrated. Fog computing is used for providing networking services, computation, and storage to the end users between the network end points and cloud servers or data centers. It encompasses the concept of a network edge, which consists of specialized devices, where the data, applications, and processes are concentrated (Negash et al. 2017). Hence, they do not exist completely in the cloud; instead, they are kept close to the end users, at the network edge, which can be considered the ultimate objective of fog computing. As mentioned, the fog is an additional layer to the cloud; thus, the fog is located beneath the cloud, like an intermediary, and delivers the services to the end users or devices faster and more efficiently. However, the computation in the fog takes place separately from and outside the cloud. Fog computing ensures that the services, including networking, computation, storage, applications, and data, are provisioned to the end users through the network edge in a distributed manner. Due to the characteristics of the fog, it enables data control in several network edge points. It also integrates the cloud services with applications and transforms traditional data centers into a distributed cloud platform (Goyal et al. 2020). Fog computing can be closely related to edge computing, which is an enhancement to the traditional computing architecture and provides a boost to fog computing as well. Edge computing enables the knowledge generation as well as analytics to take place at the data source instead of at devices and resources (Goyal et al. 2020). However, it demands resources which may or may not be always connected to a network (Yu 2016). Such resources can be tablets, sensors, mobile devices, or laptops. Studies are still going on for researching the computation near the network edge, out of which important technologies include cloudlets, and MediaBroker (Nithya et al. 2020). The need for fog computing and edge computing emerged due to increasing needs of users to get data, analytics, and services faster, conveniently, and in a distributed fashion. But it is a myth that fog computing can replace cloud computing. Fog computing is simply an increment to the cloud, which brings cloud services closer to the users. Fog computing is generally implemented over technologies which are commonly existing, such as the content delivery network (CDN), but with an adaptability to sustain complicated services and computation too (Zolfaghari et al. 2021). Overall, the characteristics and key features of fog computing can be mentioned as follows: It increases the proximity of information to the end users. It is organized hierarchically in order to support scalability and low latency. It is decentralized in nature instead of being concentrated at a geographical location. It possesses the support for mobility and constitutes a larger number of computing nodes. Fog computing enables real-time interactions among the nodes and devices. The fog forms a basis for IoT and XaaS developments as well.

3.3.4 Distributed Computing for Defense

The military possesses various challenges: the mission requirements can be unpredictable, dynamic, and complex. Distributed computing aims to overcome the limitations in the military and is the future of military missions and applications. Distributed computing has beneficial and appreciable characteristics, including rapid and faster accessibility to networks, media, storage, and data as well as applications and services. Along with these features, distributed computing can reduce costs and use optimum resources to provision the applications and services, thereby supporting and contributing to several domains. Of such domains, the military is one of them. However, the use of distributed computing has been witnessed since early times and is not a new aspect. Distributed computing is optional in military and related applications and services. Studies and research are still underway to practically implement the integration of the cloud and distributed computing logics to military applications. With this integration, military capabilities can be equipped with cost-effectiveness, flexibility, accessibility, and efficiency (Sharma et al. 2020). Distributed cloud computing can further introduce several novelties and battlefield opportunities. Distributed computing, along with the military cloud, can contribute to crucial applications in future warfare (Hossain et al. 2019). It can be firmly stated that military cloud and distributed computing are the great players in the military and possess the potential to increase situation-based awareness in the battlefields for military personnel. Moreover, it facilitates the setting up of information superiority. Distributed computing in the military domain gives rise to a new technology, as an enhancement of IoT, called "Internet of Battlefield Things" (IoBT) (Hu et al. 2019). IoBT is basically a set of interconnected and interdependent entities which can consist of storage, networks, devices, algorithms, and infrastructure. The given entities should be able to meet multiple goals, tasks, or missions. Additionally, they must operate automatically and possess adaptability and flexibility. Lastly, the entities should also be able to deal with complex services in the battlefield intelligently. The battlefield services can be processing or capturing data, predicting behavior of people and surroundings, analyzing the physical surroundings, and many more (Xi and Kamhoua 2020). The future of IoBT needs to be consistent with resilience, safety, and responsiveness. It also needs to be equipped with a short decision loop, as military operation decisions are hierarchical in nature, which requires a faster follow-up and response (Zhu and Majumdar 2020). The command chain reduces response speed, thereby affecting critical operations. Thus, a properly structured IoBT can assist military operations and functions in a highly efficient manner and enable as well as empower responsiveness and readiness. Along with computation, distributed computing also requires security and stability of military functions. Any malfunction, partial or complete loss of resources and entities, cannot be witnessed, and thus security is a factor that cannot be compromised. The modern distributed architecture of the military cloud learns from time to time along with experiences in order to improve itself and enhance its capabilities (Mutlag et al. 2019). If all the expected capabilities and characteristics are introduced in a distributed military architecture, it can largely reduce costs and improve the efficiency of missions and tasks as well.

3.4 CONSTRAINTS IN INTELLIGENT DISTRIBUTED COMPUTING

Although the distributed computing paradigm has contributed and aims to contribute a lot to the industries and technology, it encounters some unavoidable challenges and obstacles in its implementation and practice, which can be described here, one by one.

3.4.1 DISTRIBUTED SCHEDULING PROBLEMS

The demands of customers and users in real time are increasing at a rapid rate. Along with increasing demands, the competition scenario has widened in the market and has become fierce. Thus, production and manufacturing systems are emphasized to be more robust, strong, flexible, and scalable in order to cope up with the expanding globalization. Moreover, it is required that multiple production units be set up in a distributed fashion in different, separate geographical locations. In the present scenario, the distributed production systems are applied in various kinds of industries of manufacturing. The industries include automotive industries, steel production industries, food processing industries, chemical preparation industries, and many more (Wei et al. 2019). Two things are highly crucial and impact a production industry significantly: modeling and scheduling (Lu et al. 2020). Intensive studies and research are conducted in both the fields for industries. Out of the two, scheduling is a topic of concern, and it plays a crucial role in the reduction of the cost of production as well as improvement of customer satisfaction. To emphasize scheduling, various scheduling problems in production and services are focused for solutions and optimization. Some of the prominent scheduling problems include the following: job shop scheduling, flow shop scheduling, parallel machine scheduling, and several derived problems (Sonnenschein and Hinrichs 2017). Apart from the scheduling problems mentioned earlier, recent studies have devised newer kinds of scheduling problems, that is distributed scheduling (Arunarani et al. 2019). Distributed scheduling primarily emphasizes distributed intelligent systems. It has attracted wide attention, and more and more emphasis is laid on its optimization and modeling. Distributed scheduling finds its applications in many areas and fields, including distributed computation systems, intelligent systems, configuration and computation systems which are geographically distributed, etc. Specifically, in manufacturing, distributed scheduling can simultaneously schedule all the encompassing factories (Leusin et al. 2018). Comparing distributed scheduling to the traditional scheduling algorithms, they are more complex in nature and possess complicated characteristics. Along with task allocation among the associated machines, distributed scheduling can conduct task allocation or assignment to different distributed factories. On the contrary, traditional scheduling algorithms can only schedule tasks for specific factories and not all the factories in a system. Distributed scheduling also considers factors such as workload balance along with time, throughput, minimization of tardiness, etc., for the purpose of decision making (Kong et al. 2018). Considering workload balance helps in reducing total costs of production in a distributed environment of manufacturing. Factories are generally located remotely; therefore, sharing and exchange of information among the factories become difficult. This poses obstacles for several manufacturing functions such as change in delivery time, machine breakdown, etc. Due to certain constraints and its complex characteristics, distributed scheduling is a major domain for research. Driss et al. dedicatedly worked on job shop distributed scheduling techniques and summarized the techniques for optimizing and solving them (Chaouch et al. 2017). Many studies have summarized different types of scheduling problems according to shop types, solution techniques, and objective functions (Zhang et al. 2017). Optimization and approximation of scheduling problems in distributed systems are emphasized and need to be considered because they are incompatible with the traditional optimization algorithms. There exist some advanced techniques of approximation, such as evolutionary algorithms (EI) and swarm intelligence (SI) (Manne 2018). However, these algorithms may or may not guarantee optimization of distributed scheduling techniques and algorithms. Classical shop scheduling is yet another kind of scheduling problem which is generally linked to distributed scheduling to analyze distributed intelligent systems in depth (Gao et al. 2020). Other problems with which distributed scheduling problems are generally integrated include planning problems, resource allocation problems, distribution problems, and vehicle problems. Presently, there is a focus on real-life scenarios and problems where distributed scheduling can be applied. The real-life manufacturing scenarios refer to constraints which can relate to flow time, production, etc. Constraints relating to flow time include transportation time, processing time, stochastic processing time, and setup time (Bocewicz et al. 2016). On the other hand, constraints related to production consist of no wait, blocking, no idle, lot streaming, limited buffer, etc. Other than these two constraints, there exist other constraints such as task re-entrance, task heterogeneity and heterogeneity in production shops, etc. (Liu et al. 2020). One requires various techniques to solve in order to model the distributed scheduling problems, especially for intelligent systems. Mathematical programming is one of the key ingredients for modeling the problems of distributed scheduling. Distributed scheduling for intelligent systems demands mathematical modeling and programming for dispatch rules (DR), Simple Heuristics (SH), and algorithms of SI and EA (Gao et al. 2019).

Distributed scheduling is considered more complicated as compared to conventional and traditional scheduling algorithms. It is so because of the fact that distributed scheduling involves first making a decision for task assignment, followed by deciding on the allocation, as well as sequence on machines in the factories. Hence, they can be considered nondeterministic polynomial or NP-Hard problems (Azhir et al. 2019). The general approaches to solve and optimize them are based on exact heuristics and approaches, including SI and EA. A distributed scheduling algorithm may target a single or multiple objectives depending on the requirements. In recent times, more emphasis is laid on multiobjective distributed scheduling algorithms and techniques. Overall, distributed scheduling has opened the door for major advancements in the manufacturing industries.

3.4.2   Security Issues in Distributed Computing

Security is one the major aspects for any system, be it intelligent or static, and cannot be overlooked. In terms of distributed intelligent systems, security encompasses several important elements, which mainly include authentication, authorization, and availability (Sinha et al. 2017). Other important attributes to consider are integrity, encryption mechanisms, system protection, etc. In traditional systems, since they comprised single individual entities, security management was moderate. However, in the case of distributed intelligent systems, due to the distributed presence of manufacturing units, factories, and production systems, security management is a matter of concern and needs to be emphasized as a priority. It creates a need for shared responsibilities and their adherence among the authorities. In general, the attacks which take place within distributed intelligent systems include but are not limited to the following: eavesdropping, masquerading, attacks to information integrity (tampering), replay attacks, and denial of service (DoS) attacks (Jaitly et al. 2017). Eavesdropping is an attack where an illegal third party, or adversary, intercepts the secret communication between two legal communicating parties and passively listens to the messages transmitted across the secret channel. Masquerading, as the name suggests, is an attack wherein an adversary disguises himself as a legitimate user by gaining their identity and makes himself a part of secret communication. Message tampering includes modification or interception of messages or information by an attacker for fulfilling a malicious intention. It is a major threat to information integrity and consistency. A replay attack is an attack where the adversary replays the network packets and stealthily intercepts or eavesdrops on the communication. One of the most fatal attacks is the attack to compromise availability, commonly referred to as a flooding or DoS attack. According to this, the adversary floods the network with multiple anonymous network packets in order to block or deny a specific service requested by the user over the targeted network. An enhanced form of DoS attack is the distributed denial of service (DDoS) attack. In DDoS attacks, the adversary makes use of controlled entities called "bots" to individually flood the network and is more impactful as compared to DoS attacks.

These attacks should be prevented from impacting or adversely affecting a distributed intelligent system. It is very important to safeguard the reliability of the system. Security for intelligent systems encompasses both traditional security terms and privacy (Varshney et al. 2019). There are several approaches for implementing and ensuring security in distributed intelligent systems. The approaches are authentication-based, trust-based, access control–based, cryptography-based, pattern-based, policy-based, and many others (Ballardie and Crowcroft 2020).

Authentication-based security schemes include the schemes to ensure that authenticity and legality are maintained for an intelligent system; that is, no user with a fake or wrong identity is able to access the system. Some authentication-based approaches encompass remote client security and multifactor authentication schemes such as two-factor authentication (2FA) and three-factor authentication (3FA) (Haseeb et al. 2020). Trust-based schemes consist of trust-based models, which highly contribute to the development of security systems. They also include risk management and evaluate trust levels. Access control–based security approaches involve provision of roles and privileges to users in order to control and limit the access to different components of an intelligent system. Cryptography-based schemes look after encryption techniques, including public key cryptography and Extended Markup Language (XML) binding technologies (Majumder et al. 2020). Pattern-based schemes are the schemes which identify the signatures and patterns of attacks and safeguard intelligent systems. Lastly, other approaches for security include some security models proposed in various studies, e.g., mobile agent–based models, heterogeneity approaches for security, delegate object–based models, etc. (Xie et al. 2019). Mobile agent–based models possess the ability to identify and defend intelligent systems from intrusions and threats. The heterogeneity approach proposed by Xie et al. specifically ensures security in distributed scheduling systems (Weiming 2019). The delegate object–based security approach emphasizes challenges of distributed characteristic of intelligent systems and paradigms (Sengupta et al. 2020).

Some critical security challenges can be listed as follows: identifying methods for security risk assessment, integrating encryption techniques for secure communication, applying web applications in implementing security services, developing security metrics, and monitoring security and threats to it (Mohanta et al. 2020). The world is heading towards more secure distributed intelligent systems, and hence their reliability, trust, security, and safety need to be safeguarded efficiently. In the future, new approaches are finding a way to design and practically implement secure distributed intelligent systems.

3.5 CONCLUSION AND FUTURE RESEARCH DIRECTIONS

So far, this chapter has summarized the distributed paradigm and the key components to design them intelligently to practically implement an intelligent distributed paradigm for industries. This chapter has laid an emphasis especially on production and manufacturing industries to relate distributed computing and associated technologies. Additionally, the chapter has emphasized the application and association of distributed computing with intelligent systems and focused on intelligent distributed technologies. The chapter has also described the prerequisite technologies which are required for understanding and implementing the intelligent distributed systems and computing, such as IoT, AI, ML, distributed computing, and cloud computing. The chapter has also discussed the major applications of distributed computing in the domain of intelligent systems. It described the cloud-based delivery models for distributed intelligent systems, which are SaaS, PaaS, and IaaS. Additionally, a reference to XaaS has also been given in this chapter. Apart from distributed computing and cloud computing, the chapter also detailed the related and enhanced versions, including fog computing and edge computing. Fog computing is just an additional layer to the cloud, which allows faster access to networks, computation, and services. The chapter described and detailed the newer concept of computing known as jungle computing, which involves multiple devices and connectivity links among distributed systems across different geographical areas. The military also holds a scope for the contribution of intelligent distributed computing, which was mentioned in this chapter. Lastly, the chapter successfully addresses the challenges and issues which persist and create an obstacle for the implementation of intelligent systems with distributed computing. Some of the major challenges in the implementation of distributed computing are mentioned in the chapter, namely issues in distributed scheduling and modeling and security issues (attacks, threats, and intrusions) in a distributed computing architecture. Prominent distributed scheduling problems have been mentioned specifically in the chapter, along with an overview of modeling and solving the problems using mathematics and computing algorithms.

In the near future, this study can be effectively referred to for the formulation of distributed computing algorithms for intelligent systems. Moreover, this study can be extended in order to deeply study the intelligent distributed computing paradigm, which can assist with the practical implementation of intelligent distributed computing in different domains. The future prospects of this study also encompass the study of newer technologies, such as AI, IoT, ML, fog computing, and edge computing in depth with respect to distributed computing and intelligent systems. The study successfully forms a basis and aims to contribute to future studies and research in the field of distributed computing and distributed intelligent systems.

REFERENCES

Aldowah, H., Ul Rehman, S., Ghazal, S., & Naufal Umar, I. (2017). Internet of things in higher education: A study on future learning. Journal of Physics: Conference Series, 892, 012017. https://doi.org/10.1088/1742-6596/892/1/012017.

Alwakeel, A. M. (2021). An overview of fog computing and edge computing security and privacy issues. Sensors, 21(24), 8226. https://doi.org/10.3390/s21248226.

Arora, A., Kaur, A., Bhushan, B., & Saini, H. (2019). Security Concerns and Future Trends of Internet of Things. 2019 2nd International Conference on Intelligent Computing, Instrumentation and Control Technologies (ICICICT). DOI: 10.1109/icicict46008.2019.8993222.

Arunarani, A. R., Manjula, D., & Sugumaran, V. (2019). Task scheduling techniques in cloud computing: A literature survey. Future Generation Computer Systems, 91, 407–415. https://doi.org/10.1016/j.future.2018.09.014.

Azhir, E., Jafari Navimipour, N., Hosseinzadeh, M., Sharifi, A., & Darwesh, A. (2019). Deterministic and non-deterministic query optimization techniques in the cloud computing. Concurrency and Computation: Practice and Experience, 31(17). https://doi.org/10.1002/cpe.5240.

Ballardie, T., & Crowcroft, J. (2020). Multicast-specific security threats and counter-measures. Proceedings of the Symposium on Network and Distributed System Security, 2–16. https://doi.org/10.1109/ndss.1995.390649.

Bazai, S. U., Ghafoor, M. I., Aqeel, M., & Roomi, M. S. (2021). Kernel virtual machine based high performance environment for grid and jungle computing. 2021 2nd International Informatics and Software Engineering Conference (IISEC), 1–6. https://doi.org/10.1109/iisec54230.2021.9672355.

Bhushan, B., & Sahoo, G. (2017). A comprehensive survey of secure and energy efficient routing protocols and data collection approaches in wireless sensor networks. 2017 International Conference on Signal Processing and Communication (ICSPC), 294–299. DOI: 10.1109/cspc.2017.8305856.

Bocewicz, G., Nielsen, I. E., & Banaszak, Z. A. (2016). Production flows scheduling subject to fuzzy processing time constraints. International Journal of Computer Integrated Manufacturing, 29(10), 1105–1127. https://doi.org/10.1080/0951192x.2016.1145739.

Bueno, A., Godinho Filho, M., & Frank, A. G. (2020). Smart production planning and control in the industry 4.0 context: A systematic literature review. Computers & Industrial Engineering, 149, 106774. https://doi.org/10.1016/j.cie.2020.106774.

Chaouch, I., Driss, O. B., & Ghedira, K. (2017). A modified ant colony optimization algorithm for the distributed job shop scheduling problem. Procedia Computer Science, 112, 296–305. https://doi.org/10.1016/j.procs.2017.08.267.

Conti, M., & Orcioni, S. (2019). Cloud-based sustainable management of electrical and electronic equipment from production to end-of-life. International Journal of Quality & Reliability Management, 36(1), 98–119. https://doi.org/10.1108/ijqrm-02-2018-0055.

Coppolino, L., D'Antonio, S., Mazzeo, G., & Romano, L. (2019). A comparative analysis of emerging approaches for securing Java software with Intel SGX. Future Generation Computer Systems, 97, 620–633. https://doi.org/10.1016/j.future.2019.03.018.

Crocker, S. D. (2019). The ARPANET and its impact on the state of networking. Computer, 52(10), 14–23. https://doi.org/10.1109/mc.2019.2931601.

Ding, D., Han, Q.-L., Wang, Z., & Ge, X. (2019). A survey on model-based distributed control and filtering for industrial cyber-physical systems. IEEE Transactions on Industrial Informatics, 15(5), 2483–2499. https://doi.org/10.1109/tii.2019.2905295.

Ding, D., Han, Q.-L., Xiang, Y., Ge, X., & Zhang, X.-M. (2018). A survey on security control and attack detection for industrial cyber-physical systems. Neurocomputing, 275, 1674–1683. https://doi.org/10.1016/j.neucom.2017.10.009.

Dixit, S., Fallah, S., Montanaro, U., & Stevens, A. (2018). Trajectory planning and tracking for autonomous overtaking: State of the art and future prospects. Annual Reviews in Control, 45, 76–86. https://doi.org/10.1016/j.arcontrol.2018.02.001.

El Zaatari, S., Marei, M., Li, W., & Usman, Z. (2019). Cobot programming for collaborative industrial tasks: An overview. Robotics and Autonomous Systems, 116, 162–180. https://doi.org/10.1016/j.robot.2019.03.003.

Fu, Y., Hou, Y., Wang, Z., Wu, X., Gao, K., & Wang, L. (2021). Distributed scheduling problems in intelligent manufacturing systems. Tsinghua Science and Technology, 26(5), 625–645. https://doi.org/10.26599/tst.2021.9010009.

Ganapathy, A. (2020). Everything-as-a-service (xaas) in the world of technology and trade. American Journal of Trade and Policy, 7(3), 91–98. https://doi.org/10.18034/ajtp.v7i3.555.

Gao, D., Wang, G.-G., & Pedrycz, W. (2020). Solving fuzzy job-shop scheduling problem using de algorithm improved by a selection mechanism. IEEE Transactions on Fuzzy Systems, 28(12), 3265–3275. https://doi.org/10.1109/tfuzz.2020.3003506.

Gao, K., Cao, Z., Zhang, L., Chen, Z., Han, Y., & Pan, Q. (2019). A review on swarm intelligence and evolutionary algorithms for solving flexible job shop scheduling problems. IEEE/CAA Journal of Automatica Sinica, 6(4), 904–916. https://doi.org/10.1109/jas.2019.1911540.

Gebreslassie, Y., & Prasad Sharma, D. (2019). Client-side metering and monitoring of service level agreement for green cloud services. SSRN Electronic Journal. https://doi.org/10.2139/ssrn.3358354.

Gómez-Cruz, N. A., Loaiza Saa, I., & Ortega Hurtado, F. F. (2017). Agent-based simulation in management and organizational studies: A survey. European Journal of Management and Business Economics, 26(3), 313–328. https://doi.org/10.1108/ejmbe-10-2017-018.

Goyal, S., Sharma, N., Bhushan, B., Shankar, A., & Sagayam, M. (2020). IoT enabled technology in secured healthcare: Applications, challenges and future directions. Cognitive Internet of Medical Things for Smart Healthcare, 25–48. DOI: 10.1007/978-3-030-55833-8_2.

Goyal, S., Sharma, N., Kaushik, I., Bhushan, B., & Kumar, A. (2020). Precedence & issues of IoT based on edge computing. 2020 IEEE 9th International Conference on Communication Systems and Network Technologies (CSNT), 72–77. DOI: 10.1109/csnt48778.2020.9115789.

Grzonka, D., Jakóbik, A., Kołodziej, J., & Pllana, S. (2018). Using a multi-agent system and artificial intelligence for monitoring and improving the cloud performance and security. Future Generation Computer Systems, 86, 1106–1117. https://doi.org/10.1016/j.future.2017.05.046.

Haseeb, K., Almogren, A., Ud Din, I., Islam, N., & Altameem, A. (2020). SASC: Secure and authentication-based sensor cloud architecture for intelligent internet of things. Sensors, 20(9), 2468. https://doi.org/10.3390/s20092468.

Hossain, M. S., Ramli, M. R., Lee, J. M., & Kim, D.-S. (2019). Fog radio access networks in Internet of Battlefield Things (IoBT) and load balancing technology. 2019 International Conference on Information and Communication Technology Convergence (ICTC), 750–754. https://doi.org/10.1109/ictc46691.2019.8939722.

Hu, P., Dhelim, S., Ning, H., & Qiu, T. (2017). Survey on fog computing: Architecture, key technologies, applications and open issues. Journal of Network and Computer Applications, 98, 27–42. https://doi.org/10.1016/j.jnca.2017.09.002.

Hu, Y., Sanjab, A., & Saad, W. (2019). Dynamic psychological game theory for secure internet of battlefield things (IoBT) systems. IEEE Internet of Things Journal, 6(2), 3712–3726. https://doi.org/10.1109/jiot.2018.2890431.

Jaitly, S., Malhotra, H., & Bhushan, B. (2017). Security vulnerabilities and countermeasures against jamming attacks in wireless sensor networks: A survey. 2017 International Conference on Computer, Communications and Electronics (Comptelix), 559–564. DOI: 10.1109/comptelix.2017.8004033.

Jonas, E., Pu, Q., Venkataraman, S., Stoica, I., & Recht, B. (2017). Occupy the cloud. Proceedings of the 2017 Symposium on Cloud Computing, 445–451. https://doi.org/10.1145/3127479.3128601.

Kong, L., Li, H., Luo, H., Ding, L., & Zhang, X. (2018). Sustainable performance of just-in-time (JIT) management in time-dependent batch delivery scheduling of precast construction. Journal of Cleaner Production, 193, 684–701. https://doi.org/10.1016/j.jclepro.2018.05.037.

Kuan Hon, W., Millard, C., & Singh, J. (2021). Cloud technologies and services. Cloud Computing Law, 3–26. https://doi.org/10.1093/oso/9780198716662.003.0001.

Lamport, L. (2019). Time, clocks, and the ordering of events in a distributed system. Concurrency: The Works of Leslie Lamport, 21(7), 179–196. https://doi.org/10.1145/3335772.3335934.

Leusin, M., Frazzon, E., Uriona Maldonado, M., Kück, M., & Freitag, M. (2018). Solving the job-shop scheduling problem in the industry 4.0 era. Technologies, 6(4), 107. https://doi.org/10.3390/technologies6040107.

Li, W., Song, H., & Zeng, F. (2018). Policy-based secure and trustworthy sensing for internet of things in smart cities. IEEE Internet of Things Journal, 5(2), 716–723. https://doi.org/10.1109/jiot.2017.2720635.

Lin, J., Yu, W., Zhang, N., Yang, X., Zhang, H., & Zhao, W. (2017). A survey on internet of things: Architecture, enabling technologies, security and privacy, and applications. IEEE Internet of Things Journal, 4(5), 1125–1142. https://doi.org/10.1109/jiot.2017.2683200.

Liu, Y., Yu, F. R., Li, X., Ji, H., & Leung, V. C. (2020). Blockchain and machine learning for communications and networking systems. IEEE Communications Surveys & Tutorials, 22(2), 1392–1431. https://doi.org/10.1109/comst.2020.2975911.

Lu, Y., Liu, C., Wang, K. I.-K., Huang, H., & Xu, X. (2020). Digital twin-driven smart manufacturing: Connotation, reference model, applications and research issues. Robotics and Computer-Integrated Manufacturing, 61, 101837. https://doi.org/10.1016/j.rcim.2019.101837.

Mahmud, R., Kotagiri, R., & Buyya, R. (2017). Fog computing: A taxonomy, survey and future directions. Internet of Things, 103–30. https://doi.org/10.1007/978-981-10-5861-5_5.

Majumder, S., Ray, S., Sadhukhan, D., Khan, M. K., & Dasgupta, M. (2020). ECC-CoAP: Elliptic curve cryptography based constraint application protocol for internet of things. Wireless Personal Communications, 116(3), 1867–1896. https://doi.org/10.1007/s11277-020-07769-2.

Malik, A., Gautam, S., Abidin, S., & Bhushan, B. (2019). Blockchain technology-future of IoT: Including structure, limitations and various possible attacks. 2019 2nd International Conference on Intelligent Computing, Instrumentation and Control Technologies (ICICICT) 1, 1100–1104. DOI: 10.1109/icicict46008.2019.8993144.

Manne, J. R. (2018). Swarm intelligence for multi-objective optimization in engineering design. Encyclopedia of Information Science and Technology, Fourth Edition, IGI Global, US.239–250. https://doi.org/10.4018/978-1-5225-2255-3.ch022.

Mohanta, B. K., Jena, D., Satapathy, U., & Patnaik, S. (2020). Survey on IOT security: Challenges and solution using machine learning, artificial intelligence and Blockchain technology. Internet of Things, 11, 100227. https://doi.org/10.1016/j.iot.2020.100227.

Mutlag, A. A., Abd Ghani, M. K., Arunkumar, N., Mohammed, M. A., & Mohd, O. (2019). Enabling technologies for fog computing in healthcare IOT Systems. Future Generation Computer Systems, 90, 62–78. https://doi.org/10.1016/j.future.2018.07.049.

Negash, B., Rahmani, A. M., Liljeberg, P., & Jantsch, A. (2017). Fog computing fundamentals in the internet-of-things. Fog Computing in the Internet of Things, 3–13. https://doi.org/10.1007/978-3-319-57639-8_1.

Ngo, C., Demchenko, Y., & de Laat, C. (2016). Multi-tenant attribute-based access control for cloud infrastructure services. Journal of Information Security and Applications, 27–28, 65–84. https://doi.org/10.1016/j.jisa.2015.11.005.

Nguyen, T. L., Tran, Q.-T., Caire, R., Gavriluta, C., & Nguyen, V. H. (2017). Agent based distributed control of islanded microgrid — Real-time cyber-physical implementation. 2017 IEEE PES Innovative Smart Grid Technologies Conference Europe (ISGT-Europe), 1–6. https://doi.org/10.1109/isgteurope.2017.8260275.

Nithya, S., Sangeetha, M., Prethi, K. N., Sahoo, K. S., Panda, S. K., & Gandomi, A. H. (2020). SDCF: A software defined cyber foraging framework for cloudlet environment. IEEE Transactions on Network and Service Management, 17(4), 2423–2435. https://doi.org/10.1109/tnsm.2020.3015657.

Pal, D. (2021). Comparative analysis of cluster, utility, grid and cloud computing. SSRN Electronic Journal. https://doi.org/10.2139/ssrn.3849896.

Paul, P. K., Aremu, B., Aithal, P. S., Saavedra, R., & Sinha, R. R. (2020). A study on cloud computing and service market: International context with reference to India. Asian Journal of Managerial Science, 9(1), 52–56. https://doi.org/10.51983/ajms-2020.9.1.1629.

Portugal, D., Pereira, S., & Couceiro, M. S. (2017). The role of security in human-robot shared environments: A case study in ROS-based surveillance robots. 2017 26th IEEE International Symposium on Robot and Human Interactive Communication (RO-MAN), 981–986. https://doi.org/10.1109/roman.2017.8172422.

Posey, C., Kandel, A., & Langholz, G. (2020). Fuzzy hybrid systems. Hybrid Architectures for Intelligent Systems, 173–197, 1st Edition, CRC Press. https://doi.org/10.1201/9781003068075-10.

Potluri, S., & Subba Rao, K. (2017). Quality of service based task scheduling algorithms in cloud computing. International Journal of Electrical and Computer Engineering (IJECE), 7(2), 1088. https://doi.org/10.11591/ijece.v7i2.pp1088-1095.

Puri, D., & Bhushan, B. (2019). Enhancement of security and energy efficiency in WSNs: Machine learning to the rescue. 2019 International Conference on Computing, Communication, and Intelligent Systems (ICCCIS), 120–125. DOI: 10.1109/icccis48478.2019.8974465.

Sengupta, J., Ruj, S., & Das Bit, S. (2020). A comprehensive survey on attacks, security issues and blockchain solutions for IOT and IIoT. Journal of Network and Computer Applications, 149, 102481. https://doi.org/10.1016/j.jnca.2019.102481.

Sethi, R., Bhushan, B., Sharma, N., Kumar, R., & Kaushik, I. (2021). Applicability of industrial IoT in diversified sectors: Evolution, applications and challenges. In: Kumar, R., Sharma, R., Pattnaik, P.K. (eds) Multimedia Technologies in the Internet of Things Environment. Studies in Big Data, vol 79. Springer, Singapore. https://doi.org/10.1007/978-981-15-7965-3_4

Shahrestani, S. (2017). Assistive IOT: Deployment scenarios and challenges. In: Internet of Things and Smart Environments. Springer, Cham https://doi.org/10.1007/978-3-319-60164-9_5.

Sharma, P. K., Park, J. H., & Cho, K. (2020). Blockchain and federated learning-based distributed computing defence framework for sustainable society. Sustainable Cities and Society, 59, 102220. https://doi.org/10.1016/j.scs.2020.102220.

Sinha, P., Jha, V. K., Rai, A. K., & Bhushan, B. (2017). Security vulnerabilities, attacks and countermeasures in wireless sensor networks at various layers of OSI reference model: A survey. 2017 International Conference on Signal Processing and Communication (ICSPC), 288–293, doi: 10.1109/CSPC.2017.8305855.

Sonnenschein, M., & Hinrichs, C. (2017). A distributed combinatorial optimization heuristic for the scheduling of energy resources represented by self-interested agents. International Journal of Bio-Inspired Computation, 10(4), 1. https://doi.org/10.1504/ijbic.2017.10004322.

Torres, B. S., Ferreira, L. R., & Aoki, A. R. (2018). Distributed intelligent system for self-healing in smart grids. IEEE Transactions on Power Delivery, 33(5), 2394–2403. https://doi.org/10.1109/tpwrd.2018.2845695.

Turker, A., Aktepe, A., Inal, A., Ersoz, O., Das, G., & Birgoren, B. (2019). A decision support system for dynamic job-shop scheduling using real-time data with simulation. Mathematics, 7(3), 278. https://doi.org/10.3390/math7030278.

Tychalas, D., & Karatza, H. (2017). High performance system based on cloud and beyond: Jungle computing. Journal of Computational Science, 22, 131–147. https://doi.org/10.1016/j.jocs.2017.03.027.

Tzafestas, S. G. (2020). Introduction to intelligent robotic systems. Intelligent Robotic Systems, 3–42. https://doi.org/10.1201/9781003066279-2.

van Steen, M., & Tanenbaum, A. S. (2016). A brief introduction to distributed systems. Computing, 98(10), 967–1009. https://doi.org/10.1007/s00607-016-0508-7.

Varshney, T., Sharma, N., Kaushik, I., & Bhushan, B. (2019). Architectural Model of Security Threats & Their Countermeasures in IoT. 2019 International Conference on Computing, Communication, and Intelligent Systems (ICCCIS), 424–429. DOI: 10.1109/icccis48478.2019.8974544.

Wei, M., McMillan, C. A., & de la Rue du Can, S. (2019). Electrification of industry: Potential, challenges and outlook. Current Sustainable/Renewable Energy Reports, 6(4), 140–148. https://doi.org/10.1007/s40518-019-00136-1.

Weiming, S. (2019). Learning in agent-based concurrent design and Manufacturing systems. Multi-Agent Systems for Concurrent Intelligent Design and Manufacturing, Edition 1, 77–100, CRC Press https://doi.org/10.4324/9780203305607_chapter_5.

Woods, D. D., & Branlat, M. (2017). Basic patterns in how adaptive systems fail. Resilience Engineering in Practice, 127–143. CRC press https://doi.org/10.1201/9781317065265-10.

Xi, B., & Kamhoua, C. A. (2020). A hypergame-based defense strategy toward cyber deception in internet of battlefield things (IoBT). Modeling and Design of Secure Internet of Things, 59–77. https://doi.org/10.1002/9781119593386.ch3.

Xie, L., Ding, Y., Yang, H., & Wang, X. (2019). Blockchain-based secure and trustworthy internet of things in SDN-enabled 5G-vanets. IEEE Access, 7, 56656–56666. https://doi.org/10.1109/access.2019.2913682.

Yigitcanlar, T., Desouza, K. C., Butler, L., & Roozkhosh, F. (2020). Contributions and risks of Artificial Intelligence (AI) in building smarter cities: Insights from a systematic review of the literature. Energies, 13(6), 1473. https://doi.org/10.3390/en13061473.

Yu, Y. (2016). Mobile edge computing towards 5G: Vision, recent progress, and open challenges. China Communications, 13(Supplement2), 89–99. https://doi.org/10.1109/cc.2016.7833463.

Yu, Y., Cao, R. Q., & Schniederjans, D. (2016). Cloud computing and its impact on service level: A multi-agent simulation model. International Journal of Production Research, 55(15), 4341–4353. https://doi.org/10.1080/00207543.2016.1251624.

Zarrin, J., Aguiar, R. L., & Barraca, J. P. (2017). Hard: Hybrid adaptive resource discovery for jungle computing. Journal of Network and Computer Applications, 90, 42–73. https://doi.org/10.1016/j.jnca.2017.04.014.

Zhang, D., Ma, Y., Zheng, C., Zhang, Y., Hu, X. S., & Wang, D. (2018). Cooperative-competitive task allocation in edge computing for delay-sensitive social sensing. 2018 IEEE/ACM Symposium on Edge Computing (SEC), 243–259. https://doi.org/10.1109/sec.2018.00025.

Zhang, H., Tang, D., Huang, T., & Xu, C. (2016). An agent based intelligent distributed control paradigm for manufacturing systems. IFAC-PapersOnLine, 49(12), 1549–1554. https://doi.org/10.1016/j.ifacol.2016.07.800.

Zhang, J., Ding, G., Zou, Y., Qin, S., & Fu, J. (2017). Review of job shop scheduling research and its new perspectives under industry 4.0. Journal of Intelligent Manufacturing, 30(4), 1809–1830. https://doi.org/10.1007/s10845-017-1350-2.

Zheng, P., Wang, H., Sang, Z., Zhong, R. Y., Liu, Y., Liu, C., Mubarok, K., Yu, S., & Xu, X. (2018). Smart manufacturing systems for industry 4.0: Conceptual framework, scenarios, and future perspectives. Frontiers of Mechanical Engineering, 13(2), 137–150. https://doi.org/10.1007/s11465-018-0499-5.

Zhou, J. (2020). Real-time task scheduling and network device security for complex embedded systems based on deep learning networks. Microprocessors and Microsystems, 79, 103282. https://doi.org/10.1016/j.micpro.2020.103282.

Zhu, L., & Majumdar, S. (2020). Decision letter for "an invisible warfare with the internet of battlefield things: A literature review". Human behavior and emerging technologies, vol. 3(2), 255–260. https://doi.org/10.1002/hbe2.231/v2/decision1.

Zolfaghari, B., Srivastava, G., Roy, S., Nemati, H. R., Afghah, F., Koshiba, T., Razi, A., Bibak, K., Mitra, P., & Rai, B. K. (2021). Content delivery networks. ACM Computing Surveys, 53(2), 1–34. https://doi.org/10.1145/3380613.

4 Emerging Paradigm of Urban Computing

Challenges, Applications, and Future Research Directions

Saptadeepa Kalita, Avinash Kumar,
Anuj Kumar, Snigdha Kashyap,
Parma Nand, and Nandita Pokhriyal

CONTENTS

DOI: 10.1201/9781003255635-4

4.1 INTRODUCTION

Big Data has increasingly come to be a center figure in modern science and business. These data are generally generated from e-mails, click streams, multimedia data, online transactions, posts, logs, social networking data, sensors, mobile phones, and their applications (Verma and Pandey 2022). Data traditionally are stored in and managed through databases, but the sheer amount of it has grown to become so massive that it has become an incrementally difficult task to capture, store, manage, share, and analyze this data. Around 5 exabytes of data were created up to the year 2003. Today it takes but a two days' period of time to generate data of this scale. In 2012, the statistics showed an expansion in the world's digital data stats amounting to 2.72 zettabytes (Mahmud et al. 2020). It has been predicted that these stats would only continue to grow with a trend of seeing twice the amount every two years, which would reach to about 16 zettabytes (possibly more) by 2019 (Usman et al. 2020). While in the past processing information related to genomes could take up to a decade, now it hardly takes a week. Multimedia data is another facet that adds a lot of weight on the internet's data traffic and has been predicted to increase by 73% in the next few years (MacEachren 2017). Google alone has got about 9 million of these servers to keep up with the increasing data rates. The huge variety apart from quantity is what actually makes Big Data. The variety of it basically refers to the variety of sources it is generally obtained from, namely, unstructured, semi-structured, and structured (Lv et al. 2017). Structured data are the easiest to deal with—store, manage, and process—since they come already tagged. Unstructured data, on the other hand, pose a problem since they tend to be random. Semi-structured data doesn't contain formally defined fields, but the data tend to be tagged (Feasel 2020). The present volume of these data is already in the range of exabytes and zettabytes soon to reach yottabytes, outstripping the traditional data management techniques by a large extent (Tyagi et al. 2021). Velocity is required in order to work with data that have time-associated constraints; for such time-limited applications, these 4 data must be used as soon as they enter the organization's sphere of interference so as to maximize value (Bala 2022). Potential barriers to working with Big Data include privatization of business sponsorship, cost, inexperienced staff, difficult-to-design systems for analytic tasks, and lack of updated database software in analytics (Tiwari et al. 2019). Many hesitate to dabble in Big Data because of exhaustive analytics and since managing such a huge amount of data inherently poses a problem, high cost being the least of those (Mostajabi et al. 2021). The organizations that have added some sort of Big Data analyzing techniques to their modus operandi have seen benefits such as straight business insights, better aimed marketing, client-based segmentation, and better odds at identifying and implementing sales and marketing tactics. Problems are also faced while replacing or updating analytic platforms, such as updated platforms not fitting to the type of data at hand, slow loading of data, and IT being sidestepped due to enormous demand (Rawat et al. 2021). Urbanization is precisely what has led to such ginormous growth in the data trends. The rapid progress of urbanization has led to modernization of the lives of many people living in cities, as well as engendered massive challenges like environmental pollution, congestion of traffic, waste of resources, and increased energy consumption (Bhushan et al. 2020). The possibility to tackle these challenges has been brought by the most emerging interdisciplinary field: urban computing. Urban computing relates computer science with conventional city-related domains such as transportation systems, civil engineering, and environmental conditions as well as ecological conditions in the context of urban spaces (Meneguette and Boukerche 2020). Urban computing is very useful in connecting data for management and sensing data from sensors and helps to provide undisrupted services in order to persistently improve human life and operations of the smart and traditional city, as well as the environment (Haque et al. 2021). It is observed that around 50% of the world's population lives in urban areas, and it is expected that the population in cities would double in the next 40 years (Slovokhotov and Echenique 2018). Thus, effective planning of urbanization can build an intelligent ecosystem.

Urban computing is a multidimensional framework which includes key technologies like data mining techniques, processing heterogenous data, data visualization, and perception technology. The general framework for most urban computing consists of four layers: city perception layer, data management or processing layer, data analysis layer, and service layer (Manchanda et al. 2020). In the first layer, acquisition of data takes place. Generally, the type of data that is collected in this phase probes people's mobility (Mora et al. 2019). Social media data posted by people are also captured. In order to perceive data from the city, various perception devices are deployed that sense the data and upload them such as global positioning system (GPS), flow of the traffic, noise detectors, and transportation plans (Li 2021). The second layer of the framework deals in preprocessing the data by labeling them. The data on people's mobility and social media are organized in a systematic way, giving some indexing that includes spatiotemporal information as well as texts for supporting efficient data analytics (White and Clarke 2020). The third layer is concerned with analyzing the preprocessed data and finding out the anomaly of people's mobility. This anomaly can be identified by comparing the people's mobility with the original patterns. The last layer of the framework is the service provider layer, which basically allows the people to make responses in advance by issuing early warnings of anomalies based on the locations and descriptions (Ordoñez-García et al. 2017). This chapter attempts to critically analyses and address urban computing. In summary, the major contributions of this chapter are as follows:

- This work presents the need for dealing with Big Data in correlation with urban computing.
- This work outlines the complexity of data emerging in urban computing.
- The work covers vital constraints and challenges faced in the implementation of urban computing.
- The work focuses and discusses various crucial applications of urban computing The remainder of the chapter is organized as follows. Section 4.2 presents the various sources of urban computing data. Section 4.3 provides in-depth analysis of the urban computing ecosystem. Section 4.4 introduces the challenges in urban computing. Section 4.5 presents application areas of urban computing. Finally, Section 4.6 presents a conclusion, followed by future research directions.

4.2   URBAN COMPUTING DATA

The Big Data that are produced by perception devices have various utilities in different domains of urban computing. Bid Data include healthcare data, GPS data, transport data, radio frequency identification (RFID) data, Wi-Fi data, etc. (Lu et al. 2021). These Big Data are gathered, fused, and analyzed to study the activities and mobility of the people living in urban areas. This process is called urban event detection (Arora et al. 2020). The accurate analysis of these data can provide a significant amount of information about the people that can help to build up proper urban application planning (Goyal et al. 2021). There are various data sources that are used to gather in the urban event detection stage. Some of the data sources used in the urban applications are discussed next.

4.2.1   PHONE NETWORK DATA

Call record data (CRD) are generated when people exchange voice or text messages and data usages (2G/3G/4G/5G) (Ahmad et al. 2020). The attributes that are generated from this CRD can be the phone numbers of the calling as well as the receiving parties, the call duration, and the location of both the parties. Using these data, one can study the behavioral analysis of the person or set up a network between different communicators (Arora et al. 2020). Apart from this, the CRD data are more concerned about the user's location via a periodic paging signal. The cellular system uses a triangle positioning algorithm to detect a phone's location based on the nearby two or three base stations. All these data contain private information of the users, so only a portion of the data is shared after the detection of the anomaly as per the rules and regulations of that particular country. With these CDR data, analysis can be made on citywide human mobility, detection of social events, urban modeling of the city, detecting traffic jams, or tourist places and trip analysis.

4.2.2 Radio Frequency (RF)– and Infrared Spectroscopy (IR)–Based Data

Almost all the smartphones are equipped with GPS, Wi-Fi, Bluetooth, or RF sensors. The data generated from this can be utilized to analyze the crowd mobility and size of the crowd in the urban area.

4.2.2.1 Wi-Fi and Bluetooth Data

With high directional RF beams, the RF access points scan smartphones and other RF. The data obtained here are the relative spread strength (RSS) indicator, Media Access Control (MAC) address, smartphone user details, and timestamps. Many redundant data can be obtained, so the RF access point detects only the geolocation of a particular user at a particular timestamp. These data are gathered by placing Wi-Fi or Bluetooth devices with highly directional antennas in a large, crowded area and using them to find trajectories of vehicles, crowd density, and size in urban areas.

4.2.2.2 GPS Data

This type of data is gathered from GPS sensors that receive signals from a minimum of four visible satellites at a particular time. These satellites are responsible to transmit spatiotemporal information at regular intervals. These signals are captured by the GPS sensors and then determine the geolocation to track any crowd (Sundaram et al. 2020).

4.2.2.3 RFID Data

When people travel to different places, a huge volume of commuting data is generated from swiping of cards, i.e., RFID cards at ticket counters of metro, bus, or train stations as well as toll gates. When they swipe the RFID cards, the RFID reader, also known as the interrogator, scans the RFID tags using high-RF beams. Thus, these data consist of timestamps of the people entering or leaving a particular place and the identity and location of the station. From these data, an analysis can be made to monitor crowds and traffic jams around a place (Wemegah and Zhu 2017).

4.2.2.4 Optical Wireless Communication Data

This is a form of short-range optical communication that uses unguided visible IR or ultraviolet (UV) light for carrying signals between a transmitter and receiver. It is also termed visible light communication (VLC). Compared to the RF signals, this optical data transmission is faster and more accurate. But the only disadvantage of this signal is the small area coverage. The data obtained from Optical Wireless Communication (OWC) can be used for crowd-counting information.

4.2.3 Transport Data

In the present scenario, many services in the transportation system are connected through internet and information and communications technology (ICT) structures. Transport data are usually generated from several sources and sensor smart cards or fare cards and transactions made by people for ticket booking, payment at parking, or traveling through web applications or mobile applications. These central database of transport data is real time, and it contains information about the arrival and departure times, delay time, and service time (Antunes et al. 2019). Mostly these data are used for application of smart transport planning and management in urban areas. It also helps in analyzing crowd density and the estimation of mobility of people and traveler trajectories.

4.2.4   DATA FROM VARIOUS SOCIAL MEDIA

Social media data are divided into two parts: one is the social structure, generally denoted by a graph that represents the relation, dependency, and interaction among the users. The second is the user-generated social media data such as photographs, videos, and text, which represent behaviors of the user and their interests (Sinnott et al. 2018). These social media data include data generated from social platforms like Instagram, Facebook, and Twitter. Moreover, these data can be used for sentiment analysis, recommendation systems, and various other social features. Social media also helps in collecting the geotagged information by crawlers and mapped using geographic information systems (GIS) (Xu et al. 2020).

4.2.5   VIDEO DATA

With the rapid growth of urbanization, video cameras are observed to be installed almost all the important places such as train stations, banks, schools and colleges, and tourist places for better security surveillance (Kumar et al. 2021). These cameras are capable of capturing the movement of people in large gatherings. This kind of video data is important for crowd analysis, crowd tracking, crowd density estimation and managing the traffic flow, and security surveillance (Yun and Leng 2020).

4.2.6   SATELLITE AND UNMANNED AERIAL VEHICLES (UAV) IMAGES

The use of satellites and UAV for data acquisition is rapidly evolving. Satellite sensors can capture satellite images of high resolution through their sensors when they fly over a particular area (Peng et al. 2019). The images captured are converted into a mosaic image or can be captured over multiple days. Some of the best imaging satellites include Worldview, Spot, and Landsat. UAVs are popular for generating accurate data at low cost and shorter turnaround time. The UAVs record the real-time activities going on the large area from a particular altitude using the geotagged images and videos. Both the satellite and UAV images provide the information on crowds covering a broad geographic region. One of the factors that affects the data acquisition is the environmental conditions like shadow, clear sky index, and luminescence. These images are used for crowd analysis, visual surveillance in an urban area, and traffic analysis.

4.2.7   HEALTH DATA

These data are generated by hospitals and clinics that include information of the patient's health and diseases (Goyal et al. 2020). Additionally, people use health wearables to monitor the parameters of the body such as heart rate, pulse, and sleep duration. These data from the wearables are sent to the cloud for storage, with the purpose of diagnosing the disease and conducting remote medical examination of the patient (Shin and Shaban-Nejad 2018). These heath data not only help to diagnose disease but also to study the impact of the change of environmental conditions on people's health.

4.2.8 Other Hybrid Data Sources

A combination of two or more of the previously mentioned data sources can give hybrid data. Many works have been performed on these hybrid data to implement a smart urban application. Some of them are crowd monitoring using applications on a smartphone, evaluating the attractiveness of a tourist site at urban areas using CRD and Flickr data, crowd activity, and crowd size estimation are determined by the Bluetooth and GPS location data.

4.3 URBAN COMPUTING ECOSYSTEM

Urban areas are densely populated, hosting many more people than they are supposed to. The rapid growth of urbanization can bring negative impacts to people such as traffic and higher levels of environmental pollution (Zappatore et al. 2019). Due to increasing population in the urban areas, the areas become more polluted and dirtier, compromising people's health and security. In recent years, more advancements have been seen in developing and deploying miniaturized sensors in urban environments that could sense the environmental factors like humidity, air quality, temperature, noise, etc., with higher precision. In this context, concepts of the Web of Things (WoT) give rise to new ideas that can interact with the urban environment and help the people create a smart urban city (Hamdaoui et al. 2020).

The scalable as well as flexible urban computing frameworks have been developed using static artifacts, and in case there is a need for change, manual integration of new sources is performed. The main challenges in the application of smart city development using WoT are dynamicity, heterogeneity, and autonomy. The insufficiency of WoT is found in the absence of a universal, widely prevalent protocol for discovering ad hoc sensors and devices. Finally, the semantic web technologies can assure seamless integration of data.

4.3.1 Vital Elements of the Ecosystem

The input given to a WoT-based ecosystem for application of urban computing includes a combination of data produced through the sensors known as machine sensing and the data generated by people who participate to send information from mobile applications to be stored, known as human sensing. This comprises different components and provides service to build an efficient urban computing application that can discover sensors and data, analysis of real-time computation, and the mobile computing (Wazid, et al. 2021).

4.3.1.1 Sensor Data

WoT is combination of the Web and servers that are embedded on the Web, which include the wide area network (WAN) that is commonly known as the internet. The data streams from WoT include devices that are integrated with the internet and embedded with web servers or in an indirect way by connecting to the gateways. The devices help in sensing the environmental parameters generated by the sensors that can be easily reused and interoperable. Humans as sensors participate to sense data through their smart devices or install applications or by deploying perception devices at homes or other areas to sense the environment and expose their service through the internet. The contribution of data could be made more superfluous when the government bodies and nongovernment organizations contribute to the data by implanting sensor devices in urban areas.

4.3.1.2 Discovering Device, Service, and Sensor Data

This could be only achieved with the help of a wider use of a nonrigid mechanism that should be ubiquitous to the consumer or the user. These techniques should comply with the presently available internet standards, and no significant changes must be required to the technical devices and protocols.

4.3.1.3 Big Data Analysis for CEP

The data can be collected from both online and offline sensors that are deployed in government offices, social media networking, repositories storing environmental data, physical sensors, etc. (Puri and Bhushan 2019). The real-time analysis has to be made from this huge amount of data in order to provide quick recommendations to the people or input to the various smart urban real-life implementations (Satija et al. 2019).

4.3.1.4 Publish Messaging Queues

The sensor stream, which is also known as a publisher, has been abstracted from the end users, which are called subscribers, to attend asynchronous communications, including optimization of total volume of messages exchanged. Messaging queues might be from middleware or could be independent.

4.3.1.5 ICT Technologies

The basic functionality of any ICT technologies such as smartphones is to locate and interact with the environmental surroundings provided by embedded systems. ICT technologies inform the users about the ongoing environmental situation around them (Shawon et al. 2019). These mobile applications work together with the sensory services as urban mashups, integrating web services over the WAN. Some examples of these kinds of applications are time schedules of public transportation systems, weather forecasting, events in city, and notifying police for any accident or incident. Together, all these leads to assist the people of urban area to take informed decisions in their day-to-day life.

The WoT-enabled sensors that sense data in an ecosystem for urban computing can be discovered only through a scalability factor and pervasive mechanism. This discovery of WoT-enabled sensors is carried out by middleware as well as some online platforms that execute analysis on Big Data and Complex Event Processing (CEP) or by smart devices used by an urban area's citizens (Hou et al. 2021). The middleware is capable enough to make advanced reasoning of the problems, and it can even formulate real-time complex event patterns by integrating various data streams along with event services. In the case of online platforms, the mobile applications are used for locating and interacting with the sensors that are present in the nearby locations around them. They also possess the opportunity to merge these services together in order to procure an opportunistic as well as composite service known as urban mashups. A mobile application has the range of sensing only the urban environment, and therefore, these applications are only capable of making local decisions. In order to make any kind of global decision and some more advanced hypothesis, middleware can be deployed to sense the data from the whole urban infrastructure covering a wide area. This technique is the same as cloud and of fog computing. Cloud computing gives much more computational capabilities, and fog computing attempts to reduce the burden by creating local referencing.

4.3.2 REAL-TIME DISCOVERY OF THE WoT

The real challenge in WoT is the discovery of physical entities in real time. Many came up with solutions that were centralized, but they were not very successful; others were decentralized not adopted for application. The Information Retrieval System (IRS) was also not adopted because it lacks scalability in terms of the Web and an additional infrastructure is required (Bhushan et al. 2020). One of the best solutions is microformats, but it will lead to more dependency on the commercial search engines. The Domain Name System (DNS) was also not considered a perfect solution, as it was not very flexible. A novel technique was proposed for the discovery of WoT devices as well as semantic annotated data related to the Internet of Things (IoT) with the use of some web crawlers that act as agents for discovering search engines on WoT (Biswal and Bhushan 2019). First, the method starts with discovering linked data end points with the help of web crawlers that crawl over the World Wide Web (WWW). Second, the linked data points are examined one by one to see if they are associated with an IoT or WoT infrastructure. Analysis is made of the linked data end points that connect with WoT/IoT to obtain metadata as well as other information regarding the IoT/WoT equipment and services that are openly available. Lastly, the data description and services offered by the devices based on IoT/WoT are recorded.

4.3.3 BIG DATA ANALYSIS AND CEP

In urban computing, a huge amount of data is generated from middleware among smart urban infrastructure applications and data from sensors. These Big Data require analysis and processing of complex events. It seems challenging to provide assistance for quality-aware, WoT-enabled distributed systems. Almost all the applications of real-time WoT in urban computing require the capability to make decisions and determine proper handling of missing, diverse, and redundant data. To solve this issue (Khazael et al. 2021) proposed the Automated Complex Event

Implementation System (ACEIS) that serves as an adaptive CEP platform used for urban data streams. The ACEIS architecture comprises three vital elements: an application interface (AI), semantic annotation (SA), and lastly the core. The functionality of the application layer is to interact with the end users and the core module of ACEIS. This application layer also permits the end users to give input to the application as required, and the result is displayed in an intuitive way. The data can be high-level data streams from WoT devices or from a low level.

4.3.4 MOBILE IMPLEMENTATIONS FOR URBAN COMPUTING

For the purpose of creating a smart urban city, various sensors are deployed at various locations in urban areas that inform the end users about the present environmental situations where they are currently. These sensed environmental parameters are sent to the environmental services, and the mobile apps interact with the environmental services to inform about the conditions to the users. In fact, these mobile applications have a range to sense the local environment and are capable of making local decisions. In order to make global decisions, communication is required with middleware like ACEIS that helps in sensing parameters from a wide range over the city infrastructure, making reasoning and decisions at a global level. One of the mobile applications implemented based on urban computing is urban radar. It is specially designed for Android phone users. This application provides the exact user location based on the GPS of the user and Wi-Fi position. The urban radar provides a proximal radial range of meters to kilometers, with the location of the user at the center. In urban cities, a proximal radial range of 10 meters is sufficient, while in outskirts the range can be from hundreds to thousands of meters (Liu et al. 2019).

Another example of mobile applications based on urban computing is the urban mashups. These urban mashups expose the features that have been generated by the sensors as web services. Therefore, it allows the user to create real-time web mashups involving all the entities nearby. The web mashups integrate with the physical mashups, as well as sensors with the use of classic techniques based on web mashups (Cáceres et al. 2020). Physical mashups have the disadvantage of low mobility and unpredictability; therefore, physical mashups cannot be directly implemented in an urban area. Thus, urban mashups have proved to be more opportunistic than physical mashups, as they are capable of adapting in highly dynamic landscapes. The application of urban mashups is in mobile applications, where the main focus is service of the sensors.

4.3.5 SEMANTIC WEB TECHNOLOGIES

The main aim of semantic web technology is to deliver seamless integration of data and reusability of data with less effort (Guedea-Noriega and García-Sánchez 2019). To elucidate, the semantic sensor generated data streams, and lightweight data models are developed based on popular ontologies like a spatial statistical network (SSN) and Web Ontology Language (OWL), the World Wide Web Consortium (W3C), Provenance (PROV), IoT, Simulation of Urban Mobility (SUMO), and OGC GeoSPARQL. The mentioned ontologies are used for describing elements as well as at this of the urban city like pollution sensor devices, spatial information provider, cameras to monitor traffic, smart meters as well as forecasting of weather.

4.4 CHALLENGES IN URBAN COMPUTING

Urban computing is the process of data acquisition from diverse sources, integration of data, and analysis of the data in the urban space. Today, it can be seen that big cities face issues like traffic congestion, environmental pollution, and extensive power consumption. Urban computing provides the solution for this by tackling the Big Data that are generated from different sensors. To implement an ideal urban planning in an urban area, there are three basic challenges which are faced by urban computing. These challenges are discussed in the following subsections.

4.4.1 URBAN SENSING AND DATA ACQUISITION

One of the primary challenges is the collection of urban data from various sources distributed citywide at different locations in a nonintrusive and continuous manner. It seems quite easy to monitor the flow of the traffic on the roads in the city, but continuously updating the traffic flow is a difficult task, as there are not many sensors distributed over the city. Deploying new sensors or infrastructures to sense traffic data would be a burden to the city. A new concept of deploying humans as a sensor would help to tackle this issue. When a person posts any update on their social media sites or their GPS traces, it helps to understand and analyze the events going on around them. Another instance is the GPS traces of the people traveling in a city, which can reflect the patterns of the traffic and the anomalies (Sethi et al. 2020). As there are two sides of a coin, there also exist some pros and cons of deploying humans as sensors. Some of these challenges are discussed next.

4.4.1.1 Increased Energy Consumption and Privacy

The concept of nontrivial is used for applications that are based upon participatory sensing. In this context, the user continuously submits their data using smart devices (Kaiser et al. 2018). Also, they tend to optimize the energy consumption of their smart device together with protection of their private data during the process of sensing. The synchronization between energy, use of data, and privacy is optimally achieved.

4.4.1.2 Nonuniform Distribution of Sensors

Traditional sensors can be placed anywhere in an area, and these sensors are configured in such a way that they sense the data and send them at a certain frequency. But when using humans to sense the data and send it, a situation might arise when the person is unavailable at that particular moment. This may lead to missing data as well as sparsity issues. Another issue can be related to sending oversufficient data or redundant data that might affect preprocessing, making it time consuming. Moreover, this can affect the communication and storage as well. Another challenge is that some people may not share the data as well. Thus, nonlinear distribution of data takes place in the entire dataset based on the mobility of the people.

4.4.1.3 Contribution of Unstructured, Implicit, and Noisy Data

The traditional sensors generate data that are in a well-structured manner, explicit, with no noise, simple, and easier to understand, whereas the data that are provided by the humans as sensors exist in a free format like text messages or images. These data can have noise present in them and may be also in an unstructured manner and implicit. One of the examples could be the use of sensors for identifying the number of vehicles queued at a fuel station. The idea is to identify the consumption of fuel in a specified time. Here, the taxi provides the GPS trajectories that would help in the fuel consumption in specified times. Though it seems a very calculative concept, this concept could not consider other vehicles that are needed for the actual calculation of fuel consumption. The vehicles that are taking fuel from other fuel stations would not be considered in these calculations. Also, there is a possibility that the vehicles are parked in the fuel station zone or near to it for technical resolution (other than fuel filling) and for resting purposes. In both these cases, the fuel consumption based upon the GPS trajectories will degrade the quality. Therefore, the data received from these sensors must be analyzed adequately, which is received from partial, noisy, and received implicit data (Ferrara et al. 2019).

4.4.2 Computing with Heterogenous Data

Heterogenous data usually denote the datasets that contain spatial, categorical, and temporal data reflecting the dynamic state of an urban area. Heterogenous data are produced by various urban sources such as sensor devices, vehicles, apartments, or humans. Air quality data and vehicular traffic data are examples of heterogenous data.

4.4.2.1 Knowledge from Heterogeneous Data

The urban issues are very challenging and include a wide range of different types of data. Like in the case of air pollution analysis, there are various other parameters that need to be dealt with, such as uses of land, flow of the traffic, types of gas emission from households, frequency of natural disasters occurring, and meteorology (Feng et al. 2021). The current technologies deal with one particular set of data. The images are tackled using computer vision (CV), and the text data are tackled using natural language processing (NLP). The chances of achieving high accuracy are not very high when the data are extracted from different data sources. Moreover, with the increase in the number of different types of data sources, the high dimension space is reluctant, and this increases the problem of data sparsity. The performance of the model can be affected if the data are not handled in a correct way. In this case, the analytics models are trained using mutually reinforced knowledge on the various heterogenous data that are received from sensors, humans, cars and buses, and apartments (You et al. 2020).

4.4.2.2 Ability of Effective and Efficient Learning

Almost all the urban applications like detection of traffic anomalies or monitoring of environmental conditions such as air quality have the requirement of an instant response. A computing framework requires an effective as well as efficient knowledge discovery capability. For this purpose, not only does the computation need to happen faster but the there is a need for aggregation of data management and mining. Moreover, when the machine learning algorithms are used for this purpose, this helps in getting an optimal knowledge discovery ability. The earlier technologies were used for designing data sources that used a single type of data source. Therefore, there is a lack of optimal advanced methodology that could incorporate multimodal data such as geospatial data.

4.4.2.3 Visualization

In order to tackle the huge amount of data that is Big Data, it's important that the information extracted be presented in an efficient manner. The use of better visualization on the raw data could help to seed new ideas for solving the problems. Moreover, the efficient computational results will help in revealing knowledge intuitively that would enhance the decision-making capability. This will also help in establishing a correlation among the various distinct factors. The higher dimensions such as social, spatial, etc., are eminent in the case of multimodal data that are obtained in urban computational processes. The most challenging aspect of urban computing is tackling various types of distinct data in distinct views. Also, pattern detection and trend detection are challenges for urban computing. The probable solution could be a combination of an instant data mining methodology with the visualization (Ortner et al. 2017).

4.4.3 Hybrid System Blending the Physical and Virtual World

In the digital world, the data are generated from search engines and computerized games, whereas in urban computing, the data are incorporated from both worlds, like the combination of traffic data with social media. In an alternate way, the data that are generated in the physical world, such as GPS trajectories of vehicles, are directed to the digital world, i.e., the cloud system for storage. After the preprocessing of the raw data and combining them with other sources of data in the cloud system, the data are used for training the systems. The learned knowledge from the processed data is used for mobile users from the physical world, e.g., suggestions of routes while driving, ride sharing in cabs, and monitoring the quality of air. Unlike a conventional system that deals with one world only, the modern urban computing system deals with multiple worlds, making designing of the systems much more challenging. This challenge is due to the communication of the urban computing systems with a number of devices and users at the same time, as well as sending and receiving data in multiple formats and various distinct frequencies.

4.5 APPLICATION OF URBAN COMPUTING

Urban computing provides a speculative and specialized framework that integrates with the rapidly growing sophisticated innovations and applications in urban spaces. Some of the popular urban computing applications for smart urban city planning include traffic control systems, transportation management, security and safety of citizens, environment, energy resources, and social and economic reasons (Malik et al. 2019). Various challenges are faced in urban computing applications such as data sensing, data acquisition and handling Big Data that are generated by sensors, and accurate analysis of data. Therefore, urban computing can promise to provide targeted solutions to the challenges faced when building a smart urban computing application. Some of the well-known applications of urban computing are discussed in depth in the following subsections.

4.5.1 Urban Computing Application for a Smart City

City planning based on urban computing is one of the most significant applications. Most of the applications are applied in homogenous rural areas, where the sensors are placed in forest areas and glaciers or in small-scale areas like houses or small buildings. Effective planning is essential to build up an intelligent city. For this purpose, evaluation of a large number of features like people's mobility, traffic flow, and road network structures is necessary. This kind of complex factor brings challenges to urban planning applications. Previously, many works have been carried out from the surveys conducted by the people who travel from one place to another. The drawback of these labor-intensive surveys is that the data that are obtained might be insufficient and might lapse in time. In the present scenario, the data which are obtained from transport data or mobile data are much more accurate compared to the surveys, which helps in building smart urban spaces (Arora et al. 2019).

The new information as well as communication techniques have made as contribution in developing a smart city. A huge amount of data is generated from sustainable sensors placed across the city. The multimedia data such as image data and video data can be considered as the most typical form of data. With the proper implementation of CV techniques, a value-added smart city can be implemented. A smart lighting system can be one of the recent trends and applications in the field of urban computing. This smart lighting system minimizes power consumption as well as brings awareness of a sudden change in the climate (Mora et al. 2019).

4.5.2 Urban Computing for a Transportation System

A smart city needs an intelligent transportation system (ITS). In order to manage the smooth running of traffic with the urban area, the ITS proves to a vital part of urban computing. With the rapid growth of population in the urban areas, there also arises the problem of unorganized traffic. The present infrastructure of the roads is not enough to tackle the huge amount of traffic. Road traffic congestion is one of the major issues in urban areas. This congestion affects the flow of traffic in the city. This may lead to a delay in a medical emergency, causing health issues or even deaths. Therefore, a proper urban computing application on a smart transportation system is required to handle this situation. To construct a smart transportation management system, it becomes a challenge to handle the huge amount of transport data.

The aim of an intelligent traffic management system (ITMS) is to manage the urban traffic in an efficient manner in the case of emergency situations, with the usage of ICT and efficient algorithms. The ITMS plays a crucial role by minimizing the congestion of traffic as well as providing a smart parking system. Some of the technologies that support the implementation of ITMS are vehicular ad hoc networks

(WANETS), a wireless sensor network, the vehicle to vehicle (V2V) technique, and the vehicle-to-infrastructure (V2I) technique. To sense data for ITMS, sensors or closed-circuit televisions (CCTV) are placed by the roadside to sense the real-time data. An efficient architecture to control and manage the traffic congestion and transport management is developed with the use of IoT, a wireless sensor network (WSN), and cloud systems. For real-time data collection, IoT and a cloud system is used. To get an efficient and reliable traffic management system, fuzzy logic is applied. This system minimizes the travel time of the citizens of an urban area by informing them about ongoing traffic jams and their routes (Kaur and Malhotra 2017). An intelligent smart system is developed using urban computing techniques for managing emergency situations on roads by minimizing the waiting time at the traffic signals (Malik et al. 2018).

4.5.3 Urban Computing for the Environment

The rapid growth of urbanization is causing a threat to the environment. It is observed that increasing urbanization has increased the level of environmental pollution greatly. It affects the environmental aspects such as air, water, and land. Urban computing helps in the protection of the environment, as well as making people's lives modern at the same time. The air quality of urban areas is very poor due to emission of harmful gases from vehicles and industries. Urban computing offers proper monitoring of the air quality factors by implementing measurement stations in the city to check air quality (Huang et al. 2019). Air quality may vary from area to area. As a result, the mobile computing and measurement stations work in combination to overcome the nonlinearity of measurement of air quality. A project named Copenhagen Wheels has been carried out to sense environmental parameters of the city such as temperature, humidity, and CO_2 concentration by placing sensors in the bike's wheel. The advantage of this application is that it can sense the parameters of the air from different locations and can interact with the smart device user and send data to the server (Wu et al. 2021).

Similarly, lots of environmental noise is generated by millions of people. This creates noise pollution in the environment. Exposure to high levels of noise can lead to illness in people or hearing loss (Longo et al. 2018). The intensity of the noise has to be measured, and this measurement also defines people's tolerance to noises It is observed that people's tolerance to noise at day is greater as compared to night. By placing noise sensor detectors over the city at different locations, the ranking of the locations can be done based on the data at a particular time to control the noise pollution at the areas where the threshold value is exceeded (Gunatilaka 2021).

4.6 CONCLUSION AND FUTURE SCOPE

The exponential growth of urbanization has made the urban computing domain more important in application of smart city planning. With the principle of urban development, the whole urban area can be considered as integration of interconnected devices and network systems that allow access to information from the various locations within an urban area. The intelligent sensors sense the environment and generate heterogeneous data for further processing and analysis. Big Data must be tackled and made suitable for proper analysis for the application of urban computing. This chapter discusses the detailed concepts of urban computing, the various sources of sensor data, and the key challenges faced. This chapter also covers the various application areas of urban computing in terms of the environment, city planning and smart transportation systems, traffic management, and crowd analysis. The future direction of urban computing could be proper handling of data during analysis, as most of the sensors could generate unbalanced data. Some locations might have less data or might have no data at all. Application of a down-sampling method could deal with this kind of situation. Another issue to work on is the application of a learning algorithm based on data mining and machine learning to deal with the different data coming from multiple sources.

REFERENCES

Ahmad, A., Bhushan, B., Sharma, N., Kaushik, I., & Arora, S. (2020). Importunity & Evolution of IoT for 5G. 2020 IEEE 9th International Conference on Communication Systems and Network Technologies (CSNT). 102–107. DOI: 10.1109/csnt48778.2020.9115768.

Antunes, H., Figueiras, P., Costa, R., Teixeira, J., & Jardim-Gonçalves, R. (2019). Analysing Public Transport Data Through the Use of Big Data Tecnhologies for Urban Mobility. 2019 International Young Engineers Forum (YEF-ECE). 40–45. DOI: 10.1109/YEF-ECE.2019.8740816.

Arora, A., Kaur, A., Bhushan, B., & Saini, H. (2019). Security Concerns and Future Trends of Internet of Things. 2019 2nd International Conference on Intelligent Computing, Instrumentation and Control Technologies (ICICICT). 1, 891–896. DOI: 10.1109/icicict46008.2019.8993222.

Arora, S., Sharma, N., Bhushan, B., Kaushik, I., & Ahmad, A. (2020). Evolution of 5G Wireless Network in IoT. 2020 IEEE 9th International Conference on Communication Systems and Network Technologies (CSNT). 108–113. DOI: 10.1109/csnt48778.2020.9115773.

Bala, P. (2022). Introduction of Big Data with Analytics of Big Data. Research Anthology on Big Data Analytics, Architectures, and Applications, 54–66, IGI Global. https://doi.org/10.4018/978-1-6684-3662-2.ch003.

Bhushan, B., Khamparia, A., Sagayam, K. M., Sharma, S. K., Ahad, M. A., & Debnath, N. C. (2020). Blockchain for Smart Cities: A Review of Architectures, Integration Trends and Future Research Directions. Sustainable Cities and Society, 61, 102360. DOI: 10.1016/j.scs.2020.102360.

Bhushan, B., Sahoo, C., Sinha, P., & Khamparia, A. (2020). Unification of Blockchain and Internet of Things (BIoT): Requirements, Working Model, Challenges and Future Directions. Wireless Networks. 27(1), 55–90. DOI: 10.1007/s11276-020-02445-6.

Biswal, A., & Bhushan, B. (2019). Blockchain for Internet of Things: Architecture, Consensus Advancements, Challenges and Application Areas. 2019 5th International Conference on Computing, Communication, Control and Automation (ICCUBEA). 1–6. DOI: 10.1109/iccubea47591.2019.9129181.

Cáceres, P., Sierra-Alonso, A., Cuesta, C. E., Vela, B., & Cavero Barca, J. M. (2020). Improving Urban Mobility by Defining a Smart Data Integration Platform. IEEE Access, 8, 204094–204113. DOI: 10.1109/ACCESS.2020.3033584.

Feasel, K. (2020). Connecting to Hadoop. In PolyBase Revealed, 63–93. Apress, Berkeley. https://doi.org/10.1007/978-1-4842-6584-0_7.

Peng, Z., Li, H., Zeng, W., Yuan, Q., Hu, & Qu, H. (Feb. 2021). Topology Density Map for
    Urban Data Visualization and Analysis. IEEE Transactions on Visualization and
    Computer Graphics, 27(2), 828–838. DOI: 10.1109/TVCG.2020.3030469.

Ferrara, E., Fragale, L., Fortino, G., Song, W., Perra, C., Di Mauro, M., & Liotta, A. (2019).
    An AI Approach to Collecting and Analyzing Human Interactions With Urban
    Environments. IEEE Access, 7, 141476–141486. DOI: 10.1109/ACCESS.2019.2943845.

Goyal, S., Sharma, N., Bhushan, B., Shankar, A., & Sagayam, M. (2020). IoT Enabled
    Technology in Secured Healthcare: Applications, Challenges and Future Directions.
    Cognitive Internet of Medical Things for Smart Healthcare: Studies in Systems,
    Decision and Control book series (SSDC, vol. 311) Springer, 25–48. DOI: 10.1007/
    978-3-030-55833-8_2.

Goyal, S., Sharma, N., Kaushik, I., Bhushan, B., & Kumar, N. (2021). A Green 6g Network
    Era: Architecture and Propitious Technologies. Data Analytics and Management,
    part of the Lecture Notes on Data Engineering and Communications Technologies
    book series (LNDECT, vol. 54) 59–75. DOI: 10.1007/978-981-15-8335-3_7.

Guedea-Noriega, H. H., & García-Sánchez, F. (May 2019). Semantic (Big) Data Analysis:
    An Extensive Literature Review. IEEE Latin America Transactions, 17(05), 796–806.
    DOI: 10.1109/TLA.2019.8891948.

Gunatilaka, D. (2021). An IOT-enabled Acoustic Sensing Platform for Noise Pollution
    Monitoring. 2021 IEEE 12th Annual Ubiquitous Computing, Electronics & Mobile
    Communication Conference (UEMCON). 383–389. https://doi.org/10.1109/uemcon
    53757.2021.9666534.

Hamdaoui, B., Alkalbani, M., Rayes, A., & Zorba, N. (Oct. 2020). IoTShare: A Blockchain-
    Enabled IoT Resource Sharing On-Demand Protocol for Smart City Situation-
    Awareness Applications. IEEE Internet of Things Journal, 7(10), 10548–10561. DOI:
    10.1109/JIOT.2020.3004441.

Haque, A. K., Bhushan, B., & Dhiman, G. (2021). Conceptualizing Smart City Applications:
    Requirements, Architecture, Security Issues, and Emerging Trends. Expert Systems,
    39(5). https://doi.org/10.1111/exsy.12753.

Hou, C., Zhao, Q., & Başar, T. (Oct. 2021). Optimization of Web Service-Based Data-
    Collection System with Smart Sensor Nodes for Balance Between Network Traffic and
    Sensing Accuracy. IEEE Transactions on Automation Science and Engineering, 18(4),
    2022–2034. DOI: 10.1109/TASE.2020.3030835.

Huang, J., Duan, N., Ji, P., Ma, C., Ding, Y., Yu, Y., ... & Sun, W. (2018). A crowdsource-based
    sensing system for monitoring fine-grained air quality in urban environments. IEEE
    Internet of Things Journal, 6(2), 3240-3247. DOI: 10.1109/JIOT.2018.2881240

Kaiser, M. S. et al. (Oct. 2018). Advances in Crowd Analysis for Urban Applications Through
    Urban Event Detection. IEEE Transactions on Intelligent Transportation Systems,
    19(10), 3092–3112. DOI: 10.1109/TITS.2017.2771746.

Kaur, H., & Malhotra, J. (2017). An IoT-Based Smart Architecture for Traffic Management
    System. IOSR Journal of Computer Engineering, 19(04), 60–63. https://doi.org/
    10.9790/0661-1904026063.

Khazael, B., Malazi, H. T., & Clarke, S. (2021). Complex Event Processing in Smart
    City Monitoring Applications. IEEE Access, 9, 143150–143165., DOI: 10.1109/
    ACCESS.2021.3119975.

Kumar, A., Bhushan, B., Malik, A., & Kumar, R. (2021). Protocols, Solutions, and Testbeds
    for Cyber-attack Prevention in Industrial SCADA Systems. Studies in Big Data, 3,
    355–380. https://doi.org/10.1007/978-981-16-6210-2_17.

Li, H. (2021). Research on Big Data Analysis Data Acquisition and Data Analysis. 2021
    International Conference on Artificial Intelligence, Big Data and Algorithms
    (CAIBDA), 162–165. https://doi.org/10.1109/caibda53561.2021.00041.

Liu, Z., Li, Z., & Wu, K. (Aug. 2019). UniTask: A Unified Task Assignment Design for Mobile Crowdsourcing-Based Urban Sensing. IEEE Internet of Things Journal, 6(4), 6629–6641. DOI: 10.1109/JIOT.2019.2909296.

Longo, A., De Matteis, A., &Zappatore, M. (2018). Urban Pollution Monitoring Based on Mobile Crowd Sensing: An Osmotic Computing Approach. 2018 IEEE 4th International Conference on Collaboration and Internet Computing (CIC). 380–387. DOI: 10.1109/CIC.2018.00057.

Lu, K., Liu, J., Zhou, X., & Han, B. (May 2021). A Review of Big Data Applications in Urban Transit Systems. IEEE Transactions on Intelligent Transportation Systems, 22(5), 2535–2552. DOI: 10.1109/TITS.2020.2973365.

Lv, Z., Song, H., Basanta-Val, P., Steed, A. & Jo, M. (2017). Next-Generation Big Data Analytics: State of the Art, Challenges, and Future Research Topics. IEEE Transactions on Industrial Informatics, 13(4), 1891–1899. DOI: 10.1109/TII.2017.2650204.

MacEachren, A. M. (2017). Leveraging Big (GEO) Data with (GEO) Visual Analytics: Place as the Next Frontier. Spatial Data Handling in Big Data Era. Part of the Advances in Geographic Information Science book series (AGIS), 139–155. https://doi.org/10.1007/978-981-10-4424-3_10.

Mahmud, M. S., Huang, J. Z., Salloum, S., Emara, T. Z., &Sadatdiynov, K. (2020). A Survey of Data Partitioning and Sampling Methods to Support Big Data Analysis. Big Data Mining and Analytics, 3(2), 85–101. DOI: 10.26599/BDMA.2019.9020015.

Malik, A., Gautam, S., Abidin, S., & Bhushan, B. (2019). Blockchain Technology-Future Of IoT: Including Structure, Limitations And Various Possible Attacks. 2019 2nd International Conference on Intelligent Computing, Instrumentation and Control Technologies (ICICICT). 1, 1100–1104. DOI: 10.1109/icicict46008.2019.8993144.

Malik, F., Shah, M. A., & Khattak, H. A. (2018). Intelligent Transport System: An Important Aspect of Emergency Management in Smart Cities. 2018 24th International Conference on Automation and Computing (ICAC). 1–6. DOI: 10.23919/IConAC.2018.8749062.

Manchanda, C., Sharma, N., Rathi, R., Bhushan, B., & Grover, M. (2020). Neoteric Security and Privacy Sanctuary Technologies in Smart Cities. 2020 IEEE 9th International Conference on Communication Systems and Network Technologies (CSNT). 236–241. DOI: 10.1109/csnt48778.2020.9115780.

Meneguette, R. I., & Boukerche, A. (2020). Vehicular Clouds Leveraging Mobile Urban Computing Through Resource Discovery. IEEE Transactions on Intelligent Transportation Systems, 21(6), 2640–2647. https://doi.org/10.1109/tits.2019.2939249.

Mora, H., Peral, J., Ferrandez, A., Gil, D., & Szymanski, J. (2019). Distributed Architectures for Intensive Urban Computing: A Case Study on Smart Lighting for Sustainable Cities. IEEE Access, 7, 58449–58465. https://doi.org/10.1109/access.2019.2914613.

Mostajabi, F., Safaei, A. A., & Sahafi, A. (2021). A Systematic Review of Data Models for the Big Data Problem. IEEE Access, 9, 128889–128904. DOI: 10.1109/ACCESS.2021.3112880.

Ordonez-Garcia, A., Siller, M., & Begovich, O. (2017). IOT Architecture for Urban Agronomy and Precision Applications. 2017 IEEE International Autumn Meeting on Power, Electronics and Computing (ROPEC). 1–4. https://doi.org/10.1109/ropec.2017.8261582.

Ortner, T., Sorger, J., Steinlechner, H., Hesina, G., Piringer, H., & Gröller, E. (1 Feb. 2017). Vis-A-Ware: Integrating Spatial and Non-Spatial Visualization for Visibility-Aware Urban Planning. IEEE Transactions on Visualization and Computer Graphics, 23(2), 1139–1151. DOI: 10.1109/TVCG.2016.2520920.

Peng, D., Yang, W., Li, H., & Yang, X. (2019). Superpixel-Based Urban Change Detection in SAR Images Using Optimal Transport Distance. 2019 Joint Urban Remote Sensing Event (JURSE). 1–4. DOI: 10.1109/JURSE.2019.8809008.

Puri, D., & Bhushan, B. (2019). Enhancement of Security and Energy Efficiency in WSNs: Machine Learning to the Rescue. 2019 International Conference on Computing, Communication, and Intelligent Systems (ICCCIS). 120–125. DOI: 10.1109/icccis 48478.2019.8974465.

Rawat, D. B., Doku, R., & Garuba, M. (2021). Cybersecurity in Big Data Era: From Securing Big Data to Data driven Security. IEEE Transactions on Services Computing, 14(6), 2055–2072. https://doi.org/10.1109/tsc.2019.2907247.

Satija, S., Sharma, T., & Bhushan, B. (2019). Innovative Approach to Wireless Sensor Networks: SD- WSN. 2019 International Conference on Computing, Communication, and Intelligent Systems (ICCCIS). 170–175. DOI: 10.1109/icccis48478.2019.8974548.

Sethi, R., Bhushan, B., Sharma, N., Kumar, R., & Kaushik, I. (2020). Applicability of Industrial IoT in Diversified Sectors: Evolution, Applications and Challenges. Studies in Big Data Multimedia Technologies in the Internet of Things Environment, Part of the Studies in Big Data book series (SBD,vol. 79) 45–67. DOI: 10.1007/978-981-15-7965-3_4.

Shawon, M. Hasanuzzaman, Muyeen, S. M., Ghosh, A., Islam, S. M., & Baptista, M. S. (2019). Multi-Agent Systems in ICT Enabled Smart Grid: A Status Update on Technology Framework and Applications. IEEE Access, 7, 97959–97973. DOI: 10.1109/ACCESS. 2019.2929577.

Shin, E. K., & Shaban-Nejad, A. (2018). Urban Decay and Pediatric Asthma Prevalence in Memphis, Tennessee: Urban Data Integration for Efficient Population Health Surveillance. IEEE Access, 6, 46281–46289. DOI: 10.1109/ACCESS.2018.2866069.

Silva, B. N., et al. (2018). Exploiting Big Data Analytics for Urban Planning and Smart City Performance Improvement. 2018 12th International Conference on Signal Processing and Communication Systems (ICSPCS). 1–4. DOI: 10.1109/ICSPCS.2018.8631726.

Sinnott, R. O., Gong, Y., Chen, S., & Rimba, P. (2018). Urban Traffic Analysis Using Social Media Data on the Cloud. 2018 IEEE/ACM International Conference on Utility and Cloud Computing Companion (UCC Companion). 134–141. DOI: 10.1109/ UCC-Companion.2018.00047.

Slovokhotov, Y. L., & Echenique, V. X. (2018). Hyperbolic Trend of a Global Population in XVI-XX A.D. as a Mirror of Condensation in the World System. 2018 Eleventh International Conference "Management of Large-Scale System Development" (MLSD. 1–4. https://doi.org/10.1109/mlsd.2018.8551772.

Sundaram, V., Tripathy, A. K., Deshmukh, R., & Pawar, A. (2020). A Crowd Density Estimation Approach Using GPS Mobility for Its Dynamics and Predictions. 2020 3rd International Conference on Communication System, Computing and IT Applications (CSCITA). 67–72. DOI: 10.1109/CSCITA47329.2020.9137804.

Tiwari, R., Sharma, N., Kaushik, I., Tiwari, A., & Bhushan, B. (2019). Evolution of IoT & Data Analytics Using Deep Learning. 2019 International Conference on Computing, Communication, and Intelligent Systems (ICCCIS). 418–423. DOI: 10.1109/ icccis48478.2019.8974481.

Tu, H. (2020). Research on the Application of Cloud Computing Technology in Urban Rail Transit. 2020 IEEE International Conference on Advances in Electrical Engineering and Computer Applications (AEECA). 828–831. DOI: 10.1109/AEECA49918.2020.9213455.

Tyagi, A., Bhushan, B., & Singh, R. V. (2021). Big Data Analytics for Wireless Sensor Networks and Smart Grids: Applications, Design Issues, and Future Challenges. Integration of WSNs into Internet of Things, 135–163. DOI: 10.1201/9781003107521-8.

Usman, M., Jan, M. A., He, X., & Chen, J. (2020). A Mobile Multimedia Data Collection Scheme for Secured Wireless Multimedia Sensor Networks. IEEE Transactions on Network Science and Engineering, 7(1), 274–284. https://doi.org/10.1109/ tnse.2018.2863680.

Verma, C., & Pandey, R. (2022). Statistical Visualization of Big Data Through Hadoop Streaming in Rstudio. Research Anthology on Big Data Analytics, Architectures, and Applications, 758–787, IGI Global https://doi.org/10.4018/978-1-6684-3662-2.ch035.

Wazid, M., Das, A. K., Choo, K.-K. R., & Park, Y. H. (2021). SCS-WoT: Secure Communication Scheme for Web of Things Deployment. IEEE Internet of Things Journal, 9(13), 10411–10423. https://doi.org/10.1109/jiot.2021.3122007.

Wemegah, T. D., & Zhu, S. (2017). Big Data Challenges in Transportation: A Case Study of Traffic Volume Count from Massive Radio Frequency Identification (RFID) Data. 2017 International Conference on the Frontiers and Advances in Data Science (FADS). 58–63. DOI: 10.1109/FADS.2017.8253194.

White, G., & Clarke, S. (2020). Urban Intelligence with Deep Edges. IEEE Access, 8, 7518–7530. https://doi.org/10.1109/access.2020.2963912.

Wu, Y., Low, K. H., Pang, B., & Tan, Q. (2021). Swarm-based 4D Path Planning for Drone Operations in Urban Environments. IEEE Transactions on Vehicular Technology, 70(8), 7464–7479. https://doi.org/10.1109/tvt.2021.3093318.

Xu, Z., et al. (1 April-June 2020). Crowdsourcing Based Description of Urban Emergency Events Using Social Media Big Data. IEEE Transactions on Cloud Computing, 8(2), 387–397. DOI: 10.1109/TCC.2016.2517638.

You, L., Zhao, F., Cheah, L., Jeong, K., Zegras, P. C., & Ben-Akiva, M. (Oct. 2020). A Generic Future Mobility Sensing System for Travel Data Collection, Management, Fusion, and Visualization. IEEE Transactions on Intelligent Transportation Systems, 21(10), 4149–4160. DOI: 10.1109/TITS.2019.2938828.

Yun, Q., Leng, C. (2020). Intelligent Control of Urban Lighting System Based on Video Image Processing Technology. IEEE Access, 8, 155506–155518. https://doi.org/10.1109/access.2020.3019284.

Zappatore, M., Loglisci, C., Longo, A., Bochicchio, M. A., Vaira, L., & Malerba, D. (2019). Trustworthiness of Context-Aware Urban Pollution Data in Mobile Crowd Sensing. IEEE Access, 7, 154141–154156. DOI: 10.1109/ACCESS.2019.2948757.

5 Complex Event Processing Architectures for Smart City Applications

Sonal Shriti, Bharat Bhushan, Avinash Kumar,
Parma Nand, and Tanmayee Prakash Tilekar

CONTENTS

5.1 INTRODUCTION

Huge datasets that have a diverse and complicated structure are referred to as Big Data. These datasets are quite challenging to work with—they are not easy to visualize, store, and analyze to get desirable results. Extensive research is done to explore the possible patterns and correlations that might be hidden in the huge amount of data present, which is known as Big Data analytics. The information that is obtained out of this analysis is extremely helpful to acquire a much better understanding to further enhance the inner workings of organizations for maximum optimization. This prevents them from being swept away by the competition. The data are obtained from a wide array of sources such as e-mails, audios, videos, logs, search queries,

DOI: 10.1201/9781003255635-5

click streams, online transactions, social networking interactions, science data, health records, posts, sensors, etc. After they are obtained, the data are stored in databases. Due to the millions of users that surf the Web every single day, the capturing, storing, forming, managing, and sharing of data become difficult due to their growing size, and traditional database software is not enough. This makes way for the need to have some other processing techniques that will be able to handle Big Data more efficiently. Big Data play a vital role in handling events and affect them to a great extent.

An event can be defined as a change in the state of an entity or attribute. That is, when an attribute differs from its general state, such as a change in temperature, time, or any other value. An event can also relate to news feeds, messages, social media stories, traffic analysis, and much more. As the name suggests, event processing (EP) is the technique or method used to analyze as well as track data and process them in order to retrieve information and important statistics pertaining to a domain. Analyzing streams of data to acquire useful information about a real-world scenario comes under EP (Kale 2019). EP finds its use in many real-world applications, and many of these applications require information that is brought in from outside the system to be processed. It includes processing of data and then reaching a conclusion using the processed data. EP is a critical function for any organization or business and takes place across various layers of an organization. It helps in designing the key parametric indicators (KPIs), which then provide a boost to the organization's throughput, performance, and efficiency by allowing it to adhere to them. Specifically, EP is used by businesses mostly in networks and communications, which can be internal or external. This includes collecting data not only from a single source of data but from multiple, heterogeneous sources. Thus, collecting, organizing, and processing event data from multiple sources is a skill and can be rendered through EP. Investing in EP is highly beneficial and helps organizations to receive instant information on required fields. It also enables an organization to take necessary actions instantly by looking at the processed data and event analytics. There are several techniques to conduct EP. Some of the general techniques include pattern identification, event aggregation, event abstraction and event transformation, filtering and detecting of events, and identifying relationships among the data and events. Pattern detection refers to recognizing as well as analyzing the patterns of event data and identifying the regularities and irregularities within the same. Pattern identification assists in the following: signal processing, data analysis, bioinformatics, computer graphics, machine learning (ML), image processing, etc. Event abstraction refers to the technique of generalizing event data and deriving certain rules out of them (Goyal et al. 2020). Event filtering helps in selectively filtering the necessary event data to be processed and specified as well as extracting the necessary information from the event data (Bhushan et al. 2020). EP makes use of the networks and specialized devices, including sensors and actuators for data acquisition and analysis (Goyal et al. 2020). Thus, it can be stated that the Internet of Things (IoT) is a crucial contributor to EP (Sethi et al. 2020). Nowadays, EP has become a point of research and advancement and contributes greatly to academic, industrial, and research domains (Shukla and Bhushan 2021). There are various popular EP platforms, namely Apache Spark, Flink, etc. Significant developments in the field of

IoT and MI, positively affect the EP domain as well. Furthermore, the current era has witnessed significant developments in the field of event ordering, EP languages, systems scalability, and heterogeneous devices used for EP. However, there still remains a scope for focusing on certain key areas relating to EP, including system benchmarking, streaming, and graph stream processing. EP is a highly important function for the organizations to manage their operations, functions, and systems.

Complex event processing (CEP) is a kind processing of information. The primary goal of CEP is to analyze the notifications of events that are usually low-level and then detect and define scenarios that might potentially be of interest. Another way to interpret the function of CEP systems is that they detect complex events from the provided stream of simple events. This is done by making use of declarative rules (Hedtstück 2017). These rules describe the events that are derived in the form of patterns, the temporal relations between them, and what is contained in them. The term simple events is used to classify events that are primitive and low-level. On the other hand, the term complex events is used to classify events that represent and summarize a set of alternate events. Another term, derived events, is used to refer to a special kind of complex event. These are obtained by applying a technique or strategy to a particular event, or more than one for that matter. There are a number of languages and CEP systems that are used (Hedtstück 2017). However, the basic concepts, structures, and mechanisms that are employed in them stay the same. Evidently, the technology behind is CEP is nothing short of impressive. It is capable of dealing with an inordinate amount of data that are acquired from several different sources. Based on the consistency of the acquired data, CEP provides accurate results that help in real-time processing of data that is dynamic in nature. This makes the knowledge of methods and tools involved in CEP of utmost importance if the goal is to devise CEP systems that are extremely effective and robust.

This chapter addresses the approaches to explain EP as well as CEP. The major contributions in the chapter can be summarized as follows:

- This chapter defines EP and CEP in detail.
- This chapter relates technologies such as Big Data to the EP paradigm.
- This chapter provides a detailed overview of CEP.
- This chapter describes the EP architecture and various platforms for EP.
- This chapter lists the applications for EP and CEP.

The remainder of this chapter is organized as per the following scheme: Section 5.2 provides an overview of CEP. Section 5.3 addresses the architecture for event processing. Section 5.4 discusses the applications of CEP with reference to real-life scenarios. Finally, Section 5.5 concludes the chapter, followed by stating the future research directions.

5.2   CEP OVERVIEW

Some of the primary characteristics associated with CEP are efficiency, heterogeneity, timeliness, scalability, and robustness. The input for such systems is an infinite and never-ending stream of events that is gathered from different sources, which is then used in handling data in real time. The handling of data involves detecting a large number of events with low latency. CEP systems can also aggregate and filter data to discern complex events that are of semantic interest. Apart from this, there are many more tasks that CEP is capable of, and these can range from a run-of-the-mill monitoring task to tasks like algorithmic trading and detection of fraud that are extremely complicated in nature. The following subsections describe the various existing CEP engines and give a detailed definition of uncertain data.

5.2.1 CEP ENGINES

Business process management (BPM) has been the natural balancing of CEP. BPM focuses on end-to-end business processes to continuously enhance and align with the operational environment. However, a company's full implementation does not rely exclusively on its various procedures, which all come to a stop at some point. Procedures that appear to be unrelated can be extremely linked. Evaluate the following circumstance: In the aeronautical profession, it's usually advisable to keep an eye on traffic congestion to identify similarities (to determine potential weaknesses in production processes, equipment, etc.). Another strategy is to keep monitoring the life spans of currently operational vehicles and shut them down when required. Another implementation of CEP is to integrate these processes so that, in the case of the first event (deterioration), we can identify iron-based inactivity (a major event), and action is taken to implement the second event (life cycle) to send an alert to vehicles that use a pile of metal that was discovered to be problematic in the first system.

CEP and BPM should be combined on two levels: business awareness (users should understand the full favorable circumstances of their particular processes) and technical information (users should understand the full potential benefits of their processes) (there should be a way for CEP to work with BPM implementation). Table 5.1 classifies various types of CEP engines.

5.2.2 UNCERTAIN DATA

Uncertainty is a term used to describe the partial knowledge and less accuracy of data. In CEP, the decision-making process can face various issues due to incorrect, insufficient, and imprecise and unimportant information. This uncertainty is caused due to systematic errors, vagueness, lack of study, and incomplete information (Kennedy 2021). The data blocks or strings are considered uncertain when they contain incomplete data or impressive data. Various reasons such as privacy protection, signal disturbance, and weaknesses while measuring the accuracy can lead to an imprecise event. There are various types of uncertainty about which the event processing unit has to deal with, such as inconsistent event notations, inadequate event dictionary, patchy event stream, vague event patterns, etc. These factors of uncertainty tell the difference between uncertainty in event input and uncertainty in event pattern. The two major types of uncertainties are existential uncertainty and value uncertainty. Uncertainty can make uncertain data deviate

TABLE 5.1
Implementation of CEP Engines

Topic	Motive	Implementation	Limitations
Verifying systems based on Internet of Things (IoT) at runtime (Incki et al. 2017)	Failure detection at runtime by using complex event patterns and exploiting them.	IoT	Disorder related to network packets can lead to some errors left undetected.
Cognitive programming of robots by employing CEP and procedural parameters (Erich and Suzuki 2016)	Improvisation of systems of data processing in robots in terms of efficiency and power.	Robotics	Primarily focuses on processing sensor data, and the actuation is dependent on the modules that are available.
CEP for city officers (Bonino and De Russis 2018)	Providing city officers with a panel to let them implement, test, design, and create datasets accurately.	Pipelines of data processing used in smart cities.	Knowledge specific to the domain is a requirement that city officers need to have.
Consistent and knowledge-driven CEP on persistent and real-time streams (Zhou et al. 2017)	Ability to gauge dimensions like volume and velocity of the data stream.	Cyber-physical systems (CPS) and IoT	Uncertainty of events is not taken into consideration.
Simulation of detecting the violation of speed limits in real time (Rakkesh et al. 2016)	Warning responsible vehicle owners against traffic violations like exceeding the approved speed limit.	Environments that are prone to heavy traffic.	Roadblocks, crossing for pedestrians, alternate options of transportation like buses, and railway lines are not taken into consideration.
Detecting unexpected scenarios for multiple stages (Lu et al. 2016)	Tracking the process of production and processing the data stream that is highly frequent.	Workspace observation	Uncertainty of events is not taken into consideration.
Context-aware CEP on streams of RFID (Peng and He 2016)	Monitoring any trends and anomalies that might occur in the streams of events.	Mobile phones, radio frequency identification (RFID), and sensors	Not yet been tested over real-life scenarios and datasets.

from the original data because of the noise found. In the past few years, several tools have been developed for the collection and storage of data in which sometimes uncertain data are also present. Uncertainty can affect performance and efficiency of various CEP applications. Table 5.2 shows some of the implementations of CEP engines for uncertain events.

TABLE 5.2
Implementation of CEP Engines for Uncertain Events

Topic	Motive	Implementation	Limitations
Formally analyzing and modeling applications based on CEP (Debbi 2017)	Detecting discernable patterns in the events	BPM	Requires constraints so that probabilistic timed automata (PTA) can be adopted
CEP on Bayesian networks (BN) (Wang et al. 2018)	Taking pre-emptive actions by predicting possible states in the future	Online prediction done automatically	The modifying of parameters is not efficient and the issue of scalability of BN is present
Protecting privacy efficiently by making use of Markov correlations (Li et al. 2016)	Ensuring the privacy of the composite events that are modeled using Markov chains	Protecting privacy using RFID	
Performing CEP in the field of manufacturing and handling data streams that are uncertain (Mao and Tan 2015)	Ensure that the products manufactured are not of inferior quality	Automating the manufacturing process of products	

5.3 ARCHITECTURE FOR EVENT PROCESSING

An EP system architecture constitutes the tools (software and hardware) that are required to conduct EP and CEP. The tools can be categorized as following: event processing platforms (EPPs), distributed stream computing platforms (DSCPs), and libraries for CEP. Each of the following tools is discussed in the following subsections, one by one.

5.3.1 PLATFORM FOR PROCESSING EVENT PROCESSING

EPPs refer to the platforms which assist in provisioning explicit support and assistance for distributing computation resources among the nodes in a cluster of computing devices or computers. They are a critical component of an EP system or architecture. More appropriately, an EP platform can be described as a type of EP software which is responsible for provisioning high-level programming and programming models. Some of the important high-level programming models can include expressive event processing languages (EEPLs) and prebuilt or predefined functions to facilitate correlation, filtering, and abstraction (Tommasini et al. 2017). Some of the notable EPPs include the real-time streaming (RTS) platform constructed by Data Torrent, IBM Infosphere, StreamMine 3G, and Apache Apex and Flink. Each EPP possesses a specific publish pattern. A publish pattern is also known as a "subscribe pattern". According to such patterns, delivery of events to the desired parties takes place with the help of a message or data broker. A message broker can be a notification service, in general. A configuration is done for message brokers in order to extract only desired events and filter out the unwanted ones (Flouris et al. 2017). Filtering is important because the processing of events needs to be done selectively as per the requirement and in an effective manner. Each EP platform has its own features, characteristics, and specifications. For instance, platforms including RTS, TIBCO Streambase, WSO2 CEP, and Fujitsu Inter-stage CEP support temporal pattern and sequence-based queries. The SQL stream platform can support and facilitate Structured Query Language (SQL)–like queries, query composition, displaying results, database integration, and a comprehensive library of operators (Sahal et al. 2020). EPPs like Amazon Kinesis and similar ones can facilitate query composition, displaying results, distributed processing, contain a visual application debugger, and are scalable, as well as contain operator libraries. Fujitsu Inter-stage CEP and similar platforms for EP assist with geographical information systems (GIS) servers' integration, database integration, and application programming interfaces (APIs) support (Geisler 2020). Support for APIs consists of language extensions used for geoinformation processing by default. WSO2 CEP is a platform that supports maximum features and specifications, such as SQL queries and query composition, sequence queries, distributed computing, database integration, high availability of resources, scalability, support for APIs, and a wide library of operators. Almost all maximum platforms support key queries and types, including joins and windows (Zhou et al. 2017). The windows can be of given types: time windows, sliding windows, batch windows, and tuple windows. A time window is a type of window that expires the content or data contained in it when the specified period of time elapses. Second, a sliding window is a window that does not expire all the content contained in it; rather, it only expires a specific part of content or events based on the window size. Third, a batch window can be referred to as a window that expires all the events encompassed by the window at the same instance of time. Batch windows are also called "tumbling windows". Lastly, a tuple window expires the contained events based on a punctuation tuple and its receipt.

5.3.2 PLATFORMS FOR DISTRIBUTED STREAM COMPUTING

With the emergence of applications such as real-time search, high-frequency trading, and social networking apps and websites, the limits of the traditional data processing systems are being tested almost on a regular basis. Scalable stream computing solutions are therefore imperative to deal with the present continuous trend of high data rates and also to process the massive amount of data generated. For example, in order to provide a more personalized experience to users in search advertisements, the system needs to have the capability to process thousands of queries from millions and what could be even billions of users in real time—a process typically involving running a detailed analysis on user behavioral patterns across the websites and services in question. It has been found in studies that feature information of a user's session can come a long way in tremendously improving the system's overall accuracy. This improvement in the accuracy and performance would then be reflected in the content relevance improvement in the ads shown to any individual.

There are present, based on their differing architectures, many clearly separable types of DSCPs. These can be loosely classified into three broad categories, namely first-generation DSCPs, second-generation DSCPs, and geodistributed DSCPs. The first-generation DSCPs work well in facilitating a comparative view for the concerned users, primarily because of the fact that these can be inferred to have descended from the principles governing relational database systems. An example of this type is the Borealis stream processing engine (Conti et al. 2019). The project underlines three primary requirements for stream processing engines (SPEs), namely, revising the query results dynamically, modifying the results thus obtained dynamically, and making the system flexible and the scalability optimizable.

The second generation of stream systems, however, does not use the same relational-based user view methodology as used by the first generation. Some notable examples are Storm and S4. S4, an acronym for simple scalable streaming system, was a distributed processing system that could solve real-world problems in the context of search engines such as Google, Yahoo!, and Bing. The system made use of a "cost per click" billing model to generate its results. There exist so far two variations of the second-generation DSCPs, namely, micro-batching (stream processing in discrete batches) and per-event stream processing. The latter type of systems gives individual consideration to each and every event independent of the other contemporary or previously occurring events. They provide low latencies; however, the throughput also tends to be low as compared to micro-batching systems. Instances of systems that make use of per-event processing are Flink, S4, and Storm. A recently developed method of distributed data systems called batched stream processing systems has created a huge shift in the EP system's architecture. These systems introduce high latency but also come up with high throughput. Integrated later on with DryadLINQ, Comet was one of the first systems to make use of this technique. Both Storm Trident and Spark streaming are cases of systems that make use of the micro-batching technique. Some cases of the DSCPs that include stream and batch processing include Samza (Kleppmann 2019), Apache Flink (previously known as Stratosphere) (Hueske and Walther 2019), Trill, and Apache Apex (Gundabattula and Weise 2019). These have emerged as a result of a joint effort made in order to develop processing frameworks for Big Data. Most of these systems allow for a user-defined tradeoff between latency and throughput. These systems, however, do not directly fall under the category of traditional EPPs, but can instead be utilized to employ the EP functionality.

With the increasing importance and demand of cost-effective data centers around the world, geodistributed stream processing systems have come to the fore in recent times. The governing factors that have come to play an important role in the operation of an EP system installation are the general performance metrics of these systems, namely performance variability, throughput, and latency. In order to maintain a high-performing ESP across the various data centers, it is required that prominent changes be made in the traditional EP architectures. Adopting batched event transferring to facilitate an efficient event transfer is one such change. There exist, however, two key challenges in the process of creating event batches, namely determining the optimal size of the batch and determining the right time to trigger the send operation. One way to provide a dynamic adaptation for batch sizing in cloud based environments is to use the latency models; this would also allow for streaming on multiple routes across the various nodes. The evolution in this field is still underway, such as in the case of FT and declarative query languages. Recently, DSCPs such as Storm have undergone notable architectural enhancements. DSCPs that currently provide a query language that is similar to SQL are Spark Streaming, Storm, and Flink.

5.3.3 LIBRARIES FOR CEP

CEPs are stream processing systems used to correlate and analyze data about real-time events occurring in a system followed by deriving conclusions from the analysis of the aforementioned events (Hasan et al. 2019). What makes CEP different from other contemporary stream processing systems is that it allows for the defining of complex events on top of the simple ones (raw data), which helps the system in the identification of circumstances that hold some specific meaning, thereby allowing the system to generate instantaneous responses. Many different event processing languages (EPLs) that are domain specific, along with various engines for processing the events, exist, namely Apache Flink, Esper, and Microsoft Azure Stream Analytics. CEP programs are usually made up of a set of rules. Each rule in this set defines a pattern and triggers an event (complex) every time a given rule pattern makes a match for some event present in a stream.

The CEP library helps identify patterns, relationships, and abstraction of data among what seem to be unrelated events and triggers an answering response immediately. These libraries tend to be lightweight and can also be incorporated into alternate EP systems like EPPs and DSCPs. Some of the most common CPP libraries are ruleCore, Cayuga, and Siddhi. All the aforementioned libraries follow the open-source licensing scheme. Moreover, none of these libraries have support for geospatial data types. More than half of the EP systems surveyed so far have been programmed in Java (around 52% of them). About another 15% are programmed in Scala, and only a few have been written in C/C++ or C# (Cassandras 2018).

5.4 APPLICATION OF CEP IN REAL LIFE

The use of CEP in real-world applications has seen undeniable growth, especially in distributed and real-time environments. This makes it apparent that it has been widely accepted as a powerful and useful technology. It finds its use in many real-world applications. For example, the global positioning system (GPS), radio frequency identification (RFID) networks, IoT, healthcare systems, manufacturing, stock market analysis, data aggregation, etc. (Hedtstück 2017). The following subsections discuss some of the applications where CEP is employed.

5.4.1 CEP SYSTEMS IN MAUDE

Maude can essentially be defined as a language that is high-level and combines the rewriting of a system's logic specifications along with functional programming, which is done in an equational style. It provides an interpreter that is highly efficient and capable of executing specifications regardless of their type. In Maude, the use of rewriting of logic is done to provide CEP programs with semantics. Targeting Maude as the semantic domain is highly beneficial. This is because specifications based on Maude accommodate analysis being done multiple times; they are executable and facilitate simulation and the checking of several types of models (Burgueno et al. 2018). To employ this in Maude, an encoding of mechanisms is provided along with certain CEP concepts. A Maude toolkit is also provided to help analyze how such systems perform using stochastic behavior. Although CEP allows describing complex events using the help of events that are gathered from multiple sources and then taking suitable actions, when the complexity of such systems increases, it becomes more and more challenging to prove their effectiveness and to understand them as well (Nolan 2017). Formalizing CEP systems in Maude helps to mitigate these challenges, as Maude employs logic rewriting to analyze the properties of CEP formally and prove them.

5.4.2 PUBLIC MOOD TRACKING

The popularity of mobile applications based on providing social media platforms to millions of users has been increasing day by day. People all across the globe have become more and more comfortable in expressing their honest opinions and feelings on these social, digital, and public platforms. Mood tracking is one aspect of social media and the internet as a whole that has intrigued many researchers in terms of analyzing group activities. There are many reasons for this. Mood tracking can greatly facilitate medical recommendation that is personalized, enhance the analysis of polls done over such platforms, and predict the stock market. These are only a few from the long list of benefits that mood tracking can provide. Tracking and analyzing the mood of the public in real time from social platforms are challenging. This is primarily because of the sheer amount of data that is generated continuously from the data streams of social media platforms and needs to be dealt with. The data generated are mostly emotional and textual in nature (Koukaras et al. 2021). To mitigate this challenge, CEP is implemented. CEP is capable of providing an extremely efficient solution to the problem that is faced when dealing with such a large amount of constantly increasing data. Systems based on CEP are mainly used in mood tracking for the purpose of processing streams of event data and matching patterns. This is done against data generated that are very voluminous and high in speed. The data are acquired in real time from multiple sources. After necessary actions are performed, the results are found and submitted to the units that sought them. CEP employs techniques that are heavily used in real-time processing, text sentiment analysis, and mood analysis of the public (Koukaras et al. 2021). These techniques, when paired with certain statistical methods, are very beneficial. This is done by first using the sentiment analysis techniques on the available content and converting it into emotional events that can be further processed. Then a batch window technique is utilized to outline the mood events of the public. After this step is completed, methods of following trends and smoothing are used in order to alternate the points present in the time series of the public mood and identify trends that are falling and rising on an everyday basis. The usage of CEP in this task has proved to be highly efficient. Moreover, it has also been proved that it is extremely feasible as well.

5.4.3   CEP for Quantitative Finance Based on Data Processing

Quantitative finances (QF) makes use of market data and mathematical models to provide information that is quantitative in nature regarding all financial aspects (Zhang et al. 2019). The primary task of QF is modeling the available financial data, accurately pricing the assets and derivatives, and accurately predicting how the market will move in the future. Another task is to promptly identify the risks that might occur. All this is necessary, as it helps in intelligently and quickly reacting to the markets that keep changing at a very fast pace. This especially helps trading companies. The emergence of QF is very prominent and requires a wide array of technologies to perform CEP on the large amounts of diverse market-related data that currently exist. Integrating CEP with the infrastructure of Big Data is a very helpful real-world application for time-series data streaming (Munir et al. 2018). This paradigm is driven mainly by the available data. CEP is first executed in a mode that is data-driven, and in a data-parallel mode, the vastly diverse market data are managed. Processes like data aggregating and data cleaning are performed. This includes the computation of logarithmic returns, which is done on tick data, and finding its medians. Using CEP for this task results in a throughput that is very high, and in one second, the order of tick messages can go up to millions. It also leads to executing tasks in a data-driven manner, since these tasks are mostly data dependent. Another upside to using CEP is that due to the implementation approach being modular, the QF research models and the multiple methods for the crunching of data are developed rapidly to accommodate the quickly changing scenario of the market (Zhang et al. 2017). Ultimately, the combination of data-driven implementation and the infrastructure of Big Data has given engineers working in the financial field a way to incorporate many new and enhanced techniques in their work to increase its effectiveness.

5.4.4   CEP Risk Identification over the Internet

Nowadays, e-commerce has proved to be an extremely lucrative industry. This has led to it becoming a very vital part of the internet and the economy as well. Essentially, e-commerce can be referred to as online trading (Engelhardt and Magerhans 2019). This gives way to many risks that can hamper the process, as it involves customers' personal and financial information. The threat of a third party accessing this information and utilizing it for criminal and illegal activities has become more and more prominent due to the large number of loopholes that are possible in the process. For security measures, some traditional methods that are applied are encryption technology, authentication using digital certificates, verification codes, etc., but sometimes these methods do not ensure that the conduct of the user is legitimate (Kumari et al. 2019). For instance, malicious activities such as using API calls, spoofing identity, and multiple fake accounts can be participated in even if the user's identity is legal. Such cases are very common. Observing a particular user's data stream of online shopping in real time and efficiently detecting which activities can be potential risks greatly enhance the user's security. Hence, it is extremely crucial that such risks are identified and minimized. The user's behaviors are segregated according to their roles, as different kinds of users will have a different set of behavioral tasks. The contribution of CEP in risk identification is very notable. The CEP platform performs the conversion of initial streams of data into a stream of events (Ma et al. 2019). Events are essentially referred to as changes undergone in meaningful states. This conversion is carried out with the help of some very practical techniques, namely aggregation, byte filling, and filtering. After the conversion is brought to an end, the resultant stream of events that is generated is then sent to an engine for the formulation of event patterns based on the behaviors of users that can be considered a potential risk. At last, the risk events are then processed ahead so as to perform an operation involving early warning against the said risk. CEP is very effective in online risk identification because it generates a series of complex patterns of events. So, it effectively warns against possible online shopping risks in real time and in the early stages well before they occur. User risk identification is a very powerful way to ensure that the accounts and funds of the users involved are secure and safe from any undesirable threats.

5.4.5 CEP Dealing with Heart Failure Prediction

Out of all the cardiovascular diseases, heart failure is considered to be the most fatal and life-threatening. So, it is very important that the symptoms are detected early on to prevent negative consequences. Continuously checking up on the health of the patients and monitoring all the health metrics that are collected really help in such situations. An analysis such as this requires some thresholds that are predefined. The collected metrics are then compared to these thresholds. If certain metrics are lower or higher than the expected values, symptoms are detected. This demonstrates that defining accurate thresholds is extremely crucial in the process of predicting heart strokes. The more accurate they are, the better the analysis proves to be. When defined by cardiologists, these thresholds depend on the patient's many attributes like blood pressure, weight, and physical activity. Hence, they vary with differing patients. The CEP methodology finds its use in this process when combined with certain statistical approaches. This helps in finding out the required health parameters. A number of predefined rules of analysis are then executed to process the acquired data (Mdhaffar et al. 2017). These rules are heavily dependent on the patient's metric values compared to the threshold values and, hence, are modified constantly. The rules can also sometimes vary for the same patient at different points in time. If any undesirable result is generated, it is considered a symptom of possible heart failure and is communicated in the form of an alarm. The threshold values of a patient are calculated through statistical methods and updated continuously. These methods process historical data of the patient. For example, a higher body weight signifies an increase in the rate of fluid in the body. This is a symptom of heart failure. All possible symptoms are detected in a similar fashion. After the analysis, the cardiologist responsible takes the required steps necessary to reduce the risk. Some of the steps are adjusting the medication of the patient, asking the patient to stop by the emergency center nearest to them, or making an appointment for an urgent medical consultation. This approach is extremely helpful to limit the fatalities that are caused due to heart failures all across the world because of its promptness in analyzing a patient's records and updating the threshold continuously. This makes sure that the diagnosis done at any point in time is apt and is met with an appropriate response that ensures the safety of the patient.

5.5   CONCLUSION AND FUTURE RESEARCH DIRECTIONS

This chapter tried its best to cover all the important aspects associated with CEP so far. Initially, it discussed the various types of CEP engines that are put to use when working with CEP. These were briefly explained along with their motives. In addition to that, their usage in various fields and their limitations were also discussed. For example, IoT, robotics, smart cities, etc. Uncertainty in data is also one of the primary aspects of CEP. A detailed description of what uncertainty in data entails was given along with a list of CEP engines, where the main aim is to specifically target data uncertainty. Their motives, usage, and limitations were discussed as well. After that light was shed on the topic of EP system architectures. This involved looking into the various libraries and platforms that are provided to efficiently work on CEP. They were EPPs, DSCPs, and complex EPLs. These topics were elaborated, and their contributions were clearly explained. Then the role of CEP in real-world scenarios was explained. Some of the applications that were focused on were formalizing CEP systems in Maude, real-time public mood tracking, CEP for quantitative finance through data-driven execution, CEP for online risk identification, and CEP for heart failure prediction. Along with the applications, the details of their inner workings were also briefly explained. As far as future work is concerned, the study performed earlier can be referred to in order to formulate more efficient CEP systems. This study can also be extended to gain a deeper understanding of what CEP entails in order to perform more real-life practical implementations. This study has created a basis successfully, and the main aim is to make noteworthy contributions to future research and studies in the field of CEP.

REFERENCES

Bhushan, B., Sahoo, C., Sinha, P., & Khamparia, A. (2020). Unification of blockchain and internet of things (BIoT): Requirements, working model, challenges and future directions. Wireless Networks. 27, 55–90 (2021). doi:10.1007/s11276-020-02445-6

Bonino, D., & De Russis, L. (2018). Complex event processing for city officers: A filter and pipe visual approach. IEEE Internet of Things Journal, 5(2), 775–783. https://doi.org/10.1109/jiot.2017.2728089

Burgueno, L., Boubeta-Puig, J., & Vallecillo, A. (2018). Formalizing complex event processing systems in Maude. IEEE Access, 6, 23222–23241. https://doi.org/10.1109/access.2018.2831185

Conti, F., S. Colonnese, F. Cuomo, L. Chiaraviglio and I. Rubin, "Quality Of Experience Meets Operators Revenue: Dash Aware Management for Mobile Streaming," 2019 8th European Workshop on Visual Information Processing (EUVIP), 2019, pp. 64-69, doi: 10.1109/EUVIP47703.2019.8946152.

Debbi, H. (2017). Modeling and formal analysis of Probabilistic Complex Event Processing (CEP) applications. Modelling Foundations and Applications, Part of the Lecture Notes in Computer Science book series (LNPSE,vol. 10376), 248–263. https://doi.org/10.1007/978-3-319-61482-3_15

Engelhardt, J.-F., & Magerhans, A. (2019). ECommerce. Klipp & Klar, 123–177. https://doi.org/10.1007/978-3-658-26504-5_4

Erich, F., & Suzuki, K. (2016). Cognitive robot programming using procedural parameters and complex event processing. 2016 IEEE International Conference on Simulation, Modeling, and Programming for Autonomous Robots (SIMPAR) 2016, 61–66, doi: 10.1109/SIMPAR.2016.7862376. 2016, 61–66, doi: 10.1109/SIMPAR.2016.7862376.

Flouris, I., Giatrakos, N., Deligiannakis, A., Garofalakis, M., Kamp, M., & Mock, M. (2017). Issues in complex event processing: Status and prospects in the big data era. Journal of Systems and Software, 127, 217–236. https://doi.org/10.1016/j.jss.2016.06.011.

G. Cassandras, C. (2018). Event-driven control and optimization in hybrid systems. Event-Based Control and Signal Processing, 1st edition, 21–36. https://doi.org/10.1201/b19013-2

Geisler, S. (2020). Complex Event Processing (CEP). Encyclopedia of Big Data. Boca Raton: CRC Press. 1–7. https://doi.org/10.1007/978-3-319-32001-4_276-1.

Goyal, S., Sharma, N., Bhushan, B., Shankar, A., & Sagayam, M. (2020). IoT enabled technology in secured healthcare: Applications, challenges and future directions. Cognitive Internet of Medical Things for Smart Healthcare, 25–48. Switzerland: Springer International Publishing, Springer Nature. doi:10.1007/978-3-030-55833-8_2

Graham Kennedy, A. (2021). Handling uncertainty. Diagnosis, 64–86. Wiley, US. https://doi.org/10.1093/med/9780190060411.003.0005

Gundabattula, A., & Weise, T. (2019). Apache apex. Encyclopedia of Big Data Technologies, 41–51. Switzerland (AG 2019): Springer Cham, Springer Nature. https://doi.org/10.1007/978-3-319-77525-8_316

Hasan, M., Orgun, M. A., & Schwitter, R. (2019). Real-time event detection from the Twitter data stream using the twitternews+ framework. Information Processing & Management, 56(3), 1146–1165. https://doi.org/10.1016/j.ipm.2018.03.001

Hedtstück, U. (2017). Beispiele für event processing languages. Complex Event Processing, 63–73. Berlin, Heidelberg: Springer Vieweg. https://doi.org/10.1007/978-3-662-53451-9_6

Hedtstück, U. (2017). Complex event processing engines. Complex Event Processing, 75–91. Berlin, Heidelberg: Springer Vieweg. https://doi.org/10.1007/978-3-662-53451-9_7

Hedtstück, U. (2017). Regelbasiertes complex event processing. Complex Event Processing, 93–100. Berlin, Heidelberg: Springer Vieweg. https://doi.org/10.1007/978-3-662-53451-9_8

Hueske, F., & Walther, T. (2019). Apache Flink. Encyclopedia of Big Data Technologies, 51–58. Switzerland AG: Springer Cham, Springer Nature. https://doi.org/10.1007/978-3-319-77525-8_303

Incki, K., Ari, I., & Sozer, H. (2017). Runtime verification of IOT systems using complex event processing. 2017 IEEE 14th International Conference on Networking, Sensing and Control (ICNSC). 625–630. https://doi.org/10.1109/icnsc.2017.8000163

Kale, V. (2019). Big data stream processing. Parallel Computing Architectures and APIs, 1st edition, 335–360. UK: Chapman and Hall/CRC. https://doi.org/10.1201/9781351029223-25

Kleppmann, M. (2019). Apache samza. Encyclopedia of Big Data Technologies, 70–77. Switzerland AG: Springer Nature. https://doi.org/10.1007/978-3-319-77525-8_197

Koukaras, P., Tsichli, V., & Tjortjis, C. (2021). Predicting stock market movements with social media and machine learning. Proceedings of the 17th International Conference on Web Information Systems and Technologies. 436–444. https://doi.org/10.5220/0010712600003058

Kumari, A., Tanwar, S., Tyagi, S., & Kumar, N. (2019). Verification and validation techniques for streaming big data analytics in internet of things environment. IET Networks, 8(3), 155–163. https://doi.org/10.1049/iet-net.2018.5187

Li, F., Wang, N., Gu, Y., & Chen, Z. (2016). Effective privacy preservation over composite events with Markov correlations. 2016 13th Web Information Systems and Applications Conference (WISA). 215–220. https://doi.org/10.1109/wisa.2016.50

Lu, T., Zhu, Xu & Zhai, X. (2016). Multi-stage monitoring of abnormal situation based on complex event processing. Procedia Computer Science, 96, 1361-1370. https://doi.org/ 10.1016/j.procs.2016.08.181

Ma, Z., Yu, W., Zhai, X., & Jia, M. (2019). A complex event processing-based online shopping user risk identification system. IEEE Access, 7, 172088-172096. https://doi.org/10.1109/ access.2019.2955466

Mao, N., & Tan, J. (2015). Complex event processing on uncertain data streams in product manufacturing process. 2015 International Conference on Advanced Mechatronic Systems (ICAMechS), 2015, 583-588, doi: 10.1109/ICAMechS.2015.7287178.

Mdhaffar, A., Bouassida Rodriguez, I., Charfi, K., Abid, L., & Freisleben, B. (2017). CEP4HFP: Complex event processing for heart failure prediction. IEEE Transactions on NanoBioscience, 16(8), 708-717. https://doi.org/10.1109/tnb.2017.2769671

Munir, M., Baumbach, S., Gu, Y., Dengel, A., & Ahmed, S. (2018). Data analytics: Industrial perspective & solutions for streaming data. Series in Machine Perception and Artificial Intelligence. China: World Scientific. 144-168. https://doi. org/10.1142/9789813228047_0007

Nolan, B. (2017). The syntactic realisation of complex events and complex predicates in situations of Irish. Studies in Language Companion Series, 13-41. Dublin, John Benjamins Publishing Company. https://doi.org/10.1075/slcs.180.01nol

Peng, S., & He, J. (2016). Efficient context-aware nested complex event processing over RFID streams. WebAge Information Management, 125-136. Switzerland: Springer Nature. https://doi.org/10.1007/978-3-319-47121-1_11

Rakkesh, S. T., Weerasinghe, A. R., & Ranasinghe, R. A. C. (2016). Simulation of real-time vehicle speed violation detection using complex event processing. 2016 IEEE International Conference on Information and Automation for Sustainability (ICIAfS). 1-6. https://doi.org/10.1109/iciafs.2016.7946549

Sahal, R., Breslin, J. G., & Ali, M. I. (2020). Big data and stream processing platforms for Industry 4.0 requirements mapping for a predictive maintenance use case. Journal of Manufacturing Systems, 54, 138-151. https://doi.org/10.1016/j.jmsy.2019.11.004.

Sethi, R., Bhushan, B., Sharma, N., Kumar, R., & Kaushik, I. (2020). Applicability of industrial IoT in diversified sectors: Evolution, Applications and challenges. Studies in Big Data Multimedia Technologies in the Internet of Things Environment, vol. 79, 45-67. Switzerland: Springer Nature. doi:10.1007/978-981-15-7965-3_4

Shukla, N., & Bhushan, B. (2021). Attacks, vulnerabilities, and blockchain-based countermeasures in internet of things (IoT) systems. Blockchain Technology for Data Privacy Management, 295-316. doi:10.1201/9781003133391-14

Tommasini, R., Bonte, P., Della Valle, E., Mannens, E., De Turck, F., & Ongenae, F. (2017). Towards ontology-based event processing. Lecture Notes in Computer Science, Springer, Springer Nature Switzerland pp 115-127. https://doi.org/10.1007/978-3-319-54627-8_9

Wang, Y., Gao, H., & Chen, G. (2018). Predictive complex event processing based on evolving Bayesian Networks. Pattern Recognition Letters, 105, 207-216. https://doi. org/10.1016/j.patrec.2017.05.008

Zhang, P., Shi, X., & Khan, S. U. (2017). Can quantitative finance benefit from Iot? Second ACM/IEEE Symposium on Edge Computing: Workshop on Smart IoT (SmartIoT'17) at San Jose / Silicon Valley, California. https://doi.org/10.1145/3132479.3132491

Zhang, P., Shi, X., & Khan, S. U. (2019). QuantCloud: Enabling big data complex event processing for quantitative finance through a data-driven execution. IEEE Transactions on Big Data, 5(4), 564-575. https://doi.org/10.1109/tbdata.2018.2847629

Zhou, Q., Simmhan, Y., & Prasanna, V. (2017). Knowledge-infused and consistent complex event processing over real-time and persistent streams. Future Generation Computer Systems, 76, 391-406. https://doi.org/10.1016/j.future.2016.10.030

6 Portunus

Enhancing Smart City Application Connectivity with a Complex Space-Time Events Distributed System

Basmi Wadii and Azedine Boulmakoul

CONTENTS

6.1 INTRODUCTION

We are living in a world mobilized by a large swarm of smart devices that constitutes Industry 4.0. Companies worldwide alike are participating in the transformation by creating new technologies that enable new concepts to surface and bring value to societies. Among these technologies is the Internet of Things (IoT), which has become part of our daily lives discretely and overnight. IoT represents a network of devices collecting data and communicating with each other to accomplish a task depending on the context of the application they are operating within. IoT, alongside other technologies such as information and communication technology (ICT), Big Data, and artificial intelligence, plays a huge role in helping governors to make decisions in favor of improving their citizens' quality of life—such countries or cities are called "smart cities". It is a technologically modern urban area that employs electronic methods, software, and human intelligence to promote resource utilization efficiency and hasten the technological advancement and awareness of the public. Smart cities provide an ecosystem that is rich in data in terms of volume, variety, versatility, velocity, and most importantly, value. Modern machine learning technologies are not useful unless the data samples on which they operate are valid and useful. Smart cities provide quality of life for citizens but also quality data about all sorts of events that happen within its borders and beyond. For that reason, there exist concepts that help with understanding the conjunctions that occur within its event-driven parts, such as complex events.

This chapter confers a complex events framework as a system that extends the definition of sensors and events that characterize IoT to cover other sorts of events that occur in a smart city and showcases the potential of the data we can unveil using complex events. We divided this work in four separate sections. In Section 6.2, we explore what a smart city is and how IoT plays a huge role in modern societies to improve urban services and its citizens' quality of life. In Section 6.3, we present a definition of complex events processing in the literature and gradually move to how it is implemented in our work. Then in Section 6.4, we introduce the Portunus application data model that regroups all the different entities in play. Finally, in Section 6.5, we discuss the logical architecture of Portunus as a microservice-based system and the role of each of its components.

6.2   SMART CITY AND IoT APPLICATIONS

The IoT is a network of interconnected physical and virtual devices that collect and exchange data (Lo et al. 2019). They are integrated as part of systems that use cloud computing, machine learning, and data analysis to extend their range of capabilities (Panarello et al. 2018). IoT is playing the role of the physical body through which information technology can extract and analogically form a perception of the real-world properties, and in return gives an output that can be perceived by external elements, especially with the advancement in telecommunication infrastructure. The employment of IoT in day-to-day life, urban services, or in overall a large majority of an individual life brought to life the term "smart city", although its first usage dates to almost two decades ago now. A smart city is a city where governors and companies use IT, and specifically IoT technologies, to ease access to their urban services in order to promote quality of life for citizens and boost the economy (Basmi et al. 2020) (Kummitha and Crutzen 2019).

Before attaining Industry 4.0, one of the marking technologies human societies had achieved is agriculture. With the billions living on the planet, we have made revolutionary progress throughout the last centuries to be able to withstand the huge demand caused by large gas emissions and natural resources exhaustion, but succeeded in eliminating hunger in most of the modern countries. Today, we are employing IoT technologies to control the production chain without restricting ourselves to the time period or the geography to produce exotic products. Known as precision agriculture, it is the process of combining diverse information such as temperature, humidity, soil (physical-chemical soil characteristics, topography, productivity data), and historical data (Wolfert et al. 2017) to increase the efficiency and productivity in order to promote crop growth and surpass natural limits (Mazon-Olivo et al. 2018) (Hsu et al. 2020). A toll road is another example of IoT-based automation. In fact, a decade ago, a payment checkpoint was managed by full-time employees that used to work in shifts, but now, these checkpoints are totally managed by smart devices

that interact with drivers with smart dedicated tags, saving in the process time and
resources, and even reducing air pollution and improving public health (Lin et al.
2020). Therefore, IoT is a cheap technology—cheap production and small amount of
human resources are required, they take less space, are replaceable, and are designed
to consume less energy, and their output is by far unmatched and incomparable. IoT,
coupled with other modern technologies, is revolutionizing the job market, as 45
million Americans will lose their jobs by 2030 (Lund et al. 2021). Nonetheless, it
may render some jobs obsolete, but it is the driver of newer kinds of opportunities.
Accordingly, governments will adopt new politics to provide a sustainable economic
model that will push next generations to take a different leap.

Equipping a city with intelligent sensors connected to cloud-based solutions is
not the only characteristic to qualify a city as smart. In fact, the "smart" dimension
is a side effect of the collective effort in advancing modern technologies, including
Big Data, artificial intelligence, the high-speed internet that is covering the globe,
and the push toward effective clean energy (Thornbush and Golubchikov 2021). It is
also a trait that is cultivated in cities that already possess a growing economy, a well-
established evolving infrastructure, and who have continuously invested in its human
and social capital, its transport and its modern (ICT) communication infrastructure,
with the purpose of achieving a high quality of life for its citizens without repercussions on its natural resources (Caragliu and Bo 2016).

Surprisingly, there isn't a consensus on the development of smart cities, as the
actual implementations differ depending on their priorities and problems that their
governors aim to solve. For instance, Italy and more than 250 cities aim to improve
the quality of life, while reducing the energy consumption in order to decrease their
carbon footprint and move toward a sustainable approach to urban development
(Albino et al. 2015). On the other hand, the Spanish strategy focused on five different aspects: mobility, economy, environment, government, living, and people, and
over time it has converged on smart mobility, an efficient consumption of energy and
water, and reduced carbon emissions. Thus, Spanish smart cities have exceedingly
succeeded in attaining a desirable quality of life (Alet'a et al. 2017).

6.3 COMPLEX EVENT PROCESSING

Complex event processing (CEP) is the function of detecting complex patterns from
a large volume of events that are produced by event-driven computing systems when
they communicate their state changes with each other (Fardbastani and Sharifi 2019)
(Dayarathna and Perera 2018). Complexity, i.e., the variety of the events such a concept
covers, makes it impartially powerful to expand over many different types of events
originating from various event-driven systems, such as road traffic control and incident
management and security systems. CEP helps to identify the patterns in a group of
primitive events, mostly in real time but also on a higher or more abstract level (Wang
et al. 2018). By applying data mining or machine learning algorithms, it is possible to
predict new events based on the observation history and the related phenomenon attributes (Akbar et al. 2017). One of the main problems CEP aims to solve is to detect the
element of casualty between the stream of events; i.e., within a large dataset of events
from different categories, it aims to find instruments to control their occurrence

6.4   PORTUNUS APPLICATION DATA MODEL

Complex events are an abstract concept and are handled differently depending on the use case. At first, we have approached this concept by identifying it as a concept to regroup IoT events, which generally falls in the category of natural and industrial phenomenon. For example, temperature, air quality, visual data streams, traffic congestion control devices, motion detectors, etc. For such events, there exists the OGC Sensor Things application programming interface (API) data model. OGC is an organization aiming to provide an open geospatial-enabled and unified way to interconnect IoT devices and applications over the Web (OGC 2020). Inspired by their work, we designed a data model featured in Figure 6.1 that represents complex events: events that rise from event-driven systems. We define all event emitters as sensors: entities capable of observing a phenomenon, that is, an abstract concept, which is anything that generates data of absolutely any form in the real world. The "Datastream" entity is a concept in which we group the complex events that are characterized with a value, the content type of the value (numerical value, JSON, SVG, or bytes), a geolocation, and a timestamp. Additionally, it is also bound to one phenomenon, which could be natural, urban, industrial, or digital. For instance, in an urban city, the train schedule is a phenomenon that generates different types of events such as "the train arrival", "the train departure", and "delays". Each are state changes and can be attributed with values, a timestamp, and a geolocation. Another example could be a traffic accident, which can also be attributed with an enumerated

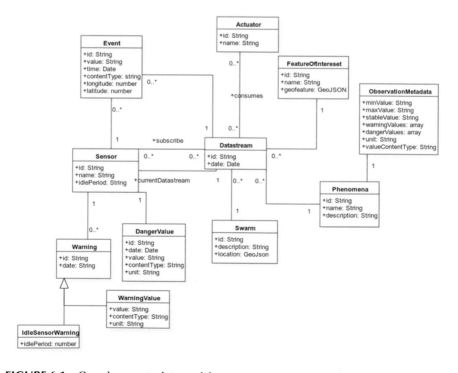

FIGURE 6.1   Complex events data model.

set of values, the time, and the geolocation where it happened. The observation metadata entity represents extra information about the generated events, such as the value unit, content type, its range, its status quo, and the values by which we can identify its warning and danger levels. Actuators are things that consume the data stream's complex events—they are entities that transform the digital events into either a work observable in the physical world or in another event and thus form a new phenomenon which can also be observed and collected.

6.5 PORTUNUS APPLICATION SERVICES ARCHITECTURE

Modern applications are now preferably designed in an architecture separated into small services that communicate with each other either synchronously or asynchronously using a message broker. The benefits are huge, starting from separation of the application domain features, which enforces the use of appropriate technologies and increases the performance of the running instances of its modules. In fact, as software grows and features are added, problems and bugs arise and hinder the speed of both the executable and the development. There is also the concept of nanoservices in which an application is separated into very small blocks of codes deployed on cloud services to perfectly optimize the cost of the application. However, it is only applicable efficiently when these functionalities are running independently over the lifetime of the application. In the context of Portunus, as illustrated in Figure 6.2, we have separated the application into four different services:

- **Data collection and distribution service:** This is the service responsible for receiving and distributing complex event things observations.
- **System things authentication:** This is the service that holds a database of all registered complex event things, and it provides an API to verify and generate tokens for connected objects.

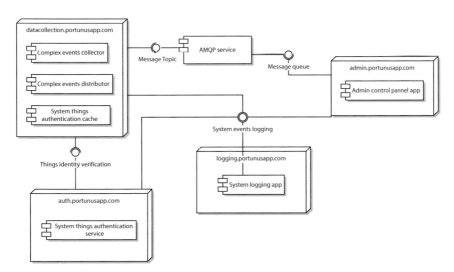

FIGURE 6.2 Portunus platform system architecture.

- **System admin service:** This is the service which gives a human interface to manage connected complex event objects, data streams, view their observations, and create reports.
- **System logging service:** This is the service that provides an API for all system services to register their system event logs.

6.5.1 DATA COLLECTION AND DISTRIBUTION SERVICE

Complex events are produced by sensors with average to high activity, i.e., they emit many observations within a short period of time. Moreover, as complex events are diverse, the total of their stream should be even bigger, and thus, it is necessary to have a distinct physical process to handle those objects. We have developed two different Node.js applications which we think is the best choice to handle the large number of incoming requests thanks to its event-driven architecture. The data collection and distribution service implement three servers:

- **HTTP:** The standard web protocol, which is the base of the World Wide Web.
- **COAP:** It's a light version of HTTP designed specifically for low-energy-consumption devices.
- **MQTT:** It's a publish/subscribe type of messaging transport protocol designed primarily for Machine-to-Machine (M2M) communication with low bandwidth and for unreliable networks with very low consumption of power (Mishra and Kertesz 2020).

The data collection and distribution service owns its own database and puts a copy of its collected complex events on the topic exchange for it to be available for the administration panel to consume asynchronously and collaborates synchronously with the authentication service to verify the identity of connected objects before accepting their incoming observations.

6.5.2 SYSTEM THINGS AUTHENTICATION

The authentication system for things is a standalone service that utilizes a private blockchain to store the system objects' (sensors and actuators) credentials. Sensors and actuators that are registered in the system generate a pair of public and private keys, which are used to encrypt and decrypt their data, respectively. Every time a sensor sends an event to the data distributor, it must encrypt it using its public key, alongside a data stream–generated subscription token. Generally, writing to a blockchain is slow, and it gets slower in correlation to the ledger size; however, reading data is as fast as reading from a regular database.

6.5.3 SYSTEM THINGS AUTHENTICATION

The system admin service consists of a web application that offers important features listed in the use case diagram in Figure 6.3 to manage the system things (sensors and actuators) and data streams with their attributed phenomena, features of interest, and all the necessary metadata to create reports with flagged values. On top of that,

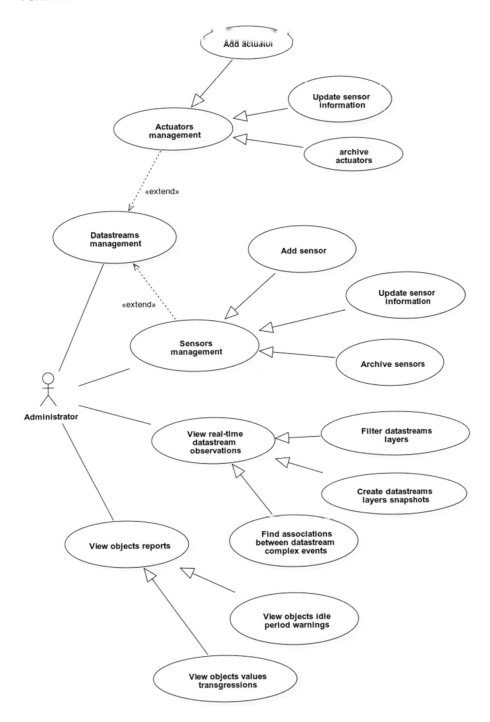

FIGURE 6.3 Portunus admin use cases.

it visualizes the data stream graphical representations on a map; actually, we mentioned previously that every data stream is associated with collected complex events, and each possess a geolocation and a timestamp. The system admin service offers a page where the administrator can create layers made of filters. These layers can be saved as snapshots and be part of reports to analyze and process later.

6.6 CONCLUSION

A smart city involves its human capital and knowledge to use information technology and a telecommunication infrastructure as a means to improve their citizens' quality of life following a specialized approach that may differ from other smart cities. The IoT is a technology that allows the optimization of resources and maximizes all its applications' performance indicators. More importantly, one of smart cities' characteristics is to combine the flux of data in order to solve very particular problems. Such a function is known as CEP, the action of identifying patterns and generalizing the concept of an event-driven system to include all sorts of state changes that occur within the perimeter of the city.

In this chapter, we propose the architecture of Portunus, a complex space-time event processing framework. Its data model was inspired by OGC Sensor Things API and directed to cover complex events as entities produced by event-driven systems experiencing a periodic series of state changes, not necessarily natural or industrial, if it has an impact on the physical world. Events are grouped in the data stream concept, an entity that relates to a phenomenon, a feature of interest, sensors that transform those observations into digital data, and actuators that acquire the data to transform it into an observable event. We presented the platform architecture as a microservice-based system composed of four services, each responsible for collecting and distributing the data, identifying connected objects, providing a human interface to analyze the reports, and logging the system events. Its data collection and distribution features support three different protocols: MQTT, HTTP, and COAP. Using small devices and connected third-party applications, its identification service uses a private blockchain to safely save its data; its admin service consists of a user interface (UI) application to manage the system objects and create and view reports using layers made of data streams.

REFERENCES

Akbar, A., A. Khan, F. Carrez, and K. Moessner (2017). Predictive analytics for complex IoT data streams. *IEEE Internet of Things Journal* 4(5), 1571–1582.
Albino, V., U. Berardi, and R. M. Dangelico (2015). Smart cities: Definitions, dimensions, performance, and initiatives. *Journal of Urban Technology* 22(1), 3–21.
Alet'a, N. B., C. M. Alonso, and R. M. A. Ruiz (2017). Smart mobility and smart environment in the Spanish cities. *Transportation Research Procedia* 24, 163–170. 3rd Conference on Sustainable Urban Mobility, 3rd CSUM 2016, 26–27 May 2016, Volos, Greece.
Basmi, W., A. Boulmakoul, L. Karim, and A. Lbath (2020). Modern approach to design a distributed and scalable platform architecture for smart cities complex events data collection. *Procedia Computer Science* 170, 43–50. The 11th International Conference on Ambient Systems, Networks and Technologies (ANT)/The 3rd International Conference on Emerging Data and Industry 4.0 (EDI40)/Affiliated Workshops, edoi: 10.1016/j.procs.2020.03.008.

Caraglu, A. and C. F. D. Du (2016). Do smart cities invest in smarter policies? learning from the past, planning for the future. *Social Science Computer Review* 34(6), 657-672.

Dayarathna, M. and S. Perera (2018). Recent advancements in event processing. *ACM Comput. Surv.* 51(2), 33.

Fardbastani, M. A. and M. Sharifi (2019). Scalable complex event processing using adaptive load balancing. *Journal of Systems and Software* 149, 305-317.

Hsu, T.-C., H. Yang, Y.-C. Chung, and C.-H. Hsu (2020). A creative IoT agriculture plat- form for cloud fog computing. *Sustainable Computing: Informatics and Systems* 28, 100285. doi: 10.1016/j.suscom.2018.10.006.

Kummitha, R. K. R. and N. Crutzen (2019). Smart cities and the citizen-driven internet of things: A qualitative inquiry into an emerging smart city. *Technological Forecasting and Social Change* 140, 44-53. doi: 10.1016/j.techfore.2018.12.001.

Lin, M.-Y., Y.-C. Chen, D.-Y. Lin, B.-F. Hwang, H.-T. Hsu, Y.-H. Cheng, Y.-T. Liu, and P.J. Tsai (2020). Effect of implementing electronic toll collection in reducing highway particulate matter pollution. *Environmental Science & Technology* 54(15), 9210-9216. PMID: 32589404.

Lo, S. K., Y. Liu, S. Y. Chia, X. Xu, Q. Lu, L. Zhu, and H. Ning (2019). Analysis of block-chain solutions for IoT: A systematic literature review. *IEEE Access* 7, 58822-58835. doi:10.1109/ACCESS.2019.2914675.

Lund, S., A. Madgavkar, J. Manyika, S. Smit, K. Ellingrud, and O. Robinson (2021). The future of work after COVID-19. Accessed March 14, 2022. https://www.weld.gov/files/sharedassets/public/departments/human-services/documents/the-future-of-work-after-covid-19-executive-summary.pdf

Mazon-Olivo, B., D. Hernández-Rojas, J. Maza-Salinas, and A. Pan (2018). Rules engine and complex event processor in the context of internet of things for precision agriculture. *Computers and Electronics in Agriculture* 154, 347-360.

Mishra, B. and A. Kertesz (2020). The use of MQTT in M2M and IoT systems: A survey. *IEEE Access* 8, 201071-201086.

OGC. (2020). OGC SensorThings API part 1: Sensing version 1.1. https://docs.ogc.org/is/18-088/18-088.html Accessed: 2022-04-09.

Panarello, A., N. Tapas, G. Merlino, F. Longo, and A. Puliafito (2018). Blockchain and IoT integration: A systematic survey. *Sensors* 18(8), doi:10.3390/s18082575.

Thornbush, M. and O. Golubchikov (2021). Smart energy cities: The evolution of the city energy sustainability nexus. *Environmental Development* 39, 100626.

Wang, Y., H. Gao, and G. Chen (2018). Predictive complex event processing based on evolv-ing Bayesian networks. *Pattern Recogn. Lett.* 105(C), 207-216.

Wolfert, S., L. Ge, C. Verdouw, and M.-J. Bogaardt (2017). Big data in smart farming – a review. *Agricultural Systems* 153, 69-80.

7 Smart Trajectories Meta-Modeling

Lamia Karim and Azedine Boulmakoul

CONTENTS

7.1 INTRODUCTION

The technological evolution of positioning devices has enabled easy, low-cost, and high-quality capture of the spatio-temporal coordinates of human beings, vehicles, and any moving object. Once these coordinates are organized as a trajectory shape, they can describe individuals' movements over a period. In addition, they hold knowledge useful for the analysis and the prediction for many services based on location.

The discovery of patterns of moving object trajectories representing high traffic density has been widely studied in various works using various approaches. These models are useful in areas such as transportation planning and traffic monitoring. Models for the discovery of high traffic flow patterns in relation to urban traffic congestion also exploit vehicle trajectories. In our previous work, we have deployed trajectories in several areas such as logistics, pedestrians, network congestion, transport of hazardous materials, etc. Trajectory management difficulties are due to several factors such as (i) very frequent updates of the geographical positions of the mobile object, (ii) conventional databases do not offer adequate management of the trajectories of moving objects from a spatio-temporal point of view, and (iii) the trajectory presentation models are specific to use cases and are part of a transactional context.

DOI: 10.1201/9781003255635-7

113

The convergence of the work we have carried out has led us to reflect on the unification of trajectory models. By unification, we express the genericity of the smart trajectory meta-model.

After the introduction, the structure of this chapter is organized as follows. Section 7.2 discusses existing models for presenting the trajectories of moving objects. It also compares existing models with the smart trajectories meta-model. Section 7.3 describes the necessary class diagrams of the smart trajectories meta-model. It presents the pattern class factory to schematize the structural process of instantiation of trajectories. This section also presents simplified instantiations of the smart trajectory model to model congestion trajectories, to model the trajectories of the ego-vehicle, to model pedestrian trajectories (a road safety project), and to model OGC® IndoorGML trajectories. Finally, the last section concludes the chapter.

7.2 COMPARISON BETWEEN THE MODELS FOR REPRESENTING TRAJECTORIES

In Table 7.1, we compare the models presented in the literature with the smart trajectories meta-model.

TABLE 7.1

Comparison of Existing Models of Trajectories with the Smart Trajectories Meta-Model

	Semantic Trajectory	Based on Ontology and Event Approach	OGC Observation Types	Hazardous Materials Transport Trajectories	Congestion Trajectories	Trajectories of Fraud	Pedestrian Risks Trajectories.	OGC® IndoorGML Trajectories
DSTTMOD (Meng and Ding 2003)	-	-	-	-	-	-	-	-
Spatio-Temporal Moving Objects (Wolfson et al. 1998)	-	-	-	-	-	-	-	-
Conceptual view on trajectories (Spaccapietra et al. 2008)	✓	-	-	-	-	-	-	-
Spatio-semantic hybrid model (Yan et al. 2010)	✓	-	-	-	-	-	-	-
Regions of interest (Giannotti et al. 2007)	-	-	-	-	-	-	-	-
Unified meta-model (Boulmakoul et al. 2012 2013)	✓	✓	✓	✓	-	-	-	-
Smart trajectories meta -model	✓	✓	✓	✓	✓	✓	✓	✓

7.3 SMART TRAJECTORIES META-MODEL

Today, ontologies are considered one of the key paradigms for semantic interoperability. They are made up of the vocabulary of terms and relationships between the
different notions of a domain. As a result, ontologies provide a unifying framework
and specify a common language, which avoid ambiguity about a given term and promote communication. This communication can be between people, between people
and the system, and between systems.

The smart trajectories meta-model in Figure 7.1 uses event ontology to provide
a new, simple, and usable way to model trajectory data and visually analyze it. For
the sake of simplicity, we briefly express the modeling elements necessary for the
understanding of the meta-model. Then, we present the pattern class factory to schematize the structural process of instantiation of trajectories. Furthermore, we present simplified instantiations of the smart trajectory meta-model to model congestion
trajectories, to model the trajectories of the ego-vehicle, to model pedestrian trajectories, to model OGC® IndoorGML trajectories, and to model trajectories of frauds.
The dominant classes in the modeling of the smart trajectories meta-model will be
described in the next subsection.

7.3.1 SEMANTICS OF DESIGN ELEMENTS OF THE SMART TRAJECTORIES META-MODEL

Advances in sensor technology make it easy to capture events with better accuracy.
Several works studied "events" and related concepts. According to Shoe's general
ontology (Heflin et al. 2022), an event is something that happens at a given place
and time. Work in (Quine 1985) describes events as bounded regions in space
and time.

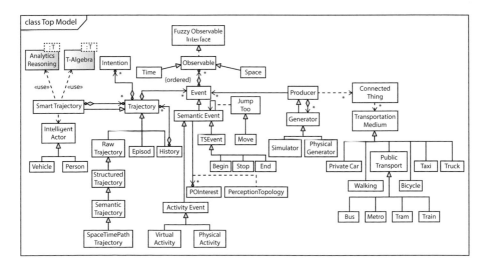

FIGURE 7.1   Smart trajectories meta-model.

In our smart trajectories meta-model illustrated in Figure 7.1, a *Spatio-Temporal Event* is composed of *Observables* of type space, time, or measurement, whereas an *Event* is produced by a *Producer*. An *Observable* can be a fuzzy physical quantity defined by an algebra. The *Producer* refers to *Connected Objects* linked to transport vehicles such as a taxi, truck, or car. It is an aggregate of *Generators* of the *Simulator* type or of the *Physical Generator Type* as sensors.

Indeed, to locate a mobile object in space and time, we have two categories of systems. The first category allows localization outside buildings called outdoor geolocation; it allows objects to be positioned using satellite systems and cellular systems. The second category, called indoor localization, makes it possible to locate mobile objects in a place in which access to satellites and global positioning system (GPS) data are not available, called indoor localization; it mainly uses Wi-Fi and Bluetooth.

However, it is better to combine different techniques to have better precision. Indeed, GPS cannot be used inside buildings, because the Global System for Mobile Communications (GSM) system is dependent on geographical coverage and the General Packet Radio Service (GPRS) network, and Wi-Fi does not work in the absence of Wi-Fi access points. For example, to track a person outside and keep his or her trajectory inside a shopping center, we will need to combine two technologies GSM and Wi-Fi. Time is equally present and important. A lot of data has a temporal component or has the potential for a related temporal component and a spatial component, such as the time and place of departure of an airplane. Most data are associated with a location, many of which are also associated with a moment in time, a time interval, or some other time component. *Spatio-Temporal* data can be used to assess the movement of the trajectory and/or understand the behavior of a moving object like a vehicle, human being, or valuable material. A *Semantic Event* is an event enriched with a location service *Toponym*, and it is associated with a point of interest and has a semantics linked to the application domain.

For trajectories, *Semantic Events* are categorized into start, stop, and end events according to the phenomenon governed by the trajectory. An *Event* is linked using a causal relationship to another event through the relationship *Jump To*. The *Jump To* relation is carried out according to the modeled phenomenon. To have a more indepth analysis on the trajectories, another type of event has been integrated—it is an *Activity Event*. This last is a *Semantic Event* materializing physical activities such as listening to music, observing a billboard, buying a product, and/or virtual activities such as sending e-mail or phoning. The *Perception-Topology* class relation denotes the neighborhood relationship of a mobile to capture nearby points of interest like restaurants, banks, administrations, pharmacies, etc.

A *Trajectory* is made up of an ordered set of *Events*; it has an intention like shopping, getting to work, going home, etc. A *History* is a trajectory made up of *Trajectories,* whereas an *Episode* is a *Trajectory* and a story is made up of *Episodes*. The trajectories are of a type presentation: raw, structured, semantic, space-time path, or smart. A *Raw Trajectory* is made up of raw events, i.e., signals composed of space, time, and measurement. The *Structured Trajectory* creates a raw trajectory by creating "topological" links between the raw events, whereas a *Semantic Trajectory* specializes a structured trajectory, and it is made up of semantic events. A *Space-Time Trajectory* creates a *Semantic Trajectory;* it consists of activity type events.

On the other hand, *Smart Trajectory* is a trajectory and itself can be composed of trajectories, and it uses the generic algebra of trajectories and exploits generic analytical primitives. In fact, *Smart Trajectory* is intended for an intelligent *Actor*, of vehicle type, or of person type, etc.

After having detailed semantics of design elements of the *Smart Trajectories* meta-model, we present in the following subsection the *Pattern Class Factory* that allows generating the different models of trajectories.

7.3.2 PATTERN CLASS FACTORY

The pattern class factory illustrated in Figure 7.2 schematizes the structural process of instantiation of different types of trajectories: congestion trajectories, trajectories of the ego-vehicle, pedestrian risks trajectories, OGC® IndoorGML trajectories, fraud trajectories, etc.

The *AbstractFactory* class encapsulates the different presentations of a smart trajectory like the raw trajectory *ConcretFactoryRAW*, structured trajectory *ConcretFactorySTR*, semantic trajectory *ConcretFactorySEMT*, trajectory based on the regions of interest, and the space-time path *ConcretFactorySTP*. The intention of using the *AbstractFactory* classes is to make our smart trajectory model extensible, i.e., we can add a new trajectory presentation without modifying the code.

In the following, we present simplified instantiations of the smart trajectory model:

* Simplified instantiation of the meta-model to model congestion trajectories
* Simplified instantiation of the meta-model to model the trajectories of the ego-vehicle
* Simplified instantiation of the meta-model to model pedestrian trajectories
* Simplified instantiation of the meta-model to model IndoorGML trajectories
* Simplified instantiation of the meta-model to model fraud trajectories

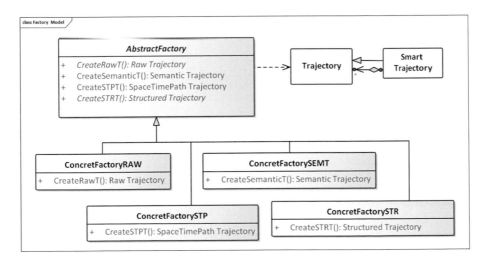

FIGURE 7.2   Factory model for trajectory instantiation.

7.3.3 MODELING CONGESTION TRAJECTORIES

Traffic congestion in the transport networks of major cities is an undeniable socio-economic problem that has arisen in many countries around the world. In this context, understanding the different traffic regulation and management operations is important to successfully manage the complexity of traffic congestion phenomena in networks (Boulmakoul et al. 2015) (Karim et al. 2017). Congestion could be defined as excessive when the marginal costs affecting society exceed the marginal costs of efforts to reduce its effects. The factors that cause congestion can be linked to micro-economic considerations relating to the road infrastructure. They may also be affected by macro-economic phenomena related to the demand for road use and depending on a set of realities related to traffic patterns and volumes. The instantiation of the meta-model makes it possible to model congestion as an illustrated trajectory. This model of trajectory improves the efficiency of road traffic and alleviates congestion, based on the analysis of traffic data.

Figure 7.3 illustrates the model of a congestion trajectory from the instantiation of the smart trajectory meta-model. The idea is to consider the *Traffic Event* as a spatio-temporal event. Indeed, the *Traffic Event* is a *Congestion Event* type, and it is composed of several physical or fuzzy quantities corresponding to observables. The first mandatory quantity is the time that allows specifying either the instant *Tm_ Instant* or the period of occurrence of the event *Tm_Period*. The second quantity

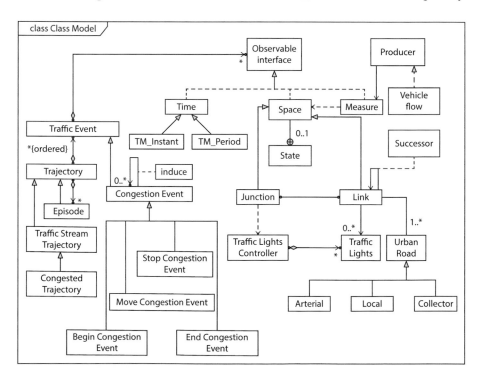

FIGURE 7.3 Simplified instantiation of the smart trajectory meta-model to model congestion trajectory.

that is also interesting for the modeling and analysis of urban traffic congestion is
space. It is characterized by a *Symbolic State* that can be either congested or free
In the case of *Traffic Congestion*, the *Space* corresponds to *Links* or *Junctions*. The
Region Of Interest is presented as an *Urban Network*. Indeed, a road network is
Arterial, Local, or *Collector* type. A *Trajectory* is a composition of links where each
link has an input and output *Junction*, and is characterized by a *State* at a specific
time. Moreover, a *Traffic Light* controller monitors it, whereas a *Vehicle Flow* type
Producer delivers these measurements.

Given a spatio-temporal network and road segments, the propagation model of
traffic congestion can be modeled by a trajectory. Each *Trajectory* is an ordered
composition of *Traffic Event*, which can correspond to a start of congestion, *Begin
congestion Event*; diffusion of a congestion, *Move congestion Event*; stop of a congestion, *Stop Congestion Event;* or the end of a congestion, *End Congestion Event*.
An event of type *Congestion Event* can induce another type *Congestion Event*; this
causality relationship is presented through the induce relationship, assuming that
congestion is a phenomenon that occurs somewhere at a given time that spreads
to other places and ends on a given date. Congestion has a *History* and *Episodes*.
The congestion propagation network characterizes the relationships between con-
gestions. As stated in the previous section, the proposed meta-model is suitable for
several situations and phenomena. The following section demonstrates the adaptability of the meta-model to present by simplified instantiation the trajectories of the
ego-vehicle model.

7.3.4 MODELING EGO-VEHICLE TRAJECTORIES

Future connected cars are equipped with computer systems with significant computing capabilities. On the one hand, these systems ensure the storage, access, management, analysis, and processing of large volumes of data, some of which will have
been produced by cameras and various sensors. These volumes of data could and
will support a range of purposes and will be processed and evaluated by the on-
board computer system. For driving in an urban transport network, knowledge of
the position and type of environment of the intelligent vehicle is essential information that allows advanced driver assistance systems or autonomous cars to perform autonomous driving maneuvers, for example, the mobile browser is assumed to
detect nearby objects. Indeed, without knowing the current location and the type of
road environment (residential area, industrial area, pedestrian area, etc.), it is very
difficult to perform driving maneuvers autonomously.

Existing location solutions are based on a combination of the global navigation
satellite system, an inertial measurement unit, and a digital map. Detection of the
type of environment is based on the geolocation system and data from laser type
sensors and/or videos. The events collected by the ego-vehicle generate usable trajectories for analysis (Boulmakoul et al. 2020). These trajectories are modeled by a
simple instantiation of smart trajectories. Figure 7.4 presents simplified instantiation
of the smart trajectories meta-model to model the trajectories of the ego-vehicle.

The smart trajectory uses the *Vehicle Interface*, which presents *Self-driving
Vehicle* and *Autonomous Driving Vehicle*. The different types of vehicles are located

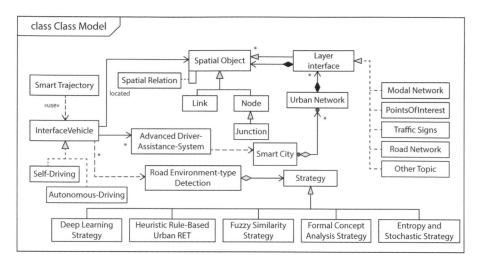

FIGURE 7.4 Simplified instantiation of the smart trajectories meta-model to model the trajectories of ego-vehicle.

in *Spatial Object,* which can be *Link, Node,* or *Junction* type objects. A *Spatial Object* is a composition of layers that can be *Road Network, Traffic Signs, Modal Network,* or *Point of Interest.* An *Urban Network* is composed of *Layers.* A *City* is made up of several urban networks depending on the mode.

An *Urban Network* consists of several layers with a geographic theme. *Spatial Objects* interact through spatial relationships that could be topological. On the other hand, connected cars will be able to exploit hardware and software resources hosted in other—static or mobile—sites so that they can exploit the extremely high performance available at those sites. In this way, advanced data analytics can be performed for real-time decision making for the connected car, for example, dynamic route planning. To detect the type of road environment, the *Interface Vehicle,* which represents the different types of vehicles, uses a strategy of road environment type detection that could be *Entropy, Stochastic Strategy, Deep Learning Strategy, Heuristic Rule Based Urban Ret, Fuzzy Similarity Strategy,* or *Formal Concept Analysis Strategy.*

After presenting in this section the modeling of ego-vehicle trajectories by an instantiation of the smart trajectories meta-model, the following section presents another area of interest. Indeed, we present the modeling of pedestrian trajectories in order to analyze and improve pedestrian safety.

7.3.5 Modeling Pedestrian Trajectories

With the aim of improving pedestrian safety, several studies analyze the traces of pedestrians in order to measure and evaluate the risks of road accidents. Spatial information on pedestrian movements is often difficult and expensive to collect. However, the trajectories produced provide concepts of temporal geography to measure pedestrian exposure. Space-time trajectories are particularly interesting and useful for measuring pedestrian exposure.

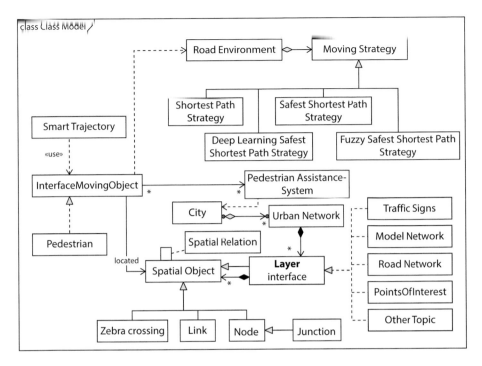

FIGURE 7.5 Simplified instantiation of the meta-model to model pedestrian trajectories.

Current research is also progressing towards a new generation of mobile applications that will allow pedestrians to find the safest way to get to the desired location (Mandar et al. 2017) (Karim et al. 2020). Other current mapping applications from major companies, such as Google Maps, offer destinations by the fastest route, without the ability to determine whether that faster route is safe for a pedestrian. Figure 7.5 illustrates a simplified instantiation of the smart meta-model for modeling pedestrian trajectories.

A pedestrian is a moving object, which has an operational type behavior like moving. It also has a tactical type behavior like choosing routes or planning and a strategic type behavior, i.e., the intention for moving. A pedestrian is located in a *Spatial Object* that could be a *Node, Junction, Link*, or *Zebra Crossing*. There is a spatial relation between spatial objects, i.e., between links, between a link and a junction, etc. *Pedestrian Assistance System*, of the artificial or cognitive system type, uses the moving object interface. This last requires the environment of the pedestrian in the road and is related to each *City*. According to the trajectory meta-model, it is made up of a set of *Urban Networks*. The latter is composed of several interfaces, which present *Road Network, Traffic Signs, Modal Network*, and *Point of Interest*. *Moving Strategy* is a behavior situated at the tactical and operational level. *Safest Shortest Path Strategy* is a kind of moving strategy. Depending on the pedestrian's road environment, there are several strategies. For example, we can use *Shortest Path Strategy, Deep Learning Safest Shortest Path Strategy, Safest Shortest Path Strategy,* and *Fuzzy Safest Shortest Path Strategy.*

The model presented in this section makes it possible to model the outdoor movement of pedestrians; however, there are several fields of application, such as the tracking of people in an airport or the understanding of customer behavior in a mall, that need to model the trajectory of moving objects inside buildings. The following section presents the modeling of IndoorGML trajectories by a simple instantiation of the smart trajectories meta-model.

7.3.6 MODELING INDOORGML TRAJECTORIES

With recent developments in spatial databases dealing with retail spaces, interior mapping, and interior positioning, several spatial information services for interior spaces have been provided.

For the purpose of modeling spatial data of indoor spaces, *IndoorGML* has been published by the OGC (Open Geospatial Consortium) as a standard data model and exchange format based on XML. The goal of *IndoorGML* is to establish a standard basis for the indoor space model. As *IndoorGML* defines a minimum data model for indoor space, more effort is needed to discover its potential aspects, which are not explicitly explained in the standard document (Daissaoui et al. 2020). Our interest lies in integrating *IndoorGML* into our meta-model and ensuring its instantiation for applications relating to interior spaces. While the *CityGML* standard covers indoor spaces and is aimed at feature modeling, the goal of our extension is to establish an agility based on the *IndoorGML* standard and use *CityGML* for an indoor space model of smart buildings.

Figure 7.6 illustrates modeling both the outdoor trajectories and the OGC® *IndoorGML* trajectories by instantiating of the meta-model. The *Moving Object*

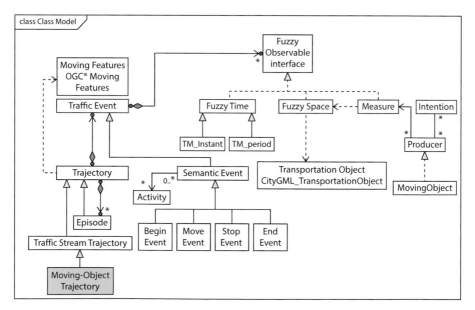

FIGURE 7.6 Simplified instantiation of the meta-model to model OGC® IndoorGML trajectories.

presents the producer. Indeed, the *Moving Object Trajectory* uses *OGC Moving Features*. We also introduced the fuzzy aspect at the *Observable Interface* level to add realness to the modeling of space and time. Fuzzy space uses the *City GML Transportation* object in order to be able to model the trajectory indoors. It is also specified that each trajectory of a mobile object has an *Intention,* for example, shopping, picking up children from school, watching a film at the cinema, etc. In addition, the activities are also listed at the level of semantic activity.

The smart trajectories meta-model allows modeling not only the spatio-temporal trajectories of moving objects such as vehicles and people, but they also make it possible to model several phenomena in a different way from what is found in the literature. In the next section, we show a new way of modeling fraud, regardless of the type of fraud, by a simple instantiation of the smart trajectory meta-model.

7.3.7 MODELING FRAUD TRAJECTORIES

Fraud is a long-standing dynamic phenomenon that occurs in many different economic activities and where fraudsters use a wide range of techniques and approaches. There is a significant number in terms of frequency and total monetary value involved. There are several types of fraud—we cite as examples credit card fraud, insurance fraud, corruption, product warranty fraud, healthcare fraud, telecommunications fraud and money laundering, etc. In almost all of the areas of fraud mentioned, information is stored electronically, but even with a deeper analysis of the data available, the fraud persists. With such amounts of data, abusive scheme transactions are hidden and difficult to detect by traditional means. These data become unwieldy when stored in relational databases, and query execution time increases as the data size increases. The trajectories of consumer behavior are mostly ignored by modern fraud detection systems. The idea of logging usage events such as credit card payment or online service to create trajectories then allows patterns to be built. A "smart" system is established to implement classifiers, allowing us to predict fraudulent trajectories in a specific domain (Karim and Boulmakoul 2021).

Regardless of the type of fraud, we model its trajectory. Figure 7.7 illustrates the instantiation of the smart trajectory meta-model to model fraud trajectories. Indeed, we assume that the trajectory of fraud is a set of fraudulent events that occur in space and time. In this case, the space component of fraudulent events can be a *Virtual Space* or *Physical Space* such as a restaurant or mall. This model describes the approach followed to catch the fraudulent patterns in different real-life applications. Since fraud is a dynamic phenomenon, it relies on the analysis of a trajectory of events depending on the application area. In this diagram, we put a focus on the consumer events as an example to illustrate and provide a more detailed overview of our model.

We distinguish between the legitimate events performed by a consumer and the disputed ones done by a fraudster. *Transaction Stream Path* is an ordered collection of *Consumer Events*. A *FraudEvent* is a *Costum Event*. A *Semantic Consumer Event* can present the beginning of fraud, continues the occurrence of a fraud named *Move*, events when fraud is silent for a period are named *Stop*, and events when fraud is solved are named *End*. A semantic fraud event has a location on a point of interest.

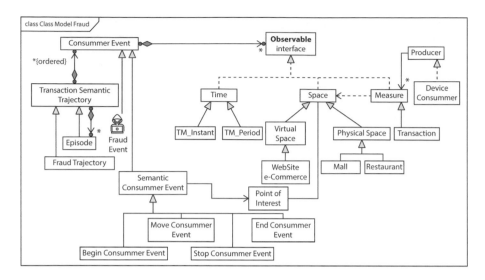

FIGURE 7.7 Simplified instantiation of the meta-model to model fraud trajectories.

7.4 CONCLUSION

In this chapter, we have described smart trajectories meta-modeling. We have demonstrated its robustness and its genericity. Indeed, smart trajectories meta-modeling is based on the ontology of space, time, and the event approach to guarantee semantic interoperability by integrating existing trajectory presentation models as well as other new models. We have also shown the classes to be instantiated to produce the model of the different instances of trajectories.

Several scopes have been reached, and the concept of the smart trajectory has interested the transport of hazardous materials, urban traffic congestion management, the Internet of Vehicles (IoV), or more generally connected objects (IoV and Internet of Things [IoT]) with their integration into edge-fog-cloud computing and Big Data analytics of trajectories in general. Obviously, other instantiations are possible and are conceivable in our perspectives.

REFERENCES

Boulmakoul, Azedine, Fazekas, Zoltán, and Karim, Lamia, Gáspár, Péter, and Cherradi, Ghyzlane. (2020). Fuzzy Similarities for Traffic Sign-based Road Environment-type Detection, Procedia Comput. Sci., vol. 170, pp 59–66.

Boulmakoul, Azedine, Karim, Lamia, and Lbath, Ahmed. (2012). Moving Object Trajectories Meta-Model and Spatio-Temporal Queries, Int. J. Database Manag. Syst., vol. 4, pp. 35–54.

Boulmakoul, Azedine, Karim, Lamia, and Mandar, Meriem. (2015). Towards Scalable Distributed Framework for Urban Congestion Traffic Patterns Warehousing, Appl. Comput. Intell. Soft Comput, vol. 2015, Artic. ID 578601. http//dx.doi.org/10.1155/2015/578601

Boulmakoul, Azedine, Karim, Lamia, Elbouziri, Adil, and Lbath, Ahmed. (2012). A System Architecture for Heterogeneous Moving Objects Trajectory Models Using Different Sensors, in *7th International Conference on System of Systems Engineering (SoSE)*, 2012, pp. 1–6, doi: 10.1109/SYSoSE.2012.6384131.

Boulmakoul, Azedine, Karim, Lamia, Elbouziri, Adil, and Lbath Ahmed (2015), A System Architecture for Heterogeneous Moving-Object Trajectory Metamodel Using Generic Sensors: Tracking Airport Security Case Study, in *IEEE Systems Journal*, vol. 9, no. 1, pp. 283–291, doi: 10.1109/JSYST.2013.2293837.

Daissaoui, Abdellah, Boulmakoul, Azedine, Karim, Lamia, and Lbath, Ahmed. (2020). IoT and Big Data Analytics for Smart Buildings: A Survey, Procedia Computer Science, vol. 170, pp. 161–168.

Fosca, Giannotti, Mirco, Nanni, Fabio, Pinelli, and Dino, Pedreschi. (2007).Trajectory pattern mining. In13th ACM SIGKDD International Conference on Knowledge Discovery and Data Mining, pp 330–339, New York, NY, ACM.

Heflin, Jeff, Hendler, James, Luke, Sean, Gasarch, Carolyn, Zhendong, Qin, Spector, Lee, and Rager, David. (2022). SHOE Base Ontology. https://www.cs.umd.edu/projects/plus/SHOE/ (Accessed March 14, 2022).

Karim, Lamia, and Boulmakoul, Azedine. (2021). Trajectory-based Modeling for Fraud Detection and Analytics: Foundation and Design, IEEE/ACS 18th International Conference on Computer Systems and Applications (AICCSA), pp. 1–7, doi: 10.1109/AICCSA53542.2021.9686920.

Karim, Lamia, Boulmakoul, Azedine, and Lbath, Ahmed. (2017). Real Time Analytics of Urban Congestion Trajectories on Hadoop-MongoDB Cloud Ecosystem, ICC '17: Proceedings of the Second International Conference on Internet of things, Data and Cloud Computing. Article No.: 29, pp 1–11, https://doi.org/10.1145/3018896.3018923.

Karim, Lamia, Boulmakoul, Azedine, and Mandar, M. (2020). ScienceDirect A New Pedestrians' Intuitionistic Fuzzy Risk Exposure Indicator and Big Data Trajectories Analytics on Spark-Hadoop Ecosystem, Procedia Comput, vol. 170, pp. 137–144 vol. ISSN 1877-0509.

Karim, Lamia, Daissaoui, Abdellah, and Boulmakoul, Azedine. (2017). Robust Routing Based on Urban Traffic Congestion Patterns, Procedia Comput. Sci., vol. 109, pp. 698–703, 2017, doi: 10.1016/j.procs.2017.05.380.

Mandar, Meriem, Karim, Lamia, Boulmakoul, Azedine, and Lbath, Ahmed. (2017). Triangular Intuitionistic Fuzzy Number Theory for Driver-pedestrians' Interactions and Risk Exposure Modeling, Procedia Comput. Sci., vol. 109, pp. 148–155, doi: 10.1016/j.procs.2017.05.309.

Meng, Xiaofeng, and Ding, Zhiming. (2003). DSTTMOD: A Discrete Spatio-Temporal Trajectory Based Moving Object Databases System, DEXA, LNCS. Berlin Heidelberg: Springer-Verlag, Vol. 2736, pp. 444–453, ISBN : 978-3-540-40806-2.

Quine, W. (1985). Events and reification. In E. Lepore & B. McLaughlin (eds.), Actions and Events: Perspectives on the Philosophy of Davidson. Oxford: Blackwell. pp. 162–171.

Spaccapietra, Stefano, Christine Parent, Maria Luisa Damiani, José Antônio Fernandes de Macêdo, Fábio André Machado Porto and Christelle Vangenot. (2008). A Conceptual View on Trajectories, Data Knowl. Eng., 65, pp 126–146.

Wolfson, O., Xu, B., Chamberlain, S., and Jiang, L. (1998). Moving Objects Databases: Issues and Solutions. Tenth International Conference on Scientific and Statistical Database Management (Cat. No.98TB100243), 1998, pp. 111–122, doi: 10.1109/SSDM.1998.688116.

Yan, Z., Parent, C., Spaccapietra, S., and Chakraborty, D. (2010). A Hybrid Model and Computing Platform for Spatio-Semantic Trajectories, The Semantic Web: Research and Applications. ESWC 2010. Lecture Notes in Computer Science, vol 6088, pp 60–75. Springer, Berlin, Heidelberg. https://doi.org/10.1007/978-3-642-13486-9_5

8 A Type-Level Trajectory Framework

*Soufiane Maguerra, Azedine Boulmakoul,
and Hassan Badir*

CONTENTS

8.1 INTRODUCTION

The trajectory model describes aspects of change across time in a very efficient manner, and the representation can be a valuable source of knowledge. For these reasons, trajectories have been used to study different attributes, including space, with different granularities, social network account activity, e-commerce, weather, and urban traffic. In each of these subjects, the trajectory has a different structure and encodes

DOI: 10.1201/9781003255635-8

different knowledge. In some applications, trajectories can be enriched with external contextual information to infer more knowledge. In these cases, the trajectory is called semantic trajectory.

In the past decades, researchers have studied trajectories intensively and proposed different models, frameworks, and ontologies. However, most trajectory models are tightly coupled with space, not leaving much flexibility for other aspects. Also, most trajectory models fail to encode the behavioral aspect of the different constructors and observations. In the literature, we could only find one paper (Ferreira et al. 2014). It provides a trajectory algebra but does not consider the semantic nature of the trajectories.

There is a need for a highly abstract, polymorphic, composable, type-safe trajectory model that enables users to construct different algebraic structures and assert that laws over them hold. The algebras can be used to build consistent, easily extensible, formal systems that behave as the integrated laws describe.

In this chapter, we tackle this matter by contributing a type-safe, functional, algebra-driven design framework for constructing algebraic trajectory data structures and checking that their algebraic laws hold. In particular, the framework covers not only spatial trajectories but can also be used to describe temporal variations of other observables, such as temperature, pressure, wind, water levels, online customer purchases, social networks activity, and account transactions, all while including the semantic side of the trajectories and the different granularities of its elements. The algebras are still, at the time of this chapter, a work in progress, and the chapter describes the latest state of the implementation.

Our framework is a library implemented with the purely functional, statically strongly typed language, Haskell. The choice of Haskell was not lightly taken. Functional programming and type-level programming are valuable assets that can build highly abstract, polymorphic, composable, type-safe software with control over side effects, clear understanding of the code, typed descriptions that can make it simple to understand by other users, and fewer bugs. Both implementations include powerful extensions such as type families that make it possible to reach a higher level of polymorphism and leverage type-level functions for compile-time validation for an additional safety layer. Singletons for dependent types make it possible to check runtime values at compile time for more type safety. Refinement types add type-level predicates for abstracting smart construction and type-level safety. Template Haskell for compile-time reflection makes meta-programming reachable within the framework. Data kinds make it possible to promote types and data constructors for the kind and type level; both new kinds and types can be used for additional type-level information, which adds more safety. As for writing mathematical proofs, QuickSpec and Small Check are used to write randomized and enumerative property tests. In general, property tests can be used to actually automatically generate, if the developer desires, thousands of tests from a single specification. Each test possibly differs from the other by the checked input, and this way the tests can cover a large number of values. Mathematical theorems can be checked using these tests on the basis that if a preposition does not fail hundreds of thousands of times, then it's less likely to fail at runtime. In particular, we use randomized property tests to check universally quantified prepositions and enumerative tests for the existentially quantified ones. The reason for this approach is because it is impossible to check an existential property by randomly choosing values; the only way is to make the gener ated values smaller and check all of them.

The remainder of this chapter is organized as follows: Section 8.2 is a state of the art on the time and trajectory subjects, Section 8.3 describes our framework, and lastly Section 8.4 serves as a conclusion and discusses the future of our work.

8.2   RELATED WORKS

In this section, we carefully describe the state of the art on the subject of trajectories. In particular, we put an emphasis on trajectory modeling. In the end, we explain how our contribution is different compared to the other studies.

8.2.1   STATE OF THE ART

The literature on trajectories can be classified into works that focus on the modeling, preprocessing, analyzing, visualization, or storing aspects (Zheng 2015). Preprocessing includes noise filtering, compression, segmentation, stay point detection, and map matching. The storage part includes studies that propose indexing approaches, integration pipelines, and schema optimization to ensure fast query processing, reads, and writes. Analytic studies aim to infer knowledge, as much as possible, from trajectories by classification, supervised or not and online or offline, or pattern detection. Visualization research has an aim to visualize trajectory data as efficiently and as much as possible to make it possible for humans to easily extract knowledge.

Another branch of studies has appeared in the last decade with a focus on Big Data (Ribeiro de Almeida et al. 2020). These studies acknowledge the vast amount of raw trajectory records that are generated instantly from various sensors and generators, either explicitly or implicitly and the limits of conventional approaches to check the integrity and ensure the availability of these data in large amounts. Therefore, they propose frameworks to process, analyze, and store these huge volumes of data by leveraging ecosystems such as Hadoop and Spark.

In our study, we have a main focus on the modeling of trajectories because modeling is an essential part in all other branches. The trajectory structure can affect the outcome and use of other approaches. In case the model is not generic enough, the study will not be useful in other fields, and there will always be a need for a new approach. Also, if the model is not correct, then it will lead to false interpretations. The next subsections describe the different trajectory models, ontologies, and the existing algebra.

8.2.1.1   Trajectory Models

As discussed in the background sections, there are different kinds of trajectories, and each domain has a focus on a specific structure. For these reasons, researchers proposed several models with an aim to find a model that can be used in as many applications as possible. In the paper (Spaccapietra et al. 2008), the authors described the concepts of stops and moves as a new effort towards semantic enrichment. The authors aimed to provide a model that can be instantiated by other domain practitioners to cover their application's needs. The semantic trajectory is richer than the raw one in terms of information and potentially also volume. Their work is limited to spatial points, and the semantics are limited to stops and moves. In (Yan et al. 2013), the semantic limit is extended to cover additional values, and the notion of structured trajectory episode was addressed. However, the semantic enrichment is limited to the episode level. Also, the scope of the model is limited to spatial points in a 2D space, and the points' time is considered as chronons. The model CONSTAnT (Bogorny et al. 2014) was proposed with a focus on semantic trajectories, and it included notions such as Object, Device, Goal, Event, Place, Environment, Behavior, and Transportation Means. The enrichment in the model covers the semantic points too, in terms of the concepts Environment and Place; at the episode level in terms of Goal, Behavior, and Transportation Means; and also at the trajectory level with Goal. The model they proposed follows a batteries-included approach, and because of this decision, it is only useful to applications that are limited to the described semantic concepts. Furthermore, the spatial domain only and chronon time limits are also present in the model. The MASTER (Mello et al. 2019) was proposed to not only cover different semantic annotations, using the concept of aspects, but also to enrich objects, trajectories, and points. An aspect can really be any semantic annotation; hence, the model can be used in different domains. Furthermore, it also models the possible relationship between objects. Nevertheless, the notion of episodes is missing, the trajectory elements are also limited to 2D spatial points with timestamps, and each aspect value is considered a nonempty array of Strings. The simple String representation is very limited, is not polymorphic, does not provide type safety, and does not offer much room for optimizations. Lastly, in (Karim et al. 2021) the authors developed a meta-model that tries to overcome the limits of the preceding ones by abstracting the notions of space and time and introducing the concept of observable, which can be any attribute. Furthermore, the model is focused on capturing pedestrian trajectories, but it can also be used in other domains by cherry-picking the relative parts. It introduces the concepts Observable, Intention, Semantic Event, and Story. However, the model does not formalize any behavioral restrictions. For instance, nothing restricts an Event to have a time observable, there is a lack of relationships between objects, and no polymorphic enrichment exists at the object or episode level. Nothing indicates that structured trajectories are composed of episodes or that semantic trajectories are possibly coupled with semantic events. No explicit, direct link exists between a story and an episode.

8.2.1.2   Ontologies

Another class of studies proposes ontologies to conceptualize trajectories. The works include datAcron (Santipantakis et al. 2017) and FrameSTEP (Nogueira et al. 2018). Both ontologies limit trajectories to the spatial domain, do not define any relationship between objects, and do not link semantics at the event or object levels. datAcron tightly links events to time intervals—we could not observe chronons in the model. FrameSTEP makes the contrast by not considering time intervals.

Furthermore, we believe that ontologies are more adequate for semantic web applications, meaning when there is a need to build a knowledge base, process semantic queries, and introduce knowledge rules. They are of little use when building backend applications, nor do they assist developers in ensuring the safety of their code, avoid breaking changes, detect bugs early, or catch incorrect statements.

8.2.1.3 Trajectory Algebra

All the works described earlier do not consider any algebraic design. In the literature, the only study proposing a trajectory algebra is (Ferreira et al. 2014). The authors followed an algebraic approach and defined algebras for time series, trajectories, and coverage. For each of these concepts they fix, control, and measure time, space, and an application theme. In detail, a time series fixes space, controls time, and measures the theme. A trajectory fixes a theme, changes time, and measures space. Lastly, a coverage fixes time, controls space, and measures a theme. Nonetheless, the clear separation between time series and trajectories goes against the definitions in (Spaccapietra et al. 2008) and (Karim et al. 2021). A time series is just a special case of a trajectory, and tightly linking space to a trajectory may restrict the use of trajectory research on time series. In addition, defining a trajectory by fixing the theme and measuring the space does not take into account the cases where the trajectory has to measure both space and theme. Lastly, the study does not consider the structured or semantic side of trajectories.

8.2.2 Our Contribution

Our research is based on all the studies noted earlier and has the same aim of defining a model that can be used by any trajectory-processing application. We carefully studied the literature and tried to propose a solution that overcomes all the drawbacks mentioned. The framework we are working on is highly abstract in the sense that it considers metaphorical, geographic, structured, and semantic trajectories, always giving the possibility to extend the algebras, define new ones, and use the existing algebraic laws. In addition, our framework is based on our past type level and algebra-driven design time framework (Maguerra et al. 2021), meaning, we are giving domain practitioners the free choice of using chronons or periods and to adopt any algebraic temporal structure they need for their application. We are heavily influenced by the algebraic approach defined in (Ferreira et al. 2014), and overcame its drawbacks by making a clear separation between semantic and nonsemantic entities and semantically enriched all of the objects, episodes, events, and trajectories. Also, we clearly made space a potential observable and not a necessity, and we tightly coupled time with events; a decision that ensures both flexibility to leverage the framework in other nonspatial fields and additional safety.

Another advantage of our framework compared to other studies is the fact that we carefully consider the technical side of applications. We acknowledge the difficulties of building reliable software; thus, we provided the framework in a purely functional, typed, and algebra-driven design setting to help build type-safe, simple, composable, highly polymorphic, and consistent software. We also acknowledge the difficulties of using, interoperating, and extending libraries; therefore, we made sure that our framework is a documentation on its own that can greatly assist developers in easily deriving the implementation relative to their domain's needs. Another issue that developers encounter is bugs and breaking changes. Type-level programming helps in detecting bugs at compile time, and also if used correctly can prevent breaking changes. The same goes for the algebraic design—the laws always ensure that the encoding is consistent.

8.3   THE TRAJECTORY FRAMEWORK

In this section, we describe the components of our framework. We detail the Haskell extensions and how they are useful. We explain how some type-level parts contribute to the safety of the code. Moreover, each of the components has both a semantic and nonsemantic side, so each subsection explains both sides.

Note that we show the code in the original language, not in a pseudo one, because Haskell, by its nature, is very concise and declarative. This means there is no need for a pseudo language, and if something is unclear in the code, we are making sure it is explained in the text code description. An additional note about the observation (===): it is meant to be identity, not equality. The framework, in the time of this writing, is still a work in progress; we did not encode all of the algebraic laws yet. Therefore, you will not find the property test laws, only the type-level encoded ones, in this chapter, but we are planning to make a new release soon. If you are interested in the source code and following the commits, please refer to our public repository (Maguerra 2021). Now, let us start with the first component: Annotation.

8.3.1   ANNOTATION

There are two different annotations in the literature: a nonsemantic one that is coupled with a nonsemantic episode and a semantic annotation that can be potentially attributed to all other components. Therefore, we separate the annotations to make the difference clear.

In the code snippet that follows, we see that both annotations are simple type classes. A type class allows ad-hoc polymorphism, meaning that the functions defined in the type class change behavior depending on the type argument. This is a pretty useful asset; however, the annotation classes defined here do not have any functions. They simply take the role of tags and add type-level information to the type variable.

```
class Annotation a
class SemanticAnnotation s
```

8.3.2   OBSERVABLE

An Observable is an observed value; it can really be of any kind. For instance, it can be longitude, latitude. temperature, pressure, or purchased products. The Observable is an element of a trajectory event. In our implementation, it is a type class tag, same as Annotations, that adds type-level information to the type parameter.

```
class Observable o
```

8.3.3 EVENT

An event is an element of a trajectory. It can be enriched with semantic contextual information or left as it is. The next two subsections explain the design of both.

8.3.3.1 Nonsemantic Event

The event can be interpreted as a list of observables and a time value. In the code snippet that follows, we describe a nonsemantic Event Algebra. The Algebra is a multiparameter type class, parameterized over a list of Observables and Time.

```
class (HObservable l, Time t) => EventAlgebra (l :: [Type]) t
where
    data Event l t
    construct :: HList l -> t -> Event l t
    observables :: Event l t -> HList l
    time :: Event l t -> t
    (===) :: Event l t -> Event l t -> Bool
```

The list of observables is a nonempty heterogeneous list (Kiselyov et al. 2004) because each observable can have a different type. Therefore, you can observe the type constructor HList, and the type parameter *l* has kind [Type]. The kind restriction ensures that only lists of types are accepted as type arguments for the event algebra class. Moreover, to ensure that only observables are accepted in the list, we used the HObservable type class constraint that can be seen next. The constraint is internal only, not public, and leverages mathematical induction to check that each type of the list is an observable. The type-level constraint showcases one side of type-level programming manifested in validating the input of the list at compile time to avoid runtime errors, meaning that the EventAlgebra accepts only lists of observables.

```
class HObservable (l :: [Type])
instance (Observable o, HObservable l) => HObservable (o ': l)
instance {-# OVERLAPPING #-} Observable o     => HObservable
(o ': '[])
```

As for time, it is completely polymorphic, and the time constraint comes from our time framework (Maguerra et al. 2021). It can be a chronon or period, and the overall algebra is an associated type family over the Event data type. Type families allow us to reach a higher level of abstractions, in the sense that they allow ad hoc polymorphism for data types. In other terms, the Event data type is abstract and changes its representation depending on the type parameters.

8.3.3.2 Semantic Event

Now for the semantic encoding. The code shown next shows the Semantic Event Algebra. Compared to the Event Algebra, there is an additional type parameter *s* to encode the semantic annotation. There is also a constructor that allows enriching an existing nonsemantic event with semantic annotation.

```
class        (        Functor (SemanticEvent l t),        HObservable
l,        SemanticAnnotation s,        Time t        ) =>
SemanticEventAlgebra l s t        where        data SemanticEvent l s t
enrich :: s -> Event l t -> SemanticEvent l s t        observables
:: SemanticEvent l s t -> HList l        semanticAnnotation ::
SemanticEvent l s t -> s        time :: SemanticEvent l s t -> t
```

The type constructor *SemanticEvent l t* is a functor over the semantic annotation. The functor abstraction allows us to transform the annotation by using the function:

```
fmap :: (s -> s') -> SemanticEvent l s t -> SemanticEvent l s' t
```

You can see that the only thing required is a function that maps s to s' and the functor will take care of mapping the *SemanticEvent* too.

8.3.3.3 Time Event Observations

In addition to the observations of each Event Algebra, we provide event time observations. The observations can be seen in the code that follows. Moreover, we provide *Event* and *SemanticEvent* instances for both *Chronon* and *Period*. Please check the source code for more information (Maguerra 2021).

```
class EventTime e where        (<) :: e -> e -> Bool        (>) :: e ->
e -> Bool        x > y = y < x        (<=) :: Eq e => e -> e -> Bool
x <= y = x < y || x == y        (>=) :: Eq e => e -> e -> Bool
x >= y = y <= x        betweenness :: e -> e -> e -> Bool
betweenness x y z = (y < x && x < z) || (z < x && x < y)
```

8.3.4 EPISODE

An episode can potentially have two annotations: one generated from the events themselves with no external contextual information, and the other inferred through an external one. Therefore, we designed two algebras: one for the nonsemantic side and the other for the semantic side. The next subsections describe the design of both.

8.3.4.1 Nonsemantic Episode

The code snippet that follows shows the Episode Algebra. It is a multiple parameter type class parameterized over Annotation, typed list of Observables, and Time.

```
class (Annotation a, Time t) => EpisodeAlgebra a l t where
data Episode a l t        construct :: a -> [Event l t] ->
Episode a l t        events :: Episode a l t -> [Event l t]
annotation :: Episode a l t -> a
```

8.3.4.2   Semantic Episode

The difference between a semantic episode and a nonsemantic one is the additional type parameter *s*. This type parameter is a semantic annotation—note that a semantic episode can have both annotations: a semantic and nonsemantic one. Another difference is that instead of parameterizing the class over a typed list of observables, we parameterize it over events, which can be either semantic or nonsemantic. To encode the restriction, at the type level, so that only events are accepted as type arguments, we leverage the closed type class *InternalEvent*. Moreover, we use the type operation *(~)* to ensure that after enriching a nonsemantic event, the semantic episode must have only nonsemantic events. This restriction comes from the fact that nonsemantic episodes contain only nonsemantic events and the semantic enrichment happens at the episode level, not the event level. The type-level equality offloads additional work from the runtime to the compiler and saves us from bugs and latency.

```
class InternalEvent e
instance InternalEvent (Event l t)
instance InternalEvent (SemanticEvent s l t)
class (Annotation a, InternalEvent e, SemanticAnnotation s)
    => SemanticEpisodeAlgebra a e s
  where
    data SemanticEpisode a e s
    construct :: a -> [e] -> s ->  SemanticEpisode a e s
    enrich
       :: e ~ Event l t
       => Episode a l t -> s ->  SemanticEpisode a e s
    elements :: SemanticEpisode a e s -> e
    annotation :: SemanticEpisode a e s -> a
    semanticAnnotation :: SemanticEpisode a e s -> s
```

8.3.5   Trajectory

Trajectories are a series of events or episodes that are all semantically annotated or not. Once a trajectory or a trajectory's elements are semantically enriched, the trajectory becomes semantic. The next subsections describe both nonsemantic and semantic trajectories.

8.3.5.1   Nonsemantic Trajectory

The nonsemantic trajectory is only composed of nonsemantic events or episodes. To make sure that this restriction is respected, we use the closed type class *InternalElement*. The type class is internal only to the module and is not exposed outside, so no additional instances can be added to it. The only instances are Event and Episode.

```
class InternalElement e
instance InternalElement (Event l t)
instance InternalElement (Episode a l t)
```

```
class InternalElement e => TrajectoryAlgebra e where
    data Trajectory e
    construct :: [e] -> Trajectory e
    elements :: Trajectory e -> [e]
```

8.3.5.2 Semantic Trajectory

The difference between a trajectory and a semantic trajectory is the additional type parameter *s*, which marks the semantic annotation of the trajectory, and the fact that the elements can also be SemanticEvent and SemanticEpisode.

```
class InternalElement e
instance InternalElement (Event l t)
instance InternalElement (SemanticEvent l s t)
instance InternalElement (Episode a l t)
instance InternalElement (SemanticEpisode a e s)
class
    (
      Functor (SemanticTrajectory e)
    , InternalElement e
    , SemanticAnnotation s) => SemanticTrajectoryAlgebra e s
    where
    data SemanticTrajectory e s
    construct :: [e] -> s -> SemanticTrajectory e s
    elements :: SemanticTrajectory e s -> [e]
    semanticAnnotation :: SemanticTrajectory e s -> s
```

8.3.6 OBJECT

An object can also be semantically enriched. Therefore, we wrote the next two subsections to describe both the nonsemantic and semantic objects. But before we get into that, each trajectory is potentially related to an object, and an object can have different types of trajectories. Therefore, in both designs, we included the function *trajectories* which evaluates to a heterogeneous list of trajectories. To ensure that the typed list includes only trajectories, we used the type family *HTrajectory*. The type family, which can be seen here, is an internal type-level function that leverages mathematical induction to prove that the types must be trajectories.

```
type family HTrajectory (l :: [Type]) :: Bool where
HTrajectory '[] = 'True    HTrajectory (Trajectory e ': t) =
HTrajectory t    HTrajectory (SemanticTrajectory e s ': t) =
HTrajectory t
```

8.3.6.1 Nonsemantic Object

The code that follows shows the design of the ObjectAlgebra.

```
class ObjectAlgebra o where    data Object o    construct :: o
-> Object o    trajectories :: HTrajectory l ~ 'True => Object
o -> HList l
```

8.3.8.2 Semantic Object

As for the semantic object, the difference is an additional type parameter o for semantic type information. Also, the Functor constraint over the *(SemanticObject o)* type constructor allows mappings for semantic annotations (check Section 8.3.3.2). Please refer to the source code for implementation details (Maguerra 2021).

8.4 CONCLUSION AND PERSPECTIVES

This chapter presents the status of our research on modeling trajectories by following an algebraic, purely functional, and typed approach. The time framework described in (Maguerra et al. 2021) is also a contribution that came from studying the trajectory domain and its needs. Both frameworks can be used in a wide range of different applications. Our design is ongoing—we will be adding more structures and laws next. We will publish the libraries in the Haskell package archive Hackage so that users can profit from them directly, as there is no need to use the repository commit hash each time. Another perspective of ours is to leverage the frameworks in an e-commerce application for data analytics. In the end, the main goal of our research is to provide programmers tools that can help them build type-safe, composable, simple, and reliable trajectory or time-related systems.

REFERENCES

Bogorny, Vania, Chiara Renso, Artur Ribeiro de Aquino, Fernando de Lucca Siqueira, and Luis Otavio Alvares. 2014. CONSTAnT—a conceptual data model for semantic trajectories of moving objects. *Transactions in GIS 18, no. 1*: 66–88.

Ferreira, Karine Reis, Gilberto Camara, and Antônio Miguel Vieira Monteiro. 2014. An algebra for spatiotemporal data: from observations to events. *Transactions in GIS 18, no. 2*: 253–269.

Karim, Lamia, Azedine Boulmakoul, and Karine Zeitouni. 2021. From raw pedestrian trajectories to semantic graph structured Model—towards an end-to-end spatiotemporal analytics framework. *Procedia Computer Science 184*, 60–67.

Kiselyov, Oleg, Ralf Lämmel, and Keean Schupke. 2004. Strongly typed heterogeneous collections. In *Proceedings of the 2004 ACM SIGPLAN Workshop on Haskell*, 96–107.

Maguerra, Soufiane, Azedine Boulmakoul, and Hassan Badir. 2021. Time framework: a type level and algebra driven design approach. In *2021 International Conference on Data Analytics for Business and Industry (ICDABI)*, 413–418.

Maguerra, Soufiane. 2021. Trajectory-framework: Type safe algebra driven design trajectory framework. https://github.com/xsoufiane/trajectory-framework (accessed March 14, 2022).

Mello, Ronaldo dos Santos, Vania Bogorny, Luis Otavio Alvares, et al. 2019. MASTER: a multiple aspect view on trajectories. *Transactions in GIS 23, no. 4*: 805–822.

Nogueira, Tales P., Reinaldo B. Braga, Carina T. de Oliveira, and Hervé Martin. 2018. FrameSTEP: a framework for annotating semantic trajectories based on episodes. *Expert Systems with Applications 92*, 533–545.

Ribeiro de Almeida, Damião, Cláudio de Souza Baptista, Fabio Gomes de Andrade, and Amilcar Soares. 2020. A survey on big data for trajectory analytics. *ISPRS International Journal of Geo-Information 9, no. 2*: 88.

Santipantakis, Georgios M., George A. Vouros, Apostolos Glenis, Christos Doulkeridis, and Akrivi Vlachou. 2017. The datAcron ontology for semantic trajectories. In *European Semantic Web Conference*, 26–30. Springer, Cham,.

Spaccapietra, Stefano, Christine Parent, Maria Luisa Damiani, Jose Antonio de Macedo, Fabio Porto, and Christelle Vangenot. 2008. A conceptual view on trajectories. *Data & Knowledge Engineering 65, no. 1*: 126–146.

Yan, Zhixian, Dipanjan Chakraborty, Christine Parent, Stefano Spaccapietra, and Karl Aberer. 2013. Semantic trajectories: mobility data computation and annotation. *ACM Transactions on Intelligent Systems and Technology (TIST) 4, no. 3*: 1–38.

Zheng, Yu. 2015. Trajectory data mining: an overview. *ACM Transactions on Intelligent Systems and Technology (TIST) 6, no. 3*: 1–41.

9 A Distributed Reactive Trajectory Framework for Nearby Event Discovery

Mohamed Nahri, Azedine Boulmakoul, and Lamia Karim

CONTENTS

9.1 INTRODUCTION

High mobility of people and goods is one of the most interesting characteristics of modern life. Studying the mobility of moving objects (human beings, vehicles, and others) has great importance, impacting different sectors such as social life, economy, smart navigation, traffic management, and others (Nahri et al. 2018) (Elbery et al. 2019) (Wu et al. 2021). In fact, discovering new knowledge related to mobility becomes more important in the era of the Internet of Things and Big Data thanks to data availability. Foremost, the study of mobility starts from tracking moving objects and representing their continued movement in the right manner. Tracking the movement of objects is promoted by advances in geolocation technologies, such as the global positioning system (GPS), and wireless technologies (Wang et al. 2021). Moreover, concepts related to the detection of activities associated with

DOI: 10.1201/9781003255635-9

139

the movement of objects have emerged thanks to advances in detection and mobile technologies. Due to the rising demand for location-based services, research topics on mobility have been extensively tackled during the past decade, especially those related to storing, querying, processing, and analyzing collected data on moving objects. Recently, real-time treatment of trajectories is rising, trying to take insights from historical data and influencing the behavior of a moving object as it happens (Taguchi, Koide, and Yoshimura 2019).

Looking at mobile objects, continuity is a principal characteristic of the movement. However, it is difficult to detect and represent this movement exactly as it is in the real world. Thus, it is approximately represented discretely via a series of spatial positions or points. Then the path of the object is corresponding usually to multiple crossed stages. Finally, the movement is represented by a trajectory comprising these stages. Every stage is generally a spatiotemporal position represented by (x, y, t), where (x, y) represents the spatial characteristic and t represents the temporal one. Obviously, periods between recording successive positions have a strong influence on the quality of a trajectory. Recently, advances in devices' computing power and bandwidth available in cities allow high-frequency recording and communicating positions and activities, producing rich, high-quality, and more accurate trajectories. Moving-object trajectory representation can have one of three principal forms, such as raw trajectory, structured trajectory, and semantic trajectory (Boulmakoul, Maguerra, and Karim 2017) (Spaccapietra et al. 2008). Thus, trjectory representation has evolved, leading to several emerging topics related to computing with trajectories. Some of those topics concern the spatiotemporal nature of trajectories, and others are related to their Big Data characteristic. Basically, considered topics are related to trajectory preprocessing, indexing, and retrieval, as well as pattern mining, location-based services, and others (Zheng and Zhou 2011). On the other hand, recent technologies allow capturing different positions and activities with interesting accuracy and communicating them with high speed as well. In fact, GPS technologies and other sensors integrated into mobile devices and modern vehicles are permitting locating moving objects with good precision, as well as sensing surrounding objects and detecting activities. Indeed, location-based services present a fundamental pillar of social networks, navigation systems, and other smart applications. Moreover, wireless technologies and cellular technologies, are now at the fifth generation, allowing a large bandwidth supporting real-time trajectory computing. Furthermore, emergence of computing concepts in smart cities such as edge, fog, and cloud (Nahri et al. 2018) and evolvement of stream processing and reactive processing ecosystems (Rinaldi et al. 2020) permit high operability with moving objects and real-time processing of generated data. Thus, trajectory processing benefits can be clearly shaped by reorienting the focus from the offline processing, corresponding to detecting spatial positions and activities, storing them, and next retrieving and mining them, to online processing treating trajectories as soon as they are constructed. The last topic is the principal problem we tend to introduce and address through this chapter.

Trajectory processing has inherited its technique from the spatial data processing field, with some specific constraints related to temporal and semantic characteristics of trajectories. Moreover, the point that trajectories are both data and patterns at the same time is adding one more complication to processing operation. Indeed, trajectory indexing and retrieval (Manolopoulos, Theodoridis, and Tsotras 2020) (He et al. 2021) and trajectory map matching (Alves Peixoto et al. 2019) are inherited from those related to spatial data. Those spatial processing concepts are helpful, especially for studying similarities between trajectories (Leal et al. 2016). Certainly, processing trajectories after their storage remains very interesting for extracting highly useful insights. This manner is representing the offline processing of trajectories. However, processing trajectories during their construction, which is rarely tackled in the state of art, seems to have great potential to shape the vision towards trajectory processing.

In this chapter, we aim to present a trajectory considering its live format and considering the history of the moving object as a past version of trajectory *Th* and the future of the trajectory *Tf* as the path to follow in the future. The border separating the history and the future versions of the trajectory represents the current location of the object. The fact of considering the future trajectories supposes that the destination of the moving object is declared beforehand. Thus, the future trajectory is declared, and sometimes predicted, with the help of navigation systems and routing algorithms (especially for a vehicle's trajectory). Furthermore, we address a problem considering permitting to a moving object to meet some events and to avoid others. The decision to meet or avoid a given event is obviously influencing *Tf*. Note that an event is supposed to have a determined period *[ti, tf]* that must be taken into account. When a moving object is deciding to meet a spatiotemporal event *Ei*, *Tf* is then split into two subtrajectories, such as the trajectory between the current location and the event position and the trajectory between the event position and the final destination. Similarly, when a moving object is deciding to avoid an event *Ej*, the future trajectory *Tf* is shaped considering avoiding the spatial area or the region where the event is happening. Moreover, we introduce the cost generated by a decision to meet or to avoid an event. The involved cost can be estimated beforehand for every potential event, before a decision is taken. This cost is represented in terms of distance or time lost or gained by a decision to meet or avoid an event.

Moreover, in this chapter, we propose designing a reactive system allowing following these online changes on trajectories and instantly suggesting events close by, as well as reacting to every meet-or-avoid decision.

The rest of this chapter is organized as follows: In Section 9.2, we present concepts related to trajectory representation, modeling, and processing, and we discuss related works to online trajectory processing. Furthermore, in Section 9.3, we formulate the problem addressed in this chapter, describing live trajectory modeling and event representation. The proposed framework for online trajectory processing is presented in Section 9.4. Finally, in Section 9.5, we conclude this chapter, presenting open perspectives.

9.2 BACKGROUND: TRAJECTORY MODELING AND PROCESSING

In this part, we introduce the latest advances on topics related to trajectory representation, modeling, and processing.

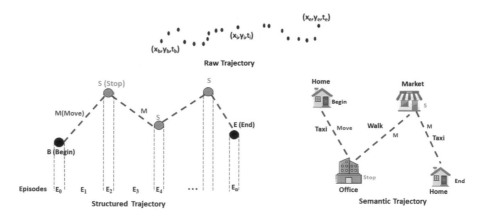

FIGURE 9.1   Raw trajectory, structured trajectory, and semantic trajectory.

9.2.1 Trajectory Representation and Modeling

The representation of a trajectory takes three basic forms, as shown in Figure 9.1:

- Raw Trajectory: Presents the basic discrete representation of the motion of a moving object. It consists of an ordered sequence of spatial recordings captured at specific times. The trajectory is represented with a tuple (x_i, y_i, t_i), knowing that (x_i, y_i) represents the geographic coordinates and t_i is the timestamp of the recording.
- Structured Trajectory: Considered a path delimited by a beginning B (Begin) and an ending E (End) and between them a sequence of M (Move) and S (Stop). M is an episode in which the object is stated with motion, and it is characterized by a traveled path and dissipated time Δt, and S is an episode in which the object is stationary, and it is characterized by position x and Δt.
- Semantic Trajectory: Also represented by a series of **Begin**, **Stop**, **Move**, and **End** episodes with more orientation towards the desired application and more enrichment by geographical information and activities carried out. These episodes in this case are considered semantic episodes.

The work (Boulmakoul 2012) presents a meta-model of trajectories based on spatial and temporal representation of the path of an object. Indeed, this meta-model takes into account the activities detected along the route and considers regions of interest (ROI). This work describes several technologies to detect and enrich trajectories with activities. Authors in (Yan et al. 2013) propose a semantic model for trajectories. They suggest a platform for processing and annotating, segmenting trajectories, and enriching them with activities. Thus, this platform is allowing trajectories to be extended to support behaviors and geographic data that are extracted from other sources of information. Indeed, the semantics of the trajectories allows the enrichment of the space-time path by detectable information (activity, any useful information, etc.) during routes. Thus, it gives a wide and holistic notion to trajectories.

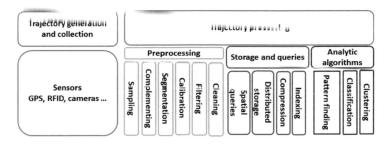

FIGURE 9.2   Trajectory processing and analytics.

9.2.2 TRAJECTORY PROCESSING AND ANALYTICS

Trajectories can be produced through varied sources. In fact, they can be constructed from GPS, vehicles, smartphones, radio frequency identification (RFID) detectors, surveillance cameras, bank cards, social networks, and others (Wang et al. 2021) (Maguerra et al. 2018). The collection of data from these different generators is followed by the stage of preparation of trajectories. This stage is called preprocessing and is essential for efficient storage and management and for further extraction of useful knowledge from these trajectories. In parallel with all these steps, the security and confidentiality aspects remain with great importance. Thus, the analytical step of the trajectories must take into consideration five major pillars (de Almeida et al. 2020) (Feng and Zhu 2016), which are preprocessing, storage, querying, knowledge extraction (data mining), and security and confidentiality. The preprocessing of the trajectories includes cleaning, filtration, complementation, calibration, segmentation, and sampling if necessary. The storage of trajectories is generally done through known spatial Big Data storage methods such as indexing, compression, and distributed storage. However, querying (Boulmakoul 2012) takes into account the spatiotemporal characteristics of the trajectories and also the behavioral parameters such as activities. Similarly, knowledge extraction part algorithms inherited from those used for Big Data analytics such as clustering (Boulmakoul et al. 2017), classification, regression, and pattern discovery. Figure 9.2 shows all these different areas related to trajectory processing and analytics.

9.2.3 TRAJECTORY QUERIES AND SIMILARITY STUDY

After the collected trajectories are finally stored in a database, querying them consists of finding spatiotemporal relationships regarding other spatial forms. The principal spatial forms of interest to the field of trajectories are points and regions and obviously trajectories (Zheng and Zhou 2011) (Zhang and Lin 2019). Indeed, spatial queries concern three types, such as trajectory-to-point, trajectory-to-region, and trajectory-to-trajectory queries.

- Trajectory-to-point: Corresponding finding trajectories regarding a fixed point in space and inversely finding points regarding a fixed trajectory, considering some spatial constraint such as distance. No temporal relationship in this case is considered.

- Trajectory-to-region: Similarly, this is correspondingly searching trajectories regarding a fixed region and inversely finding regions regarding a fixed trajectory considering spatial constraints. Also, in this case, no temporal relationship is considered.
- Trajectory-to-trajectory: Corresponding to the study of similarities between trajectories. The similarity is based on spatial relationships and might consider the temporal constraint (Leal et al. 2016).

Note that searching for trajectories through spatial relationships is time consuming. Spatial indexing of trajectories is highly required. Moreover, time indexing of trajectories is also required. Thus, a 3D indexing is performed in the case of trajectories. We note also that in Section 9.3, we consider treating relationships between trajectories and events corresponding to a point or a region with a temporal characteristic. Thus, the time constraint must be taken into account in this case.

9.2.4 Map matching

Generally speaking, map matching is used to find the exact spatial element already existing in the map or in the geographic information system relating to a given position that is represented by GPS coordinates. Usually, it consists of locating the position of a moving object relative to the road network, trying to find the nearest road segment or edge to the object location. Two kinds of algorithms are proposed for map matching: online map matching and offline map matching (Taguchi, Koide, and Yoshimura 2019) (Alves Peixoto et al. 2019). The online approach tends to retrieve or estimate the nearest segment in real time (just after the coordinate's generation), while the offline tends to find the nearest segment with high accuracy after recording the coordinates and comparing them with a large spatial dataset.

9.2.5 Spatial and Temporal Indexing for Trajectories

Spatial indexing techniques of trajectories inherit from those of normal spatial data, considering time as an additional parameter. Thus, specific indexing algorithms for trajectories are developed such as 3D R-Tree, STR-Tree, MR-Tree, HR-Tree, HR+-Tree, and others (Dur, Yigitcanlar, and Bunker 2014; Azri et al. 2013; Whitman et al. 2014; Balasubramanian and Sugumaran 2012; Manolopoulos, Theodoridis, and Tsotras 2020).

9.2.6 Online Processing of Trajectories

Online trajectory treatment consists of performing processing techniques on trajectories during their construction. In subsection 9.2.2, we describe techniques applied on trajectories after their full construction, which represent offline techniques. Authors in (Pan et al. 2020) present a framework for online trajectory similarity processing. Indeed, they propose a processing manner to fit distributed processing engines of data streams. Work cited in (Mao et al. 2018) present an online clustering technique of trajectories permeating to discover representative paths in real time. Similar works were conducted applying processing techniques (compression, filtering, etc.) on trajectories online (Zhong 2016). However, any online representation of a trajectory during its construction is not yet given. The idea in which we rely on through this work considers a live trajectory as an online version of a trajectory composed from its past and future versions. Moreover, one of the missed concepts in the state of the art is trajectory processing relative to spatiotemporal events, considering events as spatial facts or activities having starting and ending times. The point we address here consists of permitting the trajectory to meet or avoid these events. In the next section, we propose formulating this problem and related concepts. On the other hand, reactive systems are one of the key concepts that can be appealed when it comes to putting into action results of online trajectory processing.

9.3 PROBLEM FORMULATION

The problem we address in this chapter consists of online treatment of trajectories. Moreover, we assume that the movement of an object is driven basically by meeting some events while avoiding others. Furthermore, in trips, most of the time the destination is known ahead of time or declared. Thus, the trajectory from the current location to the destination can be predicted and sometimes chosen by the user. Tackling this problem has the potential to impact several activity domains such as e-commerce, smart shopping, smart navigation, traffic management, and others. Figure 9.3 presents an example of a traveler trying to join the destination D by avoiding congestion event B along with meeting event A. Obviously, every decision of meeting or avoiding has a cost to gain or to lose. This cost is evaluated basically by distance or time. In this part, we formulate this problem passing through representing live trajectories and events.

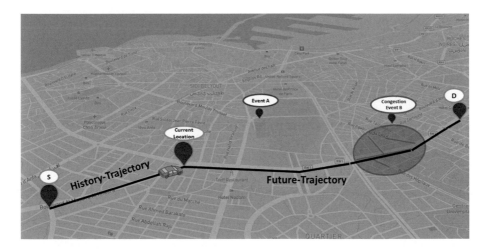

FIGURE 9.3 Trajectory-events matching.

9.3.1 Definitions and Hypothesis

Considering representations of a trajectory explained in the state of art, a given trajectory is taken with its basic representation as a raw trajectory, taking into account events met and avoided. Thus, we introduce different notions related to trajectory, event, and cost as follows:

Event: Corresponds to a spatiotemporal fact characterized by space, represented with a point or region, and period. We note:

$$E = \left(S, [ti, \, tf] \right) \tag{9.1}$$

such that S is a given point or region and ti and tf are starting and ending times of an event.

Cost: Represents the cost to gain or to lose when a decision of meeting an event is taken. We note Cl as cost to lose and Cg as cost to gain.

History-Trajectory: The path traversed by a moving object until the current position, comprising events matched. We note:

$$Th = \left((xi, \, yi, \, ti), \, M, \, A \right) \tag{9.2}$$

where (x_i, y_i, t_i) is the list of timestamped traversed positions, $M = (E_j, Cl_j)$ presents the set of met events with their timestamps, and $A = (E_k, Cg_k)$ presents the set of avoided events with their timestamps.

Future-Trajectory: Represents the intention of the moving object, and it corresponds to the path from the current position to the destination. Thus, for every version of future-trajectory Tf, a list of all potential events to meet M or to avoid is proposed A. Similarly to Th, we note:

$$Tf = \left((xi, \, yi, \, ti), \, M, \, A \right) \tag{9.3}$$

Live-Trajectory: Represents the live version of a trajectory, and it takes into consideration Th and Tf. We note:

$$Tl = Th \cup Tf \tag{9.4}$$

where $\cup$ is a concatenation operator that we explain more in the next subsection.

Final-Trajectory: Represents the final version of a trajectory Te. And it represents exactly the last version of Th. We note:

$$Te = \left((xi, \, yi, \, ti), \, M, \, A \right) \tag{9.5}$$

9.3.2 Trajectory Representation Based on History and Future

The notations presented earlier consider different kinds of used trajectories, allowing us to have a real-time situation of the moving object as well as perspectives of movement. The quest for real-time treatment of the trajectory causes us to give more importance to the live trajectory *Tl*. In fact, *Tl* corresponds to the real-time situation and perspectives of movement at every instant *t*.

Considering the time parameter, Equation 9.4 becomes:

$$Tl(t) = Th(t) \cup Tf(t) \tag{9.6}$$

Where the $\cup$ operator allows regrouping *Th* and *Tf* trajectories with all events matched by the moving object, as well as with all potential events to match in the future. The border between the *Th* and *Tf* represents the current location of the moving object.

Figure 9.4 represents an ontology describing relationships between different notions used in this section such as Trajectory, Live-Trajectory, History-Trajectory, Future-Trajectory, Event, Time, Space, Activity, and others.

As shown in Figure 9.4, a trajectory is composed of a set of spatiotemporal steps enriched with activities, corresponding to the classical representation of a trajectory that we describe in Section 9.2. The novelty in this ontology consists of adding *Live-Trajectory* as a trajectory composed of two sub-trajectories. *History-Trajectory* presents steps crossed by a moving object from the starting position until the current position. Whereas an uncertain trajectory that we call *Future-Trajectory* presents the potential path from the current position to the final destination. In fact, at every moment, a temporary *Future-Trajectory* is predicted, knowing the destination. On the other hand, we put more focus in this representation on events to meet or to avoid by a moving object. Given the *Future-Trajectory*, a set of events to meet or to avoid are proposed based on distance and time. Events that are decided to match (meet or avoid) are shaping the trajectory of a moving object. Thus, a real-time representation of the trajectory is given through this ontology, opening a wide horizon for treating trajectories as soon as they are created, as well as predicting trajectory intention and influencing its destiny.

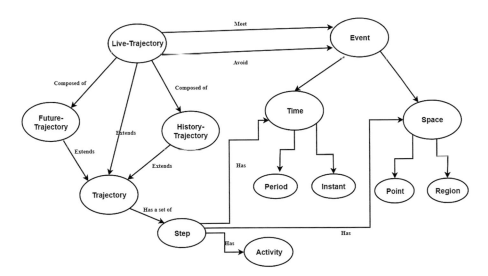

FIGURE 9.4   Ontology representing a live trajectory and its interaction with events.

9.3.3 Trajectory-Event Meeting

Knowing that the event is a spatiotemporal fact, meeting an event consists of reaching a position in space by a certain time. The first idea consists of predicting the future path of the moving object, knowing its destination. Obviously, in the case of a road network, this path can be predicted with the help of navigation systems and routing algorithms. Otherwise, we suppose that the future path is declared. Based on this future trajectory of a moving object, all potential events close by are proposed. The proposition should be based on time and distance. Finally, taking a decision to meet an event should be declared by the moving object, and the cost generated must be calculated.

Thus, as illustrated in Figure 9.5, at the instant when the object is in the position A:

$$Th = S \rightarrow A, \; Tf = A \rightarrow D \; and \; Tl = Th \cup Tf \tag{9.7}$$

We note Ei as an event happening in the position pi in the period [tis, tif].

The set of potential events to meet by the moving object is described as follows:

$$E_m = \{E_i, \, dist \, (Tf, \, p_i) \leq Dmax \; and \; t_{is} \leq t \leq t_{if}\} \tag{9.8}$$

The decision to meet an event E_i among E_m is a subject of the rerouting function.

In the case illustrated in Figure 9.5, $E_m = \{E_3, E_5, E_6\}$ and E_6 is decided to be met.

We note that the position of an event can be a point or a region. In fact, the distance calculation is related to trajectory-region and trajectory-point like queries, as explained in Section 9.2.

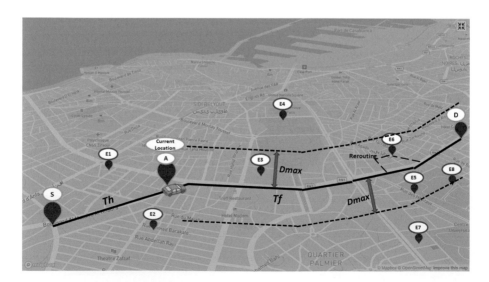

FIGURE 9.5   Trajectory-event meeting.

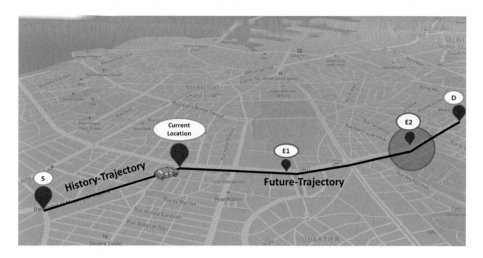

FIGURE 9.6   Trajectory-event avoiding.

9.3.4 TRAJECTORY-EVENT AVOIDING

Similarly, avoiding event operations are performed based on the future path of a moving object *Tf* considering events that can potentially be matched. However, in this case, the matching considers events positioned on the road of moving object. Thus, this is corresponding to the exact spatial matching between the trajectory and event, as described in subsection 9.2.4. Moreover, it takes into consideration the time parameter, comparing the event period and the time of accessing the position of the event by the moving object.

The set of potential events to avoid is described as follows:

$$E_a = \left\{ E_i, \ p_i \ matchs \ Tf \ and \ t_{is} \leq t \leq t_{if} \right\} \tag{9.9}$$

In the case, as illustrated in Figure 9.6, $E_a = \{E_1, E_2\}i$.

Given E_a, if E_i is decided to be avoided, a redirection or a rerouting function is called to avoid all the space concerned by E_i, generating a cost C_i.

9.4 TRAJECTORY REACTIVE SYSTEM FOR MEETING AND AVOIDING EVENTS

Real time trajectory processing, as described in previous sections, requires a reactive system responding to every change detected. Thus, the reactivity principles must be adopted early. A message-based system seems to be perfectly adapted to this kind of situation. In particular, a microservice architecture or an actor-based programming model is recommended to be used, supporting scalability and high performance (Rinaldi et al. 2020). Indeed, two essential parameters have to be taken into consideration, such as data volume and high-frequency requests. In Figure 9.7, we present a high-level design of a system for trajectory event meeting and avoiding based

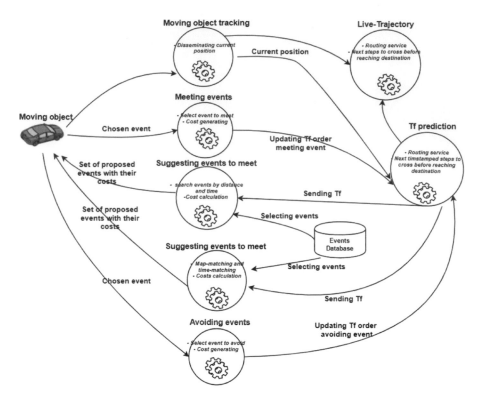

FIGURE 9.7　Reactive system for trajectory-event meeting and avoiding.

on messages and independent reactive entities. Every entity encompasses specific and coherent functionalities. Processing entities, such as moving object tracking, suggesting events to meet, suggesting events to avoid, meeting events and avoiding events, are acting directly with the live trajectory, storing its history and influencing its future. Thus, all entities satisfy the problem explained in the previous section.

The reactive system presented in Figure 9.7 remains universal for the problem of matching between trajectories and events. However, more requirements should be considered when applying it to specific fields. For instance, if we consider applying it to avoiding traffic congestion events, more compilations are added related to congestion events, space concerned, and time of occurrence. On the other hand, meeting events seems to be perfectly adapted to services like smart delivery and smart shopping.

9.5　CONCLUSION

Throughout this chapter, we presented an online trajectory processing approach. Moreover, we put more focus on real-time trajectory processing by developing a common problem consisting of discovering nearby events. Indeed, we formulated this problem based on event categories such as potential events to meet and to avoid, describing the shape of the trajectory when deciding to meet some events while avoiding others. An online representation of a trajectory is highlighted based on two subtrajectories such as History-Trajectory and Future-Trajectory. Moreover, we present a high-level design of a reactive system for meeting or avoiding events. This approach can be adopted by different fields such as e-commerce, smart navigation, traffic management, and others. Developing the system and putting it into tests in different fields might show several considerations to take into account.

REFERENCES

Alves Peixoto, Douglas, Hung Quoc Viet Nguyen, Bolong Zheng, and Xiaofang Zhou. 2019. "A Framework for Parallel Map-Matching at Scale Using Spark." *Distributed and Parallel Databases* 37 (4): 697–720. doi:10.1007/s10619-018-7254-0.

Azri, Suhaibah, Uznir Ujang, Francois Anton, Darka Mioc, and Alias Abdul Rahman. 2013. "Review of Spatial Indexing Techniques for Large Urban Data Management." In *International Symposium & Exhibition on Geoinformation (ISG)*, Malaysia, Kuala Lumpur, 24–25 September 2013.

Balasubramanian, Lakshmi, and M. Sugumaran. 2012. "A State-of-Art in R-Tree Variants for Spatial Indexing." *International Journal of Computer Applications* 42 (20): 35–41. doi:10.5120/5819-8132.

Boulmakoul, Azedine, Soufiane Maguerra, and Lamia Karim. 2017. "A Scalable, Distributed and Directed Fuzzy Relational Algorithm for Clustering Semantic Trajectories Preliminaries." In *The Sixth International Conference on Innovation and New Trends in Information Systems, Casablanca*, Morocco, 24–25 November 2017.

Boulmakoul, Azedine. 2012. "Moving Object Trajectories Meta-Model and Spatio-Temporal Queries." *International Journal of Database Management Systems* 4 (2): 35–54. doi:10.5121/ijdms.2012.4203.

de Almeida, Damião Ribeiro, Cláudio de Souza Baptista, Fabio Gomes de Andrade, and Amilcar Soares. 2020. "A Survey on Big Data for Trajectory Analytics." *ISPRS International Journal of Geo-Information* 9 (2): 1–24. doi:10.3390/ijgi9020088.

Dur, Fatih, Tan Yigitcanlar, and Jonathan Bunker. 2014. "A Spatial-Indexing Model for Measuring Neighbourhood-Level Land-Use and Transport Integration." *Environment and Planning B: Planning and Design* 41 (5): 792–812. doi:10.1068/b39028.

Elbery, Ahmed, Hossam S. Hassanein, Nizar Zorba, and Hesham A. Rakha. 2019. "VANET-Based Smart Navigation for Vehicle Crowds: FIFA World Cup 2022 Case Study." In *2019 IEEE Global Communications Conference, GLOBECOM* pp. 1–6. doi:10.1109/GLOBECOM38437.2019.9014183.

Feng, Zhenni, and Yanmin Zhu. 2016. "A Survey on Trajectory Data Mining: Techniques and Applications." *IEEE Access* 4: 2056–2067. doi:10.1109/ACCESS.2016.2553681.

He, Tianfu, Jie Bao, Yexin Li, Hui He, and Yu Zheng. 2021. "Crowd-Sensing Enhanced Parking Patrol Using Sharing Bikes' Trajectories." In IEEE Transactions on Knowledge and Data Engineering. doi:10.1109/TKDE.2021.3138195.

Leal, Eleazar, Le Gruenwald, Jianting Zhang, and Simin You. 2016. "Towards an Efficient Top-K Trajectory Similarity Query Processing Algorithm for Big Trajectory Data on GPGPUs." In *2016 IEEE International Congress on Big Data, BigData Congress*, pp. 206–213. doi:10.1109/BigDataCongress.2016.33.

Maguerra, Soufiane, Azedine Boulmakoul, Lamia Karim, and Hassan Badir. 2018. "A Survey on Solutions for Big Spatio-Temporal Data Processing and Analytics Background Spatial Data." In *Proceedings of the 7th International Conference on Innovation and New Trends in Information Systems, Marrakech*, Morocco, 21–22 December 2018, pp. 127–140.

Manolopoulos, Yannis, Yannis Theodoridis, and Vassilis J Tsotras. 2009. "Spatial Indexing Techniques." *Encyclopedia of Database Systems*, pp. 2702–2707. doi: 10.1007/978-0-387-39940-9_355.

Mao, Jiali, Qiuge Song, Cheqing Jin, Zhigang Zhang, and Aoying Zhou. 2018. "Online Clustering of Streaming Trajectories." *Frontiers of Computer Science* 12 (2): 245–263. doi:10.1007/s11704-017-6325-0.

Nahri, Mohamed, Azedine Boulmakoul, Lamia Karim, and Ahmed Lbath. 2018. "IoV Distributed Architecture for Real-Time Traffic Data Analytics." *Procedia Computer Science,* Volume 130: 480–487. doi:10.1016/j.procs.2018.04.055.

Pan, Zhicheng, Pingfu Chao, Junhua Fang, Wei Chen, Zhixu Li, and An Liu. 2020. "TraSP: A General Framework for Online Trajectory Similarity Processing." In *International Conference on Web Information Systems Engineering*, 384–397. doi:10.1007/978-3-030-62005-9_28.

Rinaldi, Luca, Massimo Torquati, Gabriele Mencagli, and Marco Danelutto. 2020. "High-Throughput Stream Processing with Actors." In *Proceedings of the 10th ACM SIGPLAN International Workshop on Programming Based on Actors, Agents, and Decentralized Control* pp. 1–10. doi:10.1145/3427760.3428338.

Spaccapietra, Stefano, Christine Parent, Maria Luisa, Jose Antonio, De Macedo, Fabio Porto, and Christelle Vangenot. 2008. "A Conceptual View on Trajectories." *Data & Knowledge Engineering* 65(1): 126–146. doi:10.1016/j.datak.2007.10.008.

Taguchi, Shun, Satoshi Koide, and Takayoshi Yoshimura. 2019. "Online Map Matching with Route Prediction." *IEEE Transactions on Intelligent Transportation Systems* 20 (1): 338–347. doi:10.1109/TITS.2018.2812147.

Wang, Sheng, Zhifeng Bao, J. Shane Culpepper, and Gao Cong. 2021. "A Survey on Trajectory Data Management, Analytics, and Learning." *ACM Computing Surveys* 54 (2): 1–36. doi:10.1145/3440207.

Whitman, Randall T., Michael B. Park, Sarah M. Ambrose, and Erik G. Hoel. 2014. "Spatial Indexing and Analytics on Hadoop." In *Proceedings of the ACM International Symposium on Advances in Geographic Information Systems*, pp. 73–82. New York, NY: ACM. doi:10.1145/2666310.2666387.

Wu, Tao, Huiqing Shen, Jianxin Qin, and Longgang Xiang. 2021. "Extracting Stops from Spatio-Temporal Trajectories within Dynamic Contextual Features." *Sustainability (Switzerland)* 13 (2): 1–25. doi:10.3390/su13020690.

Yan, Zhixian, Dipanjan Chakraborty, Christine Parent, Stefano Spaccapietra, and Karl Aberer. 2013. "Semantic Trajectories." *ACM Transactions on Intelligent Systems and Technology* 4 (3): 1–38. doi:10.1145/2483669.2483682.

Zhang, Yue, and Yaping Lin. 2019. "Quantitative Similarity Calculation Method for Trajectory-Directed Line Using Sketch Retrieval." *Journal of Visual Communication and Image Representation* 59: 448–454. doi:10.1016/j.jvcir.2019.01.040.

Zheng, Yu, and Xiaofang Zhou. 2011. *Computing with Spatial Trajectories*. New York: Springer Science & Business Media.

Zheng, Yu. 2015. "Trajectory Data Mining: An Overview." *ACM Transactions on Intelligent Systems and Technology* 6 (3): 1–41. doi:10.1145/2743025.

10 A Multidimensional Trajectory Model in the Context of Mobile Crowd Sensing

Hafsa El Hafyani, Karine Zeitouni, Yehia Taher,
Laurent Yeh, and Ahmad Ktaish

CONTENTS

DOI: 10.1201/9781003255635-10

153

10.1   INTRODUCTION

With the rapid advances of the Internet of Things (IoT), along with the widespread use of the global positioning system (GPS), and other built-in sensors, several applications have emerged to collect georeferenced data series. An application class of generating geodata series is the new paradigm called mobile crowd sensing (MCS), which empowers volunteers to contribute data acquired by a multisensor box and a mobile device. Several scenarios based on MCS exist and include but are not limited to noise sensors,[1] radiation sensors,[2] and air pollution and individual exposures such as in our context with Polluscope[3] project. For example, in an air quality (AQ) scenario, locations and periods with high pollution phenomena may be determined by mining levels of pollution over space and time. However, trajectory data collected in the context of MCS combine spatial and temporal dimensions with measurements and semantic annotations. The analysis of these data is thus not confined to the spatial and temporal dimensions but also includes information about who is moving and their contextual information. The effective representation of these four aspects of the data is a key factor for the real success of trajectory data mining and analysis in any MCS application scenario.

10.1.1   CONTEXT AND MOTIVATION

An application scenario of the MCS paradigm is the Polluscope project, in which each participant is equipped with a sensor kit and a mobile device for the transmission of the collected measurements with their GPS coordinates. The recruited participants, by the multisensor box, collects AQ measurements, including particles of different diameters (PM1.0, PM2.5, PM10), nitrogen dioxide (NO_2), and black carbon (BC), as well as climatic data, including temperature and relative humidity. The mobile device is used to collect GPS logs. In addition, a mobile application is provided so that participants can assign the context of measurements. Then, they are asked to indicate the type of place (called microenvironment) each time they change it. They also provide information on specific events that have an impact on the concentrations of pollutants and therefore on their exposure, as shown in Figure 10.1. This self-reporting of time activity is an important enrichment information for MCS. It makes it possible to interpret the observed measurements, as they are largely dependent on the type of environment (indoor, outdoor, or in transit). Without this information, the collected measurements cannot be interpreted correctly. It is worth noting that time activity is in the very heart of the field of time geography introduced by (Hägerstrand 1970). It enables insights at a higher abstraction level along the participants' trajectories.

Therefore, combining spatial location with continuous measurements and a time activity diary results in rich trajectories. Figure 10.2 shows an example of a typical rich trajectory evolving along space and time. In addition to ambient air measurements such as temperature and air pollutants, this trajectory is enriched with contextual information such as participants' microenvironments (e.g., home, office, restaurant) and air pollution–related events (e.g., smoking). Therefore, rich trajectory data do confirm the representation mentioned earlier, which can be described by a

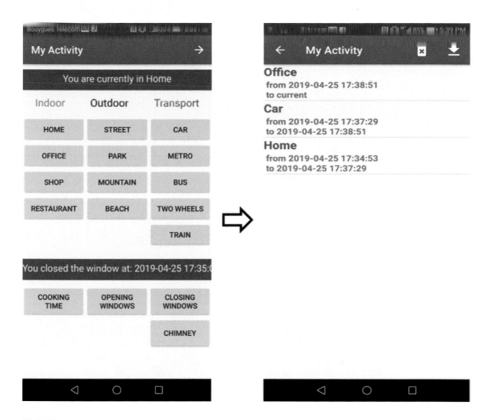

FIGURE 10.1   Annotation application on Android. Boxes on the upper side on the left are used to assign the microenvironment of the participant, such as home, office, or bus. On the bottom side, participants can declare their pollution-related events such as opening a window or cooking. The generated journal is presented on the right.

vector (user ID, timestamp, latitude, longitude, measurements, semantic). The challenge then is to perform analytical processing of rich trajectory data. For instance, to answer a query like *Which category of participant is the most exposed to NO₂ and in which microenvironment?* requires aggregating measures (here NO_2) at some space and time granularity while considering other categorical dimensions. Other queries may need not only trajectory data and contextual information but also the ability to integrate external data and perform complex analysis. For instance, answering this query: *What is the difference between the collected AQ measurements and fixed-station AQ data for the same localization and at the same time?* This query involves the integration of AQ data from fixed stations and the alignment of the spatial and temporal granularities of the two data sources in order to match and compare them. The combination and the aggregation of rich trajectories data calls for the concept of a trajectory data warehouse (TDW).

However, one particularity of MCS is its heterogeneous and imperfection properties, designating that rich trajectory data are originated from different sensors that may exhibit some issues. In fact, some sensors may be offline and do not transfer any data

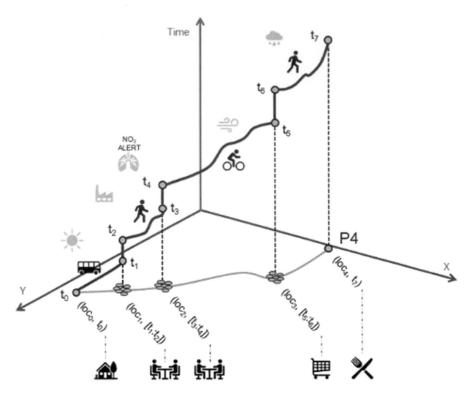

FIGURE 10.2   An example of a rich trajectory collected in the context of MCS.

for hours, which may lead to missing data problems. Furthermore, participants may not thoroughly annotate their microenvironments or may completely forget to fill in this information, resulting in trajectory data with low-confidence in self-reporting, or in worst cases, trajectory data with no semantic contextual information. With this missing data issue, analyzing rich trajectory data and extracting the *complete* exposure story of participants is not straightforward. The complete story of a subject has been addressed in (Wagner et al. 2013) to analyze the different aspects of the "mobility story" of a moving object. In our case, we intend to use this approach for extracting the complete information from rich trajectories, including interpolating missing values, inferring the context of the participants, and analyzing the exposure story of the participants.

Information about the exposure story benefits participants in triple ways: (i) Discover periods and/or locations with high pollution phenomena along the participant's trajectory, so they can change their mobility habits, if possible. (ii) Have a view on their exposure over time and detect the microenvironment with the highest/lowest levels of pollution. Users can then take actions to improve their AQ (e.g., open/close window). (iii) Provide participants with complete information on their exposure even if they did not thoroughly annotate their data or the data were not acquired. Hence, providing information from such enriched trajectory data is a fundamental issue in real-world applications.

Consequently, the multidimensional feature of rich trajectories motivates a multidimensional analysis, since it allows the exploration of these data from several perspectives (i.e., longitudinal, spatial, and temporal perspectives) and multiscale, which allows exploring the MCS data at different granularity levels.

10.1.2 PROBLEM STATEMENT

Despite extensive research efforts on modeling DTW and online analytical processing (OLAP) systems (Rivest et al. 2005, Lo et al. 2008, Wagner et al. 2013), none of them are applicable on our MCS data enriched with measurements and semantics, since it is nontrivial to adapt a generic data warehouse (DW) for spatiotemporal trajectories.

One of the strengths of MCS is the usage of different sensors designed by different manufacturers. The used sensors may differ in their sampling frequencies, which could lead to measurements at irregular, potentially asynchronous time intervals and missing values. In other words, we cannot get a complete measured state due to various times of sensor data acquisitions. For instance, GPS tracks are collected every second, while AQ data are collected every minute. The user needs to recognize the periods and locations with a high level of exposure. Combining GPS tracks and AQ time series will create a missing value issue. Plus, the time and spatial dimensions are not finite and discrete and cannot be used for aggregation, unless they are previously discretized according to a given granularity. This raises the following questions: *How do we manage the diversity of the granularity of these data? How do we ensure the usage of grouping and aggregating of measurements when some dimensions such as time and space do not present finite and discrete domains?* Since the data captured by the sensors are provided with a given accuracy, for comparable measurements, aggregation or grouping these data may not be possible. For example, the coordinates of two participant trajectories walking together may not be numerically identical even though they are acquired at the same time. Thus, a query that requires the grouping of these similar trajectories cannot be computed. In addition to this, another question that arises is: *How do we associate a concept hierarchy with these two continuous dimensions?* And lastly: *How do we handle the missing values in a multigranular DW?*

Furthermore, the semantic of events reporting the time activity (i.e., context) is also different from the sensor updates because it is categorical and relates to large intervals or sporadic events, while sensor updates are numerical and reported continuously. For instance, microenvironments depict participants' contexts for a period of time (possibly a large interval), whereas air pollution–related events report temporary and sporadic activities for a brief period. Therefore, a natural question arises: *How do we model event reporting with respect to their semantics?* Additionally, in the context of MCS, sensors continuously collect huge amounts of rich trajectory data. Thus, we need an efficient implementation of the data model to handle the volume and the velocity of this large-scale data.

10.1.3 STATE OF THE ART

Following the huge generation of spatiotemporal data, it became commonly known that nonspatial DWs are not sufficient to fully exploit the spatial dimension of geolocated data (Rivest et al. 2005, Jensen et al. 2017). A rethinking of the traditional solutions was needed. In this section, we review the main research on rich trajectory modeling. Depending on whether the model requires spatiotemporal data enriched with measurements and events, we discuss trajectory-based and enriched trajectory models.

10.1.3.1 Trajectory-Based Models

Solutions such as spatial OLAP (SOLAP) (Rivest et al. 2005) offer OLAP functionalities coupled with geographic information system (GIS) functionalities for the analysis of geolocated data. They have the ability to perform multidimensional exploration of the data, which can be presented in both detailed and aggregated forms (Bimonte et al. 2007). However, in a SOLAP model, the spatial attribute is represented as a cartographic object (i.e., points, lines, and polygons). This opens the issue of drawing the spatial dimension's hierarchy in the SOLAP model. Typically, a spatial hierarchy is depicted by the topological relations (i.e., inclusion, overlap) between members of the same and/or different spatial levels. This affects the accuracy of the aggregation process (Bimonte et al. 2007). The Open Geospatial Consortium (OGC) seeks to take inspiration from OLAP cubes to apply them on a multidimensional ("n-D") array of values. Data cubes for geospatial information provide a way to integrate observations and geospatial data for efficient data analytics, using the geospatial coverage (e.g., rasters and imagery) data structure. Unlike conventional data cubes in OLAP, the dimensions refer to metrics and not categorical or semantic data and cover entirely some spatial region. In MCS, only visited locations are materialized.

In the context of moving objects, a tailor-made data model has been proposed in (Wan et al. 2007), where the concepts of continuous dimension and continuous fact make it possible to capture the spatiotemporal fact of mobility in a predefined network. An adapted indexing method makes it possible to respond effectively to spatiotemporal aggregate queries. The advantage of this model is to allow spatial and temporal queries on the fly, without being limited to a prior division of space or time. The downside is that this model is more difficult to implement. Defining a granularity and a spatial and temporal frame of reference for the dimensions is a solution often adopted in practice. This is the case in the model proposed for the analysis of spatiotemporal activities in (Savary et al. 2004). These works are the most similar to our context, but they were limited to trajectories without associated measures.

(Leonardi et al. 2014) propose a DW model that includes both temporal and spatial dimensions for modeling a TDW and offer visual OLAP operations. However, the authors claim that their proposed model does not concern semantically annotated trajectories and complex measures. (Iftikhar and Pedersen 2010) proposed a multigranular DW by adding a granularity tag to each row. Their work focused on time dimension and did not consider disaggregation.

10.1.3.2 Enriched Trajectory Models

Sequential OLAP was proposed by (Lo et al. 2008) to support OLAP operations for sequences (S-OLAP). An event in an S-OLAP system consists of a number of **dimensions** and **measures**, and each dimension may be associated with a **concept hierarchy**. If there is a logical ordering among a set of events, the events can form a sequence. A logical ordering can be based on another attribute (e.g., time attribute). Yet in the MCS context, events are not as dense and regular as measurements and do not necessarily indicate a logical ordering. Built upon sequential OLAP, interval OLAP, or I-OLAP (Koncilia et al. 2014) was proposed to analyze and process efficiently data organized as intervals in an OLAP way. (Koncilia et al. 2014)) define an interval as the time between two consecutive events. Compared to our model, we introduce the spatial dimension in the data analysis and exploration.

(Wagner et al. 2013) tackle the problem of semantic trajectory data enriched with domain knowledge such as transportation means and propose a TDW model for mobility called *Mob-Warehouse*. The raised questions of who, where, when, what, why, and how address the trajectory's features and analyze the different aspects of the "mobility story". Indeed, the spatial dimension is an important one in the analysis, yet it is not the only focus of our analysis. One of the important facets in our analysis is to address the temporal facet with any dimensions, and not only on the spatial one, which is not possible with *Mob-Warehouse*. We are interested in the analysis of the desired phenomena during specific time periods with regard to the spatial dimension. In addition to the addressed works noted earlier, (Vaisman and Zimányi 2019), based on their moving object database (MOD) called MobilityDB, extend an existing proposal on TDW and integrate relational warehouse data with moving object data to realize the notion of spatiotemporal queries as defined in (Vaisman and Zimányi 2009).

The authors introduce mobility DW, which gives the possibility to define moving objects as measures in a DW fact table. While the authors claim that their proposal does not exclude semantic information about the moving object, it can be included as dimensions in the mobility DW. However, the authors do not include continuous measurements in their proposal, and it remains an important facet in MCS data mining and analysis.

In another work, (Karim and Boulmakoul 2021) design a dedicated trajectory-based fraud detection framework based on graph techniques and semantic search. The authors define the basis of the fraud as a story that has both meaning and intention and consider the events trajectory to detect the fraud. Indeed, the semantic facet is an important view in fraud analytics. However, other facets such as the continuous measurements and the spatial dimension are important in our analytics.

Motivated by these limitations, in this work, we propose a multidimensional data model for rich trajectory data modeling and analysis, which takes into consideration the particularity of MCS data, as well as the specific semantic and nature of self-reporting (i.e., microenvironments and events declared by participants) to capture every facet of the data.

10.1.4 CONTRIBUTIONS

To the best of our knowledge, this is the first contribution that adopts a multidimensional model to meet the requirements of the complete exposure and mobility stories. We investigate the capability of OLAP systems in handling MCS data and identify their limitations. In particular, we adopt the discretization method of the spatial and temporal dimensions so they can be used in an OLAP system. Based on machine learning, we propose new operators for spatial and temporal disaggregation to deal

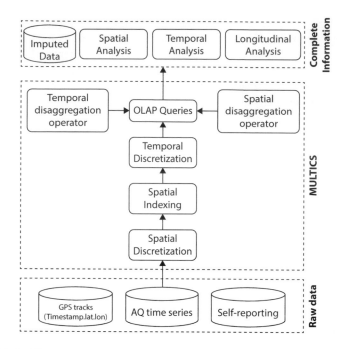

FIGURE 10.3 The conceptual architecture of the proposed solution for rich trajectories.

with missing values issues in a multigranular DW. Figure 10.3 illustrates the conceptual architecture of the proposed solution. Specifically, the main contributions of this work are:

- We propose *MULTICS*, a **MULTI**-dimensional model for **C**rowd **S**ensing, to deal with the multidimensional feature of MCS data and capture every facet.
- We adopt the discretization method of the spatial and temporal dimensions by setting a minimal granularity.
- Time is also discretized (in our application, the granularity of the minute was chosen). For the spatial dimension, the study area is divided into pixels of predefined size.[4] Trajectories are converted to pixel references combined with time units (i.e., members of spatial and time dimensions), which adapts to the exploration at different scales in a hierarchical manner. We also adopt a spatial index to speed up the spatial queries.
- Besides the embedded operators, we introduce two operators of spatial and temporal disaggregation based on machine learning to handle the problem of missing values by producing finer resolution data from coarse data.
- Along with the collected data, the proposed model has the ability to store external facts and/or dimensions, as well as data derived by, e.g., a machine learning process.
- Finally, extensive experiments on real-world data collected by participants demonstrate the usefulness of our proposed model in solving real application scenarios.

The rest of the chapter is organized as follows: The following section presents the main challenging characteristics for multidimensional rich trajectory modeling. Section 10.3 introduces the general schema and our application scenarios. Section 10.4 presents the implementation of MULTICS. Experiments are conducted on real environmental data from MCS campaigns. Finally, the last section summarizes our main contributions and draws our perspectives.

10.2   REQUIREMENTS OF MULTIDIMENSIONAL DATA MODELING IN MCS

This section introduces the main characteristics of rich trajectory data collected in MCS, which are subject to many limitations. Indeed, data measured by mobile sensors can be represented by multivariate time series, which are characterized by the presence of a spatial dimension forming trajectories. Equivalently, we can use these data as spatiotemporal trajectories enriched by additional measurements throughout the collection period. Such type of data exhibits a number of challenging characteristics.

Spatial and Temporal Autocorrelation. From the modeling view, a distinctive aspect of such data series is the spatial autocorrelation (Miller 2004). The same holds for consecutive observations, as the variation of physical phenomena is usually smooth. This means that collected data are not independent, and so, the spatial and the temporal dimensions should be organized and indexed accordingly.

Multigranularity. Another fundamental characteristic of mobile sensor data is the diversity of their granularity, under both the temporal and spatial dimensions. The temporal domain is typically represented at different time granularities. The spatial entity can be represented at different scales within a hierarchy of regions or cells. Combining multiple datasets with several granularities and changing the granularity of a dataset are important analysis tasks that we intend to deal with. Thus, we need to define a framework that takes into account spatial and temporal granularities and allows the shifting from one granularity to another. The passage from a finer resolution to a higher resolution is motivated by temporal/spatial aggregation, while spatial/temporal disaggregation is advocated when the lower-granularity data is missing.

Data Volume. Huge amounts of data are being collected continuously in MCS campaigns (as many as the number of equipped holders) in different geographical areas. Big Data processing techniques are necessary to allow an efficient interactive data analysis.

10.3   CONCEPTS FOR THE MULTIDIMENSIONAL DATA MODEL

As discussed in Section 10.1, the key insight of this work is to adopt a multidimensional model to mine rich trajectory data and extract users' complete story of exposure to pollution so that they possess the ability to take appropriate actions. In this section, we define the proper steps taken to achieve the ultimate goal. First, we introduce an overview of MCS data used for modeling rich trajectories. Afterwards, we discretize the spatial and temporal dimensions along with introducing two operators: *spatial and temporal disaggregation* (cf. Figure 10.3). Thereafter, we demonstrate the proposed model for rich trajectories, MULTICS, including the overall framework and the details of each component.

10.3.1 Overview

Data collected in the context of MCS combines geolocation with observations and measurements over time, resulting in "rich trajectories". As a running application example, we consider a database obtained from the Polluscope project. A cohort of volunteers have been equipped with individual sensors collecting several pollution measurements along with GPS data. In Polluscope, three data collection campaigns have been conducted. Each campaign is characterized by a start date, an end date, and a person in charge (i.e. responsible). Each campaign was spread over 12 weeks, with a collection generally carried out every other week (in order to check and requalify the sensors). More than 103 volunteers participated in the campaigns. These participants were equipped with kits that contain air pollution sensors and tablets empowered with GPS chipsets. The sensors collect, every minute, time-annotated measurements of particulate matter (PM1.0, PM10, PM2.5), NO_2, BC, temperature, and relative humidity. The tablet was used to geolocate the participants (GPS tracks are collected every one or three seconds) and to fill in their time activity via a mobile app developed for this purpose. Activities last a certain time and represent microenvironments which can be indoor environments (e.g., home, office, restaurant), outdoor environments (e.g., park, street), or even transportation modes (e.g., car, bus). In addition, the participant fills in the events related to air pollution, designating temporary actions over a short period (e.g., opening a window, cooking, smoking, lighting the fireplace). It becomes obvious that the collected data show properties of auto-correlation and multigranularity. The proposed solution should maintain the locality of the spatially close data and to take into account the diversity of granularities. We introduce *spatial discretization and indexing* in order to keep spatially close data together and guarantee a hierarchical spatial representation.

10.3.2 Spatial Discretization

In a multidimensional model and OLAP systems, dimensions have finite and generally known values in advance, so as to be used for grouping and aggregating the measures reported in the fact table. However, spatial and temporal data represent a continuous domain. They cannot be used for aggregation as they are represented. Subsequently, and in order to allow the representation of the spatial dimension, we transform the reported positions (i.e., latitude and longitude) into discrete values referencing a pixel of a rectangular grid with a spatial resolution (here of 50 m). The center subplot of Figure 10.4 illustrates the partition of the Paris region into a rectangular grid along the longitude and latitude dimensions. Assume the minimum latitude and longitude of the region are, respectively, Lat_{min} and Lon_{min}, and the maximum values are Lat_{max} and Lon_{max}. We split the region into $cw*ch$ cells along the 2D axes with a grid side length of λ (here $\lambda = 50$), where cw and ch are, respectively, the number of vertical and horizontal splits of the grid. The finer granularity of the spatial dimension is a cell of 50 m (the choice of this value is discussed in Section 10.1.4). The spatial hierarchy can then be represented by grouping cells with a grid side length of $k*\lambda$, where 2^k is the number of cells in the grouping (cf. Figure 10.5).

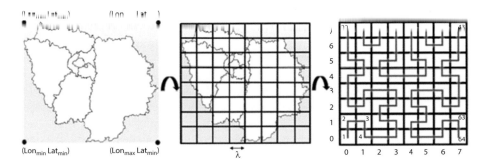

FIGURE 10.4   Spatial dimension representation

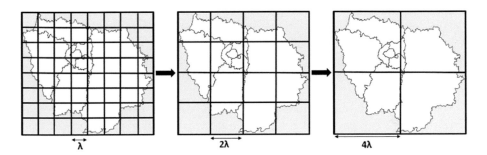

FIGURE 10.5   Spatial hierarchy representation.

10.3.3   TEMPORAL DISCRETIZATION

Likewise, the temporal data are brought back to a minimum threshold. A careful choice of the lowest level of granularity in time is needed in order to provide a good trade-off between precision and storage costs. In fact, while AQ measurements are acquired every one minute, GPS data are collected every one or three seconds. However, participants do not change their location very frequently within one minute inside a cell of 50 m spatial resolution. Hence, the temporal minimum threshold of one minute was chosen here. In this way, spatial and temporal dimensions can be supported by OLAP systems, unlike the original representation of infinite space and time.

10.3.4   SPATIAL INDEXING

There remains the question of maintaining locality in the organization of the spatial dimension. For this, we adopt the spatial indexing using 2D Hilbert Space-Filling Curves (SFCs), which provides a grouping feature per proximity (Moon et al. 2001). In other words, neighboring cells are likely to be assigned to a close Hilbert index. Moreover, Hilbert SFC shows a fractal property that eases the exploration at different levels of the spatial hierarchy, leaving the way for the "roll-up" and "drill-down" within the spatial dimension like zooming in images. By dividing the Hilbert index, we systematically move to a higher level of hierarchy with a grouping of 2^{2n} cells. Figure 10.4 shows our spatial data indexation and representation using Hilbert space-filling curves. The spatial extent is defined to cover the study area (i.e., Paris region). It is worth noting that we only maintain cells corresponding to the locations with GPS data, which is much more compact than solutions like geocubes.

The adopted rasterization approach is gainful in different ways. It allows us to derive the stay areas (often indoors) of participants. For example, we can discover the places where a participant spends time the most based on cells' densities, which can be calculated by counting the number of measurements associated with the cell (the pixel) per participant in the period. This approach also makes it possible to detect spatial outliers and spatial noise. Furthermore, it provides an equal-area pixelization that makes it easy to share the spatial dimension with external sources provided in raster formats (tiff, geoTIFF, etc.) such as AQ maps.[5]

10.3.5 TEMPORAL DISAGGREGATION

One of the main contributions of our MULTICS model is the introduction of new operators, namely, spatial and temporal disaggregation, to derive finer-grained data from coarse data.

Temporal disaggregation methods aim at deriving the data from low-frequency time series to higher-frequency time series. These methods can be categorized as models that:

- Do not rely on any indicator series. These models purely rely on mathematical criteria or time-series models such as ARIMA to obtain data at a higher level.
- Use one or more indicator series observed at higher frequency as a proxy to derive the finer-level time series.

In this chapter, we offer an illustrative example of temporal disaggregation using the R package "tempdisagg" (Sax and Steiner 2013) to perform the optimal procedures of Chow-Lin-Maxlog (Chow and Lin 1971). This example illustrates how we can descend from a low-granularity time series to a higher-granularity time series by applying one of the temporal disaggregation methods with the help of ancillary time-series data (see Section 10.4).

10.3.6 SPATIAL DISAGGREGATION

Spatial disaggregation, on the other hand, refers to the process of converting low-resolution spatial data to higher-resolution data. The most basic approaches for spatial disaggregation are mass-preserving areal weighting (Goodchild and Lam 1980) and pycnophylactic interpolation (Tobler 1979).

In recent research, methods exploring machine learning techniques for combining pycnophylactic interpolation and dasymetric weighting were proposed. (Monteiro et al. 2019) present a hybrid disaggregation procedure to historical data (e.g., estimate population in one census year within the units of another year). (Stevens et al. 2015) combine widely available, remotely sensed, and geospatial data (also referred to as covariates), such as the presence of hospitals, road networks, land cover, etc., that contribute to the modeled dasymetric weights to disaggregate census counts at a country level. Each covariate is projected on a grid pixel of 100 m spatial resolution and then aggregated by census units or villages. Their key contribution is to extract a training set at village's level to learn a random forest (RF) regression model. The RF model is then able to predict the population density at a finer level (100 m spatial resolution). An illustrative example of spatial disaggregation is discussed in Section 10.4 in which we combine different geospatial data such as road networks and the presence of parks to disaggregate NO_2 values.

10.3.7 MULTICS GENERAL SCHEMA

With the transformation of the spatial dimension, the use of OLAP functionalities is obviously possible via our presentation. In this subsection, the schema of MULTICS is described in detail. It adopts a snowflake multidimensional model.

The multidimensional model in MULTICS is generic and can be used for many MCS applications. We use the context of Polluscope in this subsection for illustration purposes.

The general schema is presented in Figure 10.6, which illustrates how we define the dimensions and fact tables. It contains six fact tables. The first table measurement stores the sensors' readings. The fact table measurement relates air pollutant measurements values, depicted by the attribute measurement_value, to five dimensions:

1. Users assigns participants their demographic data.
2. Campaign assigns to each campaign a specific campaign_id and gives information about the start_date, the end_date, and the person in charge, i.e. the responsible.
3. Location is the spatial dimensions, which give information about the Hilbert SFC indices assigned to the grid cells.

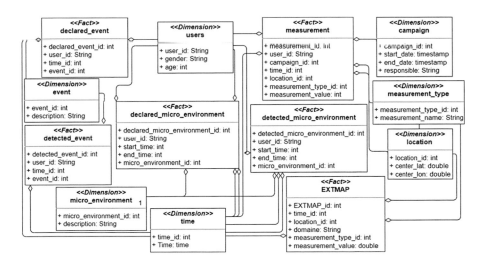

FIGURE 10.6   MULTICS conceptual schema.

4. Time is the time dimension.
5. measurement_type depicts a dictionary of the type of measurements (e.g., PM2.5, NO_2). This could be any observation or measurement, for instance, noise, ozone, pollen, etc. The schema is therefore generic and applicable to any MCS application context.

Thereafter, MULTICS defines five additional fact tables: declared_micro_environment_record, detected_micro_environment_record, declared_event_record, detected_event_record, and EXTMAP. The first table links the participants' information to their declared self-reporting. It describes the microenvironment with a start and end time of presence. The dimension micro_environment contains descriptions about indoor and outdoor microenvironments as well as the exhaustive list of transportation means. The fact table declared_event_record is similar to the declared_micro_environment_ record, except the events are characterized by a temporary timestamp because they are brief. The dimension event depicts the information about the exhaustive dictionary of air pollution–related events.

However, not all the participants thoroughly annotate their microenvironment. Therefore, there is a great interest in automatically detecting the context of the participants without burdening them. Thus, in a previous work (Abboud et al. 2021), we have proposed to learn the microenvironment of participants from multivariate time series based on a multiview stacking generalization approach. We emphasize that MULTICS tackles the particularity of rich trajectories collected in the context of MCS as location-based data series. Nevertheless, MULTICS can integrate external information as dimensions or fact tables. We extend the traditional data warehouse to support external geographic and temporal information from other sources. For instance, we can enrich the data warehouse with external geographic sources such as cartographic layers (e.g., roads, city boundaries) and points of interest (PoIs). External temporal sources can be, for instance, temporary events related to the observed phenomena, such as a fire that emits pollutants, or a confinement leading to a significant drop in traffic, which is a source of pollution. The EXTMAP fact table represents spatiotemporal external sources, such as AQ maps, that can be compared with the collected MCS data.

10.3.8 APPLICATION SCENARIOS

The multidimensional model thus proposed allows analyzing data at different scales and hierarchies. Besides, it enables data to be viewed and modeled in different views, more generally from different facets of dimensions, and more precisely at different locations and periods. We emphasize its usefulness in analyzing and exploring rich annotated trajectories in the context of MCS by introducing some use cases:

*   **Longitudinal Analysis**, which refers to the assessment of individual exposure over time. It allows us to follow the evolution of individual exposure to pollution while detecting periods of high and low levels of pollution. Data can be aggregated over time periods, such as rush hours, weekends, weekdays, etc. The analysis can also be broken into periods spent by

microenvironment which is valuable for understanding the exposure contexts and the differences between them.

- **Spatial Analysis** consists of detecting locations with high-level pollution phenomena. It allows us to emphasize pollution phenomena in different locations. For one participant, the spatial analysis permits us to follow the level of pollution throughout the participant's trajectory, and therefore discover locations with high levels of pollution. Likewise, for all participants combined, we can identify on the map the locations with high-level pollution phenomena. Spatial analysis can be generalized to types of micro-environments as reported by participants, which opens the way to the traceability of the microenvironment with the highest and lowest exposure for each participant or for all participants combined.

- **Temporal Analysis**, which refers to the analysis of measurements over time. In addition to the aforementioned longitudinal analysis, which focuses on the individual dimension, we are interested in other temporal analysis, which combines data from several participants. One example is to analyze the set of measurements reported for different time periods (e.g., peak hours, day of the week, weekend, month, season, year). Another is to focus on a specific microenvironment or area to assess the impact of certain policies on the level of pollution over time.

- **Temporal Disaggregation**, also known as temporal distribution, is the process of converting a low-frequency time series (e.g., annual time series) to a higher-frequency time series (e.g., monthly time series). This might be based on mathematical criteria or time-series model such as ARIMA, or on other consistent temporal data used as a proxy. For instance, if we know the average value of particulate matter (for which the sampling frequency is lower), we can use a consistent time series such as NO_2 (which is sampled every minute) to disaggregate or interpolate *Particulate Matters* from low-frequency to higher-frequency time series.

- **Spatial Disaggregation** refers to the process of producing high-resolution estimates of data distribution (e.g., 25 m^2) from coarse geospatial data (e.g., 1 km^2). The generation of fine-grained data can be done by utilizing techniques that are often in conjunction with ancillary data or statistical modeling methods. For instance, disaggregating NO_2 values from a coarse resolution, such as 20 0m^2, to a finer resolution, such as 50 m^2, using the distribution of buildings and roads as ancillary data in a machine learning model.

10.4   IMPLEMENTATION AND EXPERIMENTATION

10.4.1   Experimental Design

MULTICS is implemented under *Spark*[6] 3.1.2, *Hadoop*[7] 3.2.2, and *Python* 3.9.2. We take advantage of *Spark SQL* OLAP-like querying capabilities. *Spark SQL* employs Hadoop HDFS for data distribution and can answer simple queries; plus it provides optimized OLAP operators.

As mentioned earlier, more than 103 volunteers have participated in the data collection phase, which lasts one week for each participant. GPS data are collected at a frequency of 1 to 30 seconds, while pollutants' measurements are collected every minute, thus resulting in approximately 10 million rows of time-annotated measurements, plus a few annotations of data by the type of microenvironment and pollution-related events.

The spatial dimension is defined by the pixels (finite set and easiest to compare) rather than the exact position and is organized according to the Hilbert index order. This spatial organization is useful for dimensional analysis. Indeed, by referring to longitudinal analysis, one individual can discover the pixels with high pollution and the time spent within, and thus generate heat maps of their exposure. Likewise, this comparison can be performed for different participants combined at the same location (at the fine pixel level or at any level of the spatial hierarchy) in order to identify the participants with the highest exposure. As we can see, there are many different facets for exploring the data. All the possible combinations need to be reachable. In the next section, we introduce some possible combinations of the dimensions, i.e., location, time, and pollutant types.

10.4.2 Longitudinal Analysis

Longitudinal analysis consists of analyzing participant exposure over time. It captures the individual exposure view. We illustrate the perspective of the individual exposure along the daily activities (i.e., for each time interval that the individual has spent in each microenvironment) in Figure 10.7, which depicts the evolution of the collected pollutants throughout the whole period of the campaign, as well as the correlation between the concentrations of pollutants and the detected microenvironments.

10.4.3 Spatial Analysis

Spatial analysis addresses the problem of detecting locations with high-level pollution phenomena. It consists of expressing this phenomena spread in the geographical

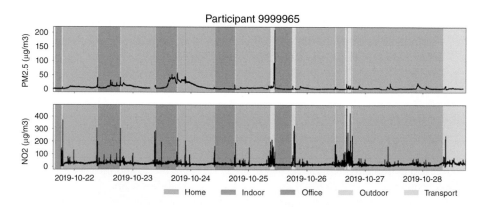

FIGURE 10.7   Longitudinal analysis per microenvironment.

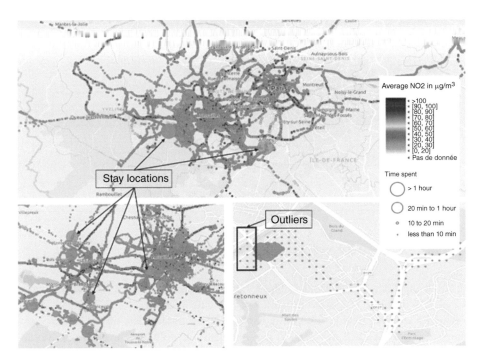

FIGURE 10.8 Concentrations of NO$_2$ for all participants combined in Paris and Versailles regions.

localization and selecting the most visited locations by all participants with high or low levels of pollution. As Hilbert SFC is, by definition, hierarchical, moving to a coarse level of hierarchy only needs to divide the Hilbert index by 2^{2n}, where n is a given hierarchy level. Moreover, this harmonized presentation of the spatial dimension allows is to maintain proximate objects close. In Figure 10.8, the upper map indicates the aggregated trajectories of all our participants combined in Paris region, the lower-left map consists of a zoom of the upper map on the Versailles region, while the lower-right map depicts a parcel of one participant's trajectory over the whole period of the campaign. On the one hand, this spatial presentation allows us to discover the stay areas of participants based on cell density. The circles on the maps depict the places where participants spend most of their time; the radius of the circle is proportional to the time spent in the cell. On the other hand, by setting a threshold, this approach allows the detection of spatial outliers and spatial noise that are close to stay areas; below this threshold, the points will be considered outliers. Moreover, because neighboring cells are likely to be assigned a close Hilbert index, finding the cells that are close to the stay areas is not problematic.

10.4.4 TEMPORAL ANALYSIS

Temporal analysis consists of analyzing the exposure to pollution over time. It permits us to get insight about the phenomena during specific periods while navigating

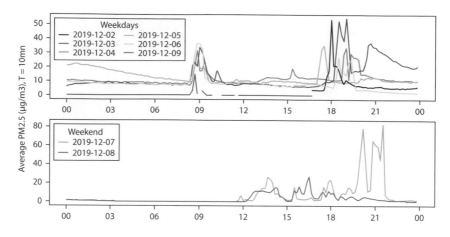

FIGURE 10.9   Ten-minute average concentrations of each day.

through the temporal hierarchy. The temporal hierarchy can be defined in many ways. For instance, it can be illustrated by minute → hour → weekdays/weekend → month → year. Besides obtaining each pollutant time series, one of the applications of our MULTICS model is to get aggregates of pollutants over a time hierarchy (e.g., every 10 minutes, every one hour, every weekend). Figure 10.9 shows the average concentrations of NO_2 every 10 minutes for each day, including weekdays and weekends, for one participant without any restriction on their whereabouts. The graph shows a recurrent behavior of the PM2.5 depicted by high concentration values every weekday around 9 a.m. (i.e., morning rush hour) and between 5 p.m. and 8 p.m. (i.e., evening rush hours). The analysis can go further by adding a GROUP BY clause on participants' microenvironments and computing the hourly average per microenvironment for all participants combined to uncover the microenvironment with the richest pollution phenomena. Table 10.1 depicts the maximums of hourly averages per microenvironment for all participants combined. That says participants are mostly exposed to the highest hourly average of NO_2 and particulate matters in the microenvironment "Office". As for BC, participants are mostly exposed to the highest hourly average in indoor spaces.

TABLE 10.1

Maximum of Hourly Averages per Microenvironment for All Participants Combined

	PM1.0 (Max)	PM2.5 (Max)	PM10 (Max)	NO_2 (Max)	BC (Max)
Office	122.5	216.5	250.5	401.5	14,089.2
Home	69.6	115.0	128.0	371.4	9943.2
Indoor	109.3	153.2	198.5	153.5	20,917.8
Outdoor	130.1	213.0	236.4	147.9	12,240.1
Transport	109.3	196.3	226.7	142.1	11,246.6

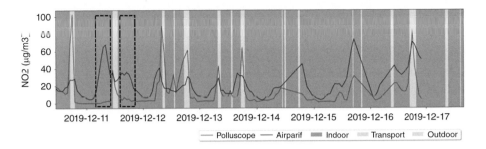

FIGURE 10.10 Airparif data versus the collected data for NO_2 in one trajectory.

As stated earlier, MULTICS can integrate external information that can be used in order to enrich the data with AQ information and/or to compare it to the collected data. In our context, and as an example, we added Airparif[8] AW data as an external source into the model. Airparif provides the same area pixelization as ours in the Paris region. It is set to 12.5×12.5 m² in Paris, 25×25 m² in the inner suburbs, and 50×50 m² elsewhere in the rest of the region. Therefore, by matching the Hilbert index and the timestamp, we can compare Airparif values with the collected data. Figure 10.10 depicts such comparison of NO_2 values throughout one participant trajectory. The two graphs preserve the same tendency, indicating the existence of a correlation between the two sources of data. However, we can remark that for some periods (denoted by the black dashed squares in Figure 10.10), the dark gray (i.e., Airparif) and the light gray (i.e., Polluscope) lines do not share the same behavior. After verifying the participant-declared microenvironment, this difference is due to an indoor space where the participant was during these periods.

10.4.5 TEMPORAL DISAGGREGATION

In this section, we illustrate an example of temporal disaggregation of particulate matter PM10 using NO_2 time-series data as a proxy. As a matter of fact, particulate matters are available every minute. So, for the sake of the experiment, we compute the average of PM10 every hour to get the low-frequency time series, and then disaggregate these data into a higher frequency using NO_2 time series as a proxy and the maximum log likelihood estimates of the Chow-Lin (Chow-Lin-Maxlog) method. Therefore, we can compare the results to the ground truth.

Figure 10.11 expresses the results of the temporal disaggregation process. NO2_1minute, illustrated by the light gray line, denotes the time series of NO_2 at one-minute frequency. PM10_1hour, illustrated by the dark gray curve, denotes the hourly average of the PM10 time series. PM10_1minute, illustrated by the lighter gray curve, indicates the ground truth of the collected measurements of PM10 at the frequency of one minute. Lastly, PM_predicted_1minute, illustrated by the darker gray curve, indicates the predicted values of the disaggregation process on PM10 data from a low frequency (i.e., every hour) to a higher frequency (i.e., every minute).

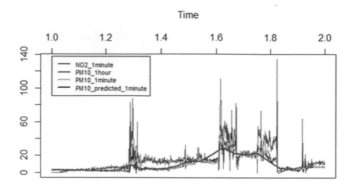

FIGURE 10.11   Example of the disaggregation of PM10 time series using NO_2 data as a proxy with the Chow-Lin-Maxlog method.

As shown in Figure 10.11, the estimates of PM10 follow the trend of NO_2 with respect to the aggregated values of PM10. More particularly, extreme values in NO_2 pull the estimate values to their levels, which explains the occurrence of peaks in the estimated values of PM10. The results of temporal disaggregation show a clear match between the darker gray curve (prediction) and the lighter gray (ground truth) PM10 value. The root mean square error (RMSE) was found to be equal to 4.079.

10.4.6   SPATIAL DISAGGREGATION

In our context, we train an RF regression model on a resolution of 200 m² to estimate the values of NO^2 at a finer level of 50 m². For this purpose, the average values of NO_2 collected over one week by 11 participants are computed over pixels of 200 m spatial resolution. These values are used as a response variable in the model.

As for the model features, pollution level is highly correlated with land use categories, so we integrate this information using Open Street Map (OSM) datasets. Table 10.2 expresses the land cover information types used in the RF regression model as explanatory variables.

TABLE 10.2

Description of the Covariates Used for the Spatial Disaggregation of NO_2 Values

Type	Feature	Description
Traffic	Point	Turning circle, parking, mini roundabout, crossing, traffic signals fuel
Transport	Point	Tram stop, bus stop, railways halt, bus station, taxi, railways station
Land use	Polygon	Forest park
Railways	Line	Subway rail
Roads	Line	Motorway trunk, secondary tertiary, primary, max speed = 30 km/h, max speed = 50 km/h, max speed = 70 km/h, max speed = 100 km/h, max speed = 130 km/h

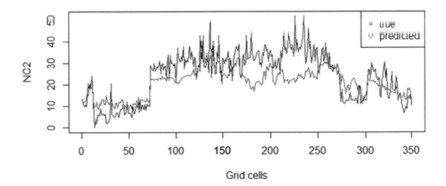

FIGURE 10.12   Real versus estimated values of NO$_2$ after the spatial disaggregation.

The distance from pixel center to covariates included in the model is added to the data, so that it can be incorporated in the model. We conducted a collinearity test between covariates to ensure the validity of the model. The RF model is then trained at a coarse level (i.e., 200 m^2). Thus, to ensure the validity of our model, we use the data over a whole week of one participant in the testing phase. As inputs, we feed to the model the distance to covariates from a finer level (i.e., 50 m^2 pixels). The model returns as output the estimates values of NO$_2$, as shown in Figure 10.12. The dark gray line denotes the real values, while the light gray line indicates the predicted values. RMSE was calculated to report the performance of the disaggregation model. Compared to the ground truth, the NO$_2$ model returns an RMSE of 8.29. We notice that the estimates of NO$_2$ follow the trend of the real values. Figure 10.12 also shows a reasonable fit in the predicted values (light gray) versus the real values (dark gray). Notice that the x-axis is the Hilbert index, which captures the variation among neighboring pixels.

10.4.7 COMPUTATIONAL COSTS

In order to evaluate the performance of the MULTICS model in terms of execution time, we have used a distributed system of five nodes, each node with a capacity of 32 Go.

As the Polluscope deployment remains limited due to its exploratory aspect, the DW does not reach the expected volume in the MCS context in spite of its 4.4 million tuples in the fact table measurement. Therefore, we opt for synthetic data to augment the data and achieve the desired volume. Furthermore, we selected six different queries with different complexity, either by adding a ROLLUP clause to the queries, or computing different aggregated values with different conditions, or performing a LEFT JOIN operator on two fact tables and then computing the different aggregates. Therefore, we compare the performance of the model with respect to the data volume and query complexity. The results of the performance tests are shown in Figure 10.13. The curves show that the execution time seems to follow a linear behavior along data volume. It can also be seen that the difference between execution times of the six queries is not uncanny, and it varies between 10 seconds and 20 seconds.

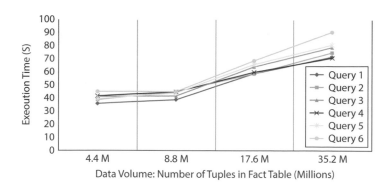

FIGURE 10.13   Execution time varying the data volume.

10.5   CONCLUSION AND PERSPECTIVES

This chapter tackles the exploration and analysis of a georeferenced data series collected in the context of MCS. Several works attempted to deal with the complex nature of such data. This chapter tries to fill the gap between raw data and usable information by providing a multidimensional view on the data.

After analyzing the requirements of multidimensional data modeling in MCS, this chapter introduces such a multidimensional data model designed for processing and querying the different aspects of individual trajectories together with underlying pollution measures. The implementation of the model was based on the Spark SQL and Hadoop ecosystem for data analysis in order to consider all the aspects of the data. The core data model and the methodology considered is applied to urban mobility and pollution data but is generic enough to act as a reference model for other applications.

We intend to use the OLAP model, on one hand, to detect anomalies by exploring the data and correcting it by applying, for instance, statistical smoothing methods such as moving average and exponential moving average. On the other hand, the OLAP model can be utilized for data enrichment such as the derivation of stops (i.e., stay locations) according to the density of points per cell, or according to the sparsity of GPS readings over time. Finally, in the case of real-life use, the volume increases continuously, which could be a bottleneck. The question that arises is: *With the incurred data volume, is it reasonable to store all the data in the data warehouse, or store only the fresh measurement stream, or use a hybrid model where historical data are aggregated to some extent to achieve a trade-off between utility and efficiency?* This leads to new challenges in terms of model and maintenance operations (to trigger the aggregation). The suggested disaggregation operator will be useful to estimate the original facts in this context. For more information on the proposed framework, refer to (El Hafyani 2022).

ACKNOWLEDGMENTS

This work has been supported by the French National Research Agency (ANR) project Polluscope, funded under the grant agreement ANR-15-CE22-0018. We are thankful to VGP (Thomas Bonhoure) for facilitating the campaign. We would like to thank all the members of the Polluscope consortia who contributed in one way or another to this work: Salim Srairi and Jean-Marc Naude (CEREMA), who conducted the campaign; Boris Dessimond and Isabella Annesi-Maesano (Sorbonne University), for their contribution to the campaign; Valérie Gros and Nicolas Bonnaire (LSCE) and Anne Kauffman and Christophe Debert (Airparif), for their contribution in the periodic qualification of the sensors and their active involvement in the project. Finally, we would like to thank the participants for their great effort in carrying the sensors, without whom this work would not be possible.

NOTES

1. http://ambiciti.io
2. https://openradiation.org
3. http://polluscope.uvsq.fr
4. The chosen division follows the same division as the Air Quality Monitoring Association in the Paris region AirParif (https://www.airparif.asso.fr/). It is set to 12.5×12.5 m^2 in Paris, 25×25 m^2 in the inner suburbs, and 50×50 m^2 elsewhere in the rest of the region.
5. https://www.breezometer.com/
6. https://spark.apache.org
7. https://hadoop.apache.org
8. Air quality observatory in Paris region: https://www.airparif.asso.fr/

REFERENCES

Abboud, M., H. El Hafyani, J. Zuo, K. Zeitouni, and Y. Taher. "Micro-Environment Recognition in the Context of Environmental Crowdsensing." Workshops of the EDBT/ICDT Joint Conference, EDBT/ICDT-WS, BMDA, vol. 2841, 2021.

Bimonte, S.O., A. Tchounikine, and M. Miquel. "Spatial OLAP: Open Issues and a Web Based Prototype." 10th AGILE International Conference on Geographic Information Science, p. 11. 2007.

Chow, G. C., and A. Lin. "Best Linear Unbiased Interpolation, Distribution, and Extrapolation of Time Series by Related Series." The Review of Economics and Statistics, vol. 53, no. 4 (1971): 372–375.

El Hafyani, H. "Spatio-temporal Data Analytics in the Context of Environmental Crowdsensing." PhD diss., Versailles, Université Paris-Saclay, 2022.

Goodchild, M. F., and N. S. N. Lam. "Areal Interpolation: A Variant of the Traditional Spatial Problem." Geo-processing 1, no. 3 (1980): 297–312.

Hägerstrand, T. "What About People in Regional Science?." Papers of the Regional Science Association 24, (1970): 7–24.

Iftikhar, N., and T. B. Pedersen. "Schema Design Alternatives for Multi-Granular Data Warehousing." International Conference on Database and Expert Systems Applications, vol. 6262, pp. 111–125. 2010.

Jensen, S. K., T. B. Pedersen, and C. Thomsen. "Time Series Management Systems: A Survey." IEEE Transactions on Knowledge and Data Engineering 29, no. 11 (2017): 2581–2600.

Karim, L., and A. Boulmakoul. "Trajectory-Based Modeling for Fraud Detection and Analytics: Foundation and Design." 2021 IEEE/ACS 18th International Conference on Computer Systems and Applications (AICCSA), pp. 1–7. 2021. doi: 10.1109/AICCSA53542.2021.9686920.

Koncilia, C., T. Morzy, R. Wrembel, and J. Eder. "Interval OLAP: Analyzing Interval Data." International Conference on Data Warehousing and Knowledge Discovery, pp. 233–244. 2014.

Leonardi, L., S. Orlando, A. Raffaetà, A. Roncato, C. Silvestri, G. Andrienko, and N. Andrienko. "A General Framework for Trajectory Data Warehousing and Visual OLAP." GeoInformatica 18, no. 2 (2014): 273–312.

Lo, E., B. Kao, W. S. Ho, S. D. Lee, C. K. Chui, and D. W. Cheung. "OLAP on Sequence Data." Proceedings of the 2008 ACM SIGMOD International Conference on Management of Data, pp. 649–660. 2008.

Miller, H. J. "Tobler's First Law and Spatial Analysis." Annals of the Association of American Geographers 94, no. 2 (2004): 284–289.

Monteiro, J., B. Martins, P. Murrieta-Flores, and J. M. Pires. "Spatial Disaggregation of Historical Census Data Leveraging Multiple Sources of Ancillary Information." ISPRS International Journal of Geo-Information 8, no. 8 (2019): 327.

Moon, B., H. V. Jagadish, C. Faloutsos, and J. H. Saltz. "Analysis of the Clustering Properties of the Hilbert Space-filling Curve." IEEE Transactions on Knowledge and Data Engineering 13, no. 1 (2001): 124–141.

Rivest, S., Y. Bédard, M. J. Proulx, M. Nadeau, F. Hubert, and J. Pastor. "SOLAP Technology: Merging Business Intelligence with Geospatial Technology for Interactive Spatio-temporal Exploration and Analysis of Data." ISPRS Journal of Photogrammetry and Remote Sensing 60, no. 1 (2005): 17–33.

Savary, L., T. Wan, and K. Zeitouni. "Spatio-temporal Data Warehouse Design for Human Activity Pattern Analysis." Proceedings. 15th International Workshop on Database and Expert Systems Applications, pp. 814–818. 2004.

Sax, C., and P. Steiner. "Tempdisagg: Methods for Temporal Disaggregation and Interpolation of Time Series". The R Journal 5/2 (2013): 88–87.

Stevens, F. R., A. E. Gaughan, C. Linard, and A. J. Tatem. "Disaggregating Census Data for Population Mapping Using Random Forests with Remotely-sensed and Ancillary Data." PloS One 10, no. 2 (2015): e0107042.

Tobler, W. R. "Smooth Pycnophylactic Interpolation for Geographical Regions." Journal of the American Statistical Association 74, no. 367 (1979): 519–530.

Vaisman, A, and E. Zimányi. "What Is Spatio-Temporal Data Warehousing?." International Conference on Data Warehousing and Knowledge Discovery, vol. 5691, pp. 9–23. 2009.

Vaisman, A., and E. Zimányi. "Mobility Data Warehouses." ISPRS International Journal of Geo-Information 8, no. 4 (2019): 170.

Wagner, R., J. A. F. de Macedo, A Raffaetà, C. Renso, A. Roncato, and R. Trasarti. "Mob-Warehouse: A Semantic Approach for Mobility Analysis with a Trajectory Data Warehouse." International Conference on Conceptual Modeling, vol. 8697, pp. 127–136. 2013.

Wan, T., K. Zeitouni, and X. Meng. "An OLAP System for Network-Constrained Moving Objects." Proceedings of the 2007 ACM Symposium on Applied Computing, pp. 13–18. 2007.

11 Trajectory Mining Based on Process Mining in RORO Terminals

Performance-Driven Analysis to Support Trajectory Redesign

*Mouna Amrou Mhand, Azedine
Boulmakoul, and Hassan Badir*

CONTENTS

11.1 INTRODUCTION

Roll On/Roll Off (RORO) transportation represents an efficient and competitive tool used for the large-scale transshipment of rolling cargo. It is characterized by its fast speed, low cost, and environmental safety. However, its low-efficient resources used for operations have constrained its development. Moreover, with the use of advanced technologies in the operational business processes in RORO terminals, the amount of data generated is important. Therefore, exploiting this immense volume of data

DOI: 10.1201/9781003255635-11

would be of great benefit in terms of resource reuse and infrastructure management that would directly impact service quality. Furthermore, process mining techniques have become a crucial component for supporting systems redesign and enhancements, as they aim to utilize the data recorded in event logs to extract hidden insights into the operational process. Also, they rely on the real data produced. Process mining is a potent technique for enabling us to have a transparent view of how the system is working, rather than how the manager or the logistician thinks it is functioning.

Process mining is widely used for process model discovery, conformance checking, identifying mismatches and performing process enhancements. On this basis, this work presents the applicability of process mining on RORO terminal event logs using data generated from the export process of one of the biggest ports in Africa. The objective is to mine trajectories in this terminal to assist the process redesign based on the performance analysis results. The idea is to increase transport performance and achieve traffic efficiency by minimizing service time and resource consumption, detecting bottlenecks, and addressing them.

The authors were motivated by the following questions to carry out this study: (M1) How can these immense data generated from the running processes of RORO terminals be exploited? (M2) How can these data be used to support decision-making and process redesign? (M3) What are the most adequate process mining algorithms that we can exploit to extract valuable insights to mine trajectories in RORO terminals from the collected logs?

The objectives of this research are to (RO1) Mine RORO terminal trajectories using data generated by the operational processes. (RO2) Apply process mining using the inductive visual miner (IVM) algorithm and the dotted chart to see the distribution of events over time and extract performance measures about RORO terminal service areas. (RO3) Identify potential bottlenecks and areas that consume time and resources and focus more on them.

This chapter has the following research questions (RQs): (RQ1): How can the immense volume of operational processes in RORO terminal data be exploited to mine trajectories to assist process redesign by managers and logisticians? (RQ2) How can automatic process mining algorithms support mining RORO terminal trajectories to discover process models and help in performance analysis to extract fact-based insights to support process redesign? (RQ3) How can we contribute to this research area?

The remainder of this chapter is organized as follows. Section 11.2 presents the research background consisting of process mining and RORO terminals' operational processes. Section 11.3 covers related work on process mining and the RORO terminal transportation field. Section 11.4 introduces the case study of trajectory mining in RORO terminals' logistics processes and presents the dataset used, tools and technologies employed for the experimental analysis and results as well as the discussion of the experimentation and analysis. Finally, the chapter concludes with a scope for future research in Section 11.5.

11.2   RESEARCH BACKGROUND

This work is founded on two main research areas, RORO terminals as the application domain and process mining as a data technique used for investigating and mining RORO terminals trajectories.

This section first defines process mining and its importance. Second, RORO terminals' key components involved in the operational logistics processes are described.

11.2.1 Process Mining

Process mining is an emerging concept that bridges the gap between data science and process science (van der Aalst 2016). This technique consists of discovering process models, performing conformance checking, and performing analysis that serves to detect bottlenecks and discover hidden patterns to propose enhancements that can decrease resource overuse and increase profitability. Process mining was proposed in 1999 by the Dutch scientist Wil van der Aalst in the Netherlands. The entrance door of any process mining technique is an event log representing one or multiple business processes. It consists of events that have occurred listed with their attributes, which primarily contain the case identifier, activity name, and timestamps of the start and end of this activity. These data are stored in information systems adopted by the organization. We distinguish between three categories of process mining techniques. The first is process model discovery, which takes an event log and produces a process model without using a priori information about the log. Second, conformance checking aims at checking if what is registered in the log is the same as what is taking place in reality through the comparison of an existent process model to an event log retrieved from the same process. Third, process enhancement intends to either improve an existing process model or extend the existing process model to reflect new perspectives based on performance results. Process mining is a powerful practice that has numerous benefits for improving processes based on fact-based decisions and supporting optimization measures through the analysis of process traces reflecting the real-life scenario happening. Using this powerful technique, deviations and bottlenecks can be detected and avoided, the cost can be lowered and the gain increased.

11.2.2 Logistics Processes Related to Transportation in RORO Terminals

The concept of RORO refers to ships specializing in the transshipment of rolling cargo. It is a very popular means of transportation in the automotive world. RORO ships are particularly adapted for short-distance transport. RORO terminals are port terminals that are used for all the operations related to RORO transportation. These terminals involve logistics processes organizing the ordering of each task. We describe two major types involved, the export and the import. Each one generates a massive volume of data at each terminal area in addition to every operation performed that we consider an event. The processes in RORO terminals are uncertain, dynamic, and complex due to the interdependency between the components constituting them. And this impacts the global performance of the system. Therefore, an impeccable interaction between the port/terminal and the ship is crucial for the favorable outcome of the RORO mode of transport.

11.3 RELATED WORK

In this section, we will cover the most recent research carried out on process mining in the first part. For the second part, we will cover research on RORO terminals.

Recently, we observed an important orientation to process mining and its usage in multiple domains such as healthcare, e-learning, robotics, etc. Some of these contributions mainly focused on proposing literature reviews, others introduced case studies applied in various domains, while the remaining addressed taxonomy, and algorithm enhancements. (Cerezo et al. 2019) proposed a work in the educational domain, where the authors applied an inductive miner algorithm over the interaction traces from 101 university students in a course given over one semester on the Moodle 2.0 platform. (Zerbino et al., 2021) presented a literature review on process mining in business management. (Pika et al., 2020) proposed a work in the healthcare domain where they used process mining to build a framework for privacy-preserving support for healthcare analysis. (Pika et al., 2021) suggested an application of process mining to enhance Big Data visualization. (Leemans et al., 2021) showed that stochastic conformance checking enables detailed diagnostics projected on both model and log. The authors also extended the reallocation matrix to consider paths, and the proposed approach has been implemented in PROM. In robotics, (Leno et al., 2020) defined a set of basic concepts of robotic process mining and presented a pipeline of processing steps that would allow a robotic process mining tool to generate robotic process automation scripts from user interface (UI) logs. In the work of (Sato et al., 2022), the authors conducted a systematic literature review on the intersection of these areas, and thus, reviewed concept drift in process mining and brought forward a taxonomy of existent techniques for drift detection in addition to online process mining. (Julian et al., 2022) used process mining and deep learning architecture for Diabetes intensive care unit patients to predict in-hospital mortality of diabetes. In regard to publications on RORO terminals, we note that the existing literature is scarce and does not have much focus. Among the recent works on RORO, we include (Görçün and Küçükönder, 2021), where the authors proposed an approach for the process of port RORO terminal selection based on the criteria importance through an inter-criteria correlation technique and the evaluation based on distance from the average solution method to evaluate the RORO marine ports selected. (Chen et al., 2021) investigated the storage location assignment problem for the arrival of cars at the yard, aiming to improve the ship-loading efficiency and contribute to efficient storage at RORO terminals. (Shen et al., 2021) focused on the operation flow of a commercial automobile RORO terminal by designing the specific import operation flow and export operation flow and summarized the loading and unloading operation plan. (Ventura et al., 2020) proposed a chapter where they provided a characterization of the European RORO ship fleet based on a comprehensive ship database.

Based on these recent works that we have included here, no work has previously addressed logistics trajectories in RORO terminals from a process mining perspective. Most of the solutions rely on simulation systems for any enhancement or trajectory optimization. These operational terminals generate numerous amounts of data that can be utilized to provide fact-based insights about these operational processes to accomplish transparent decision-making. After reviewing all this research related to process mining and its field of application, we haven't found any work addressing the case of logistics in port terminals —more specifically in RORO terminals. Hence, we tackled this research gap. In this work, we focus on port logistics as a field of application of process mining. Here, we explore process mining to mine trajectories in RORO port terminals to assist in process redesign. In this next section, we will be discussing in detail the case study chosen in this work.

11.4 CASE STUDY: TRAJECTORY MINING OF RORO TERMINAL LOGISTICS

This section is devoted to the case study, which consists of trajectory mining of RORO terminals' logistics processes. The objective here is to assist RORO terminal trajectory redesign and enhancement based on the findings of process mining results. The data used here originated from one of the biggest ports in Africa consisting specifically of data related to the export process. We will present in this section the research objectives, the tools, and the algorithms used. Also, we will illustrate the results and discussion. We note that the work presented in this chapter is a continuation of a previous work already published (Mhand et al., 2019), where we focused on the process model discovery phase using heuristics and IVM algorithm.

11.4.1 Tools and Algorithms Used

To perform mining trajectories, we used an open-source tool, the PROM6 framework. It is an extensible framework that supports a large variety of plugins for process mining. The tool is developed in Java as an independent platform. We suggest referring to the website[1] for in-depth documentation.

In this case study, we used two plugins from the PROM6 framework to perform trajectory mining. We referred to the IVM and the dotted chart. These two tools are detailed in the next sections.

11.4.1.1 Inductive Visual Miner Algorithm

The IVM is an algorithm that can be used to perform business process discovery, conformance checking, and performance analysis. This algorithm takes an event log, discovers the process model, compares it to the event log, and highlights potential enhancements through the possibility of examining performance measures such as service time, sojourn time, etc., all in an automated manner. This work provides a complete guide on IVM.[2] We chose this algorithm because it has proven its efficiency. It enables us to do the three process mining techniques from process discovery to enhancement. IVM is a strong tool to model process trees that are efficient in dealing with noisy and incomplete data.

We note that there is no previous research that has used IVM to mine trajectories for the case of the transportation domain, specifically in RORO port terminals.

11.4.1.2 Dotted Chart

The dotted chart analysis demonstrates the distribution of events over time. Every dot represents an event, the color of the dots is the action type, every horizontal line is the case, and every vertical line is the moment in time. Notice that there are empty vertical lines, which means no actions or fewer actions happen in these moments. In this case, empty vertical lines mean weekends or holidays. Dotted charts help in viewing process event distribution over time.

11.4.2 Dataset Source and Preprocessing

11.4.2.1 Data Source

The dataset has been retrieved from the information system used by the RORO terminal port management for the period April 2 to 15, 2019. It is related to rolling cargo export process traceability.

11.4.2.2 Dataset Filtering and Noise Removal

To prepare the raw data to construct an event log that can be used as the entry point of the analysis, we implemented a program that takes a CSV file containing the data that was stored in the database. and as an output, it returns the event log that can be analyzed using the PROM6 tool to obtain interpretable results during mining. The program has been implemented using the multi-paradigm language Scala. The steps are illustrated in the algorithm illustrated in Figure 11.1.

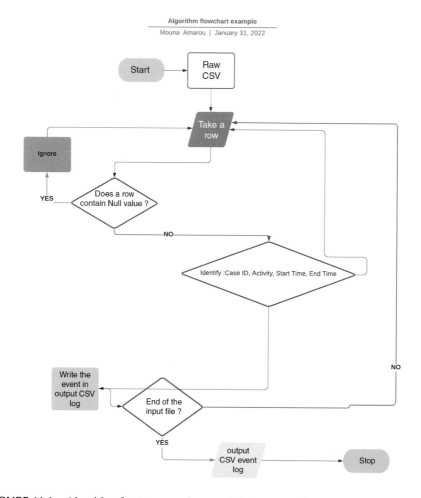

FIGURE 11.1   Algorithm for preprocessing raw data to an event log.

11.4.3 TRAJECTORY MINING AND EXPERIMENTAL RESULTS

The experimental results have been performed in a Windows environment. The laptop's characteristics are as follows:

- 8 GB memory, 256 GB SSD
- Core i7 fifth generation
- 2.1 GHz processor

The dataset used is generated from the export process of a port terminal in Africa to build a process mining model to mine trajectories and extract hidden patterns to assist the trajectory redesign. We highlight that the results have been found using the IVM algorithm.

- Experiment 1: IVM process model deviation identification

The examination here aims to identify process deviation, also referred to as trajectory deviation, in the case of RORO terminals. First of all, deviations mean that process discovery models can leave a behavior expressed in the log out of the model, as it can model the behavior. In Figure 11.2, we identify where the model and log disagree. We note that these results are obtained using the optimal alignment, as mentioned in Table 11.1.

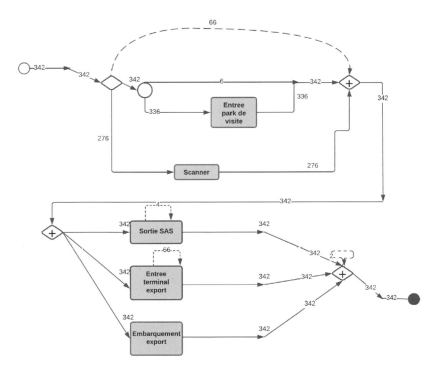

FIGURE 11.2   A close look at a deviation.

TABLE 11.1
Optimal Alignment Measures Used in IVM

Activity Frequency	Path Frequency
93%	99%

For instance, we observe a model move 66 times for *Scanner*, which means that it was expected but was not expressed in the event log. For activities (*Sortie SAS* and *Entree terminal export*), we identify a log move that refers to indicating that four times in the event log, after the execution of *Sortie SAS*, an event happened in the event log, while this should not happen according to the model. Having an in-depth look at the traces, we can see the events where the disagreement happened. In Figure 11.3, we provide an example of deviation.

- Experiment 2: Process performance analysis using IVM

Now we take an interest in process performance in this second experiment. To acquire a clear vision of the *average service time,* Figure 11.4 provides the results obtained from the analysis. We notice that *Sortie SAS activity* is taking the longest period, with *05:29:16:36*, followed by *entrée terminal export*, with *04:16:25:31.*

In regard to the *average sojourn time* (see Figure 11.5), *Embarquement Export* takes the longest time in the process, with *06:00:01:087.* This could be explained by the dependency on other factors, namely vessel arrival, berth, etc. By taking a look at the details of the performance of *Sortie SAS* (Figure. 11.5), we observe that the sojourn time for some cases may sometimes be up to a day and 10 hours. Therefore, we detect signs of bottlenecks. We present a summary in Figure 11.6.

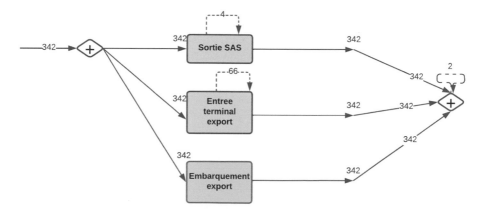

FIGURE 11.3   Paths and deviations: Inductive visual miner – Model compliance checking.

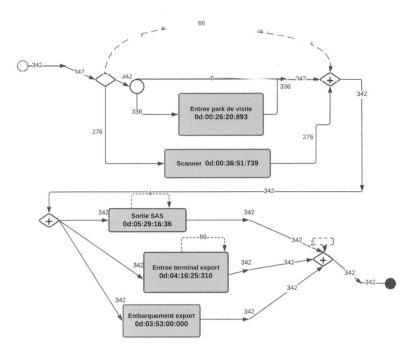

FIGURE 11.4   Performance analysis IVM: Service time per activity.

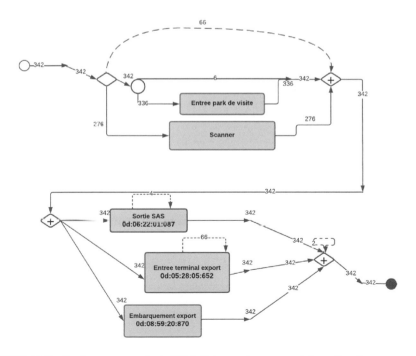

FIGURE 11.5   Performance analysis IVM: Sojourn time per activity – Result.

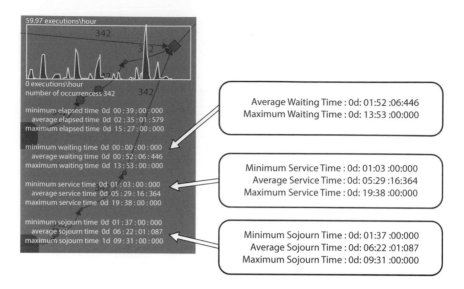

FIGURE 11.6   A summary of performance measures.

- 3: Event distribution over time using the dotted chart

In this third experiment, we will be using the dotted chart for performance analysis. It does not focus on the routing of work or the activities, but on the case to show overall event performance information of the log. In the chart, events are expressed as dots, and the time is measured along the horizontal axis of the chart. The vertical axis represents case IDs, and events are illustrated according to their activity IDs. To view the distribution of events by case duration, the horizontal axis is fixed at the starting time of cases. The results are illustrated in Figure 11.7.

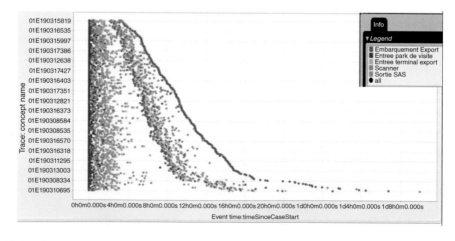

FIGURE 11.7   Distribution of cases according to trace duration.

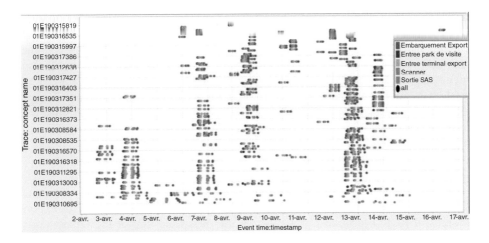

FIGURE 11.8 Distribution of cargo flow in the terminal per day of the week.

Here we can see a long tail towards the end, showing that the activity duration is not constant—the average task duration is about 18 hours, whereas some can last up to one day and 10 hours approximately. Another pattern we used to identify the days of the week, where cargo flow is at its peak, is illustrated in Figure 11.8. We can see that freight flow was high on days 4, 7, 9, 13, and 14.

11.5 RESULTS DISCUSSION AND ANALYSIS

Currently, there is very little research regarding trajectory mining in RORO terminals using process mining techniques. This is a novel approach applied in the field of port terminals, more specifically, in RORO terminals. As we can see based on these results, *SAS* activity takes approximately 8h in total, and in some cases, it may take up to a day. This can be due to multiple factors, namely not enough resources (personal as well as organizational) and also the location of this *SAS* station. This issue can be addressed, whether by increasing the personnel in this specific stop or perhaps offering multiple *SAS* stations. That way, we can serve more vehicles in fewer periods. Another approach would be to interfere at the level of process redesign by changing the location of this service. Regarding sojourn time, we observe that the longest duration was 6h approximately in the *Embarquement export area*. This highlights signs of bottlenecks, which has a direct impact on the performance of the rest of the services as well. For instance, if the terminal has a capacity of 200 vehicles, it will not be possible to exceed this capacity once it is reached, resulting in major financial loss. This issue can be addressed by increasing the terminal's capacity and collaborating with the involved parties, namely the maritime side, to attribute a minimal berthing time as well as service time in this area. This gives more benefits to every party involved. Also, during the days of the week when the cargo reaches its peak, it is recommended that all the parties involved should be ready to manage this load. As we noticed from the performance analysis carried out, we pulled out real valuable insights from the data generated from the logistics operational export process in this RORO terminal. This information provides the opportunity to identify and comprehend hidden patterns that lead over time to resource overuse, financial and human, as well as equipment. Having a clear vision of what happens in the terminal in real-time would help managers to discover the areas that need more focus and enhancement, predicting bottlenecks would make them have a prepared strategy to address them before they can happen, which has valuable gains in time and money, especially in such critical domains. On a final note, we performed the mining of trajectories in this terminal. All of this can assist managers in supporting trajectory redesign.

11.6   CONCLUSION AND FUTURE WORK

In conclusion, this chapter presents the applicability of process mining to the export process of RORO terminals. For this, we use data from one of the largest ports in Africa. The objective is to mine trajectories and extract hidden insights regarding performance and bottlenecks that can be taken into consideration in process redesign.

In this work, we identified the deviations in the process through conformance checking that has been done on the event log using IVM. Also, we analyzed the distribution of events over time using the dotted chart. Moreover, we carried out performance analysis and identified the areas in the terminal that present resource overuse and demand more attention and improvements. Based on the results and the discussion, we illustrated how this competent tool can be used to increase gain in time and resources. As a future exploration, we intend to develop a simulation model based on the process model discovered and use it to test the different changes required and how they impact the terminal's performance before making any change in real life, as these domains demand a lot of costs.

NOTES

1. https://www.promtools.org/doku.php
2. http://www.leemans.ch/publications/ivm.pdf

REFERENCES

Cerezo, R., Bogarín, A., Esteban, M. and Romero, C., 2019. Process Mining for Self-regulated Learning Assessment in E-learning. *Journal of Computing in Higher Education*, 32(1), pp. 74–88.

Chen, X., Li, F., Jia, B., Wu, J., Gao, Z. and Liu, R., 2021. Optimizing Storage Location Assignment in an Automotive Ro-Ro Terminal. *Transportation Research Part B: Methodological*, 143, pp. 249–281.

Görçün, Ö. F. and Küçükönder, H., 2021. An Integrated MCDM Approach for Evaluating the Ro-Ro Marine Port Selection Process: A Case Study in Black Sea Region. *Australian Journal of Maritime & Ocean Affairs*, 13 (3): 203–223.

Leemans, S., van der Aalst, W., Brockhoff, T. and Polyvyanyy, A., 2021. Stochastic Process Mining: Earth Movers' Stochastic Conformance. *Information Systems*, 102, p. 101724.

Leno, V., Polyvyanyy, A., Dumas, M., La Rosa, M. and Maggi, F., 2020. Robotic Process Mining: Vision and Challenges. *Business & Information Systems Engineering*, 63(3), pp. 301–314.

Mhand, M. A., Boulmakoul, A., Badir, H. and Eddoujaji., 2019. "Process mining driven analysis Model discovery: Case study of RoRo terminals." *INTIS*

Pika, A., ter Hofstede, A., Perrons, R., Grossmann, G., Stumptner, M. and Cooley, J., 2021. Using Big Data to Improve Safety Performance: An Application of Process Mining to Enhance Data Visualisation. *Big Data Research*, 25, p. 100210.

Pika, A., Wynn, M., Budiono, S., ter Hofstede, A., van der Aalst, W. and Reijers, H., 2020. Privacy-Preserving Process Mining in Healthcare. *International Journal of Environmental Research and Public Health*, 17(5), p. 1612.

Sato, D., De Freitas, S., Barddal, J. and Scalabrin, E., 2022. A Survey on Concept Drift in Process Mining. *ACM Computing Surveys*, 54(9), pp. 1–38.

Shen, M., Yao, Y., Jiang, S., Ma, R. and Wei, Y., 2021. Research on the Operation Flow of Commercial Automobile Ro-Ro Terminal. *IOP Conference Series: Earth and Environmental Science,* 692(2).

Theis, J., Galanter, W., Boyd, A. and Darabi, H., 2022. Improving the In-Hospital Mortality Prediction of Diabetes ICU Patients Using a Process Mining/Deep Learning Architecture. *IEEE Journal of Biomedical and Health Informatics*, 26(1), pp. 388–399.

van der Aalst, Wil M. 2016. Data Science in Action. Springer, pp. 3–23.

Ventura, M., Santos, T. A. and Soares, C. G., 2020. "Ro-Ro Ships and Dedicated Short Sea Shipping Terminals." *Short Sea Shipping in the Age of Sustainable Development and Information Technology*, 22–57.

Zerbino, P., Stefanini, A. and Aloini, D., 2021. Process Science in Action: A Literature Review on Process Mining in Business Management. *Technological Forecasting and Social Change*, 172, p. 121021.

12 Aspects of Mobility Data in the Fog/Cloud Era

Directions from a Pilot Case Study of Hazmat Transportation Telemonitoring in an Urban Area

Adil El Bouziri and Azedine Boulmakoul

CONTENTS

DOI: 10.1201/9781003255635-12

191

12.1 INTRODUCTION: MOBILITY DATA AND DIRECTION FROM HAZMAT TRANSPORTATION

12.1.1 HAZMAT TRANSPORTATION MONITORING

Hazmat (hazardous materials) transportation, even with a low accident probability in some areas, can cause catastrophic consequences when the event affects large numbers of people, infrastructure, ecosystems, buildings, and other resources. This kind of transport has increased significantly and affects various modes like roads, boats, trains, planes, pipelines, etc. (Garbolino et al. 2012). In particular, the transport of dangerous goods by road can be the cause of accidents in urban areas. It therefore represents one of the factors that increase the vulnerability of the crossed territories. Thus, there is a need to find and improve methods and tools to help decision makers and the private sector assess the vulnerability of areas under their control, with particular emphasis on explosions. Among the challenges of this case study is to find a reactive ecosystem for an open architecture system with broad interoperability, designed to provide risk and environment monitoring, and to give specific services to mobile users. Figure 12.1 gives an example of services that can be proposed by this type of system (Cherradi et al. 2018)—for example, the simulation of the explosion of the critical point of high risk and the surveillance of the prohibited regions that cannot crossed by the trucks of dangerous goods. Moreover, combining the Internet of Things (IoT) paradigm and the fog/cloud environment offers new computing

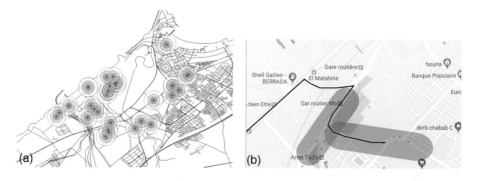

FIGURE 12.1 (a) Simulation of explosion of high-risk points. (b) Regions prohibited from hazmat transportation.

possibilities aimed to handle the new processing needs of IoT applications in order to enhance service quality and minimize latency (Lan et al 2010, Yousefpour et al 2019). In fact, this work adopts a microservice-based architecture with highly scalable applications within a fog/cloud environment.

12.1.2 Mobility Data: Management of Moving Object Trajectories

The paradigm shift regarding mobility data is emerging in many areas, in particular in critical systems. This needs advanced approaches that support the manipulation, detection, and prediction of relevant data, events, and trajectories of the mobile entities. These data themselves could possibly be noisy and heterogeneous and probably require processing on several levels of the ecosystem that has to be recommended. Thus, we can meet the demanded requirements with emerging technologies ranging from simple sensors, intelligent or not, to an arsenal of computing resources, storage and component software at the data center of cloud systems. Furthermore, there is an explosion of applications using on-board systems in healthcare, the automotive industry, robotics, smart homes, and smart cities, reflecting the rapid development of the IoT. The smart city is an interesting example where context awareness applied in smart city environments introduces new ways of deploying user-centric services.

There was in this context a real need to understand mobile data and to perform advanced modeling for mobile objects to implement efficient systems. The evolution of this work ranges from mobile object modeling and mobile queries to mobile and spatiotemporal (including NoSQL) databases capable of handling, storing, and analyzing the trajectories of moving objects with the rich specific annotations in the era of Big Data. Thus, the contribution has focused on spatiotemporal data, in particular those related to mobile objects, databases of moving objects, and spatiotemporal queries (Güting and Schneider 2005, Güting et al. 2000). The work of the authors in (Boulmakoul and Bouziri 2012) presented a global object-oriented modeling of moving objects on the multimodal transport network. In addition, other research works processed and modeled the semantic trajectories of moving objects. Interesting details are given in (Boulmakoul et al. 2015, Zheng and Zhou 2011). In addition, a specific smart trajectory modeling is detailed in the work (Karim et al. 2021) for a specific domain like a smart city.

12.1.3 Mobility Data and a Secured Transportation System

The risk associated with the transport of dangerous goods is a complex risk. It depends on several factors: traffic concentration, weather conditions, etc. An evaluation methodology is also needed to analyze the consequences for the environment and the population in the event of an accident, without forgetting the importance of optimization and planning of routes in order to reduce the occurrence of accidents and mitigate the serious consequences. Thus, we need in this critical system a spatial (or eventually spatiotemporal) decision support system (SDSS) that must be implemented within a modern architecture allowing efficient monitoring and surveillance. The work has to deal with the management of risk and routing in hazmat transportation to increase the performance of the overall system and to realize the secure transportation of hazardous materials (Boulmakoul et al. 2015, Boulmakoul 2006).

Finally, this chapter is organized as follows: Section 12.2 gives a short overview of mobility data and modeling of moving objects; Section 12.3 presents the main elements of fuzzy routing to integrate in this system; Section 12.4 introduces an overview of fog/cloud computing, and Section 12.5 briefly presents a pilot case study about hazmat transport telemonitoring.

12.2 MOBILITY DATA: REPRESENTATION AND MANAGEMENT OF MOVING OBJECT TRAJECTORIES FOR EXPLORATION

This section proposes an overview of the work on modeling moving objects that constitutes a basic "framework" for the development of the context-based services and applications. It is based on several specifications which relate to the spatial model, the temporal model, and that of the transportation network. In the same way, this model is based on a set of research works related to moving objects and to spatiotemporal queries (Güting and Schneider 2005, Güting et al. 2000).

12.2.1   THE GLOBAL OBJECT-ORIENTED MODELING OF MOVING OBJECTS

The model is represented by spatiotemporal classes with mobility aspects. This is a result of the work on mobile object modeling and location-based services. The context-based services are concerned with the moving point objects over a predefined network infrastructure. The emphasis is put on the relationships with the main classes which represent the multimodal transportation network. As a simple view, Figure 12.2 shows the diagram of packages. The package *pk_MobileObject* is the core of the data model of moving objects. It contains the main classes and operations

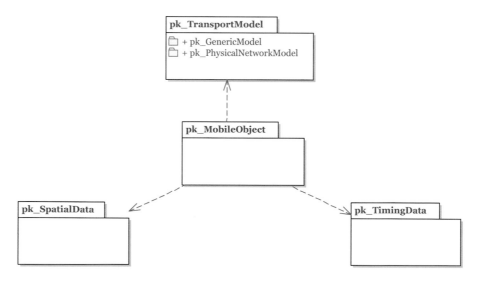

FIGURE 12.2   Package diagram of mobile object data model.

essential for the interrogation of the mobile data. The package *pk TransportModel* represents the modeling of the transportation network. It has at least two subpackages concerning a multimodal generic model and a physical network model.

12.2.2 THE MULTIMODAL TRANSPORTATION NETWORK MODEL

The representation of the multimodal transportation network is one of the most important aspects. The multimodal transportation network is modeled as an oriented graph. The focus is specifically on the definitions and semantics of these two entity types and the relationships between them. A *node* (in terms of generic topology) is the smallest identified location in space. It can play many different roles in the transportation network. It can represent the location of bus stops, parking places, or other types of nodes. The hypernode and hyperlink entities are introduced in this model. A *hypernode* is a node composed of one or more nodes; i.e., a hypernode is an intermodal station where people can change their mode of transport. The *route node* is an important entity that depends on the physical infrastructure and represents the way taken by a mobile object in a transport network. The class *Route* represents an abstract concept. Its purpose is to describe a path independently of the infrastructure pattern (cf. Figure 12.3).

12.2.3 SPATIOTEMPORAL QUERIES

12.2.3.1 New Abstract Data Types and Spatiotemporal Predicates (Basic Algebra for Moving Objects)

Some operations within the data model use the concept of spatiotemporal data types introduced into the following work (Güting and Schneider 2005). The latter is based on a set of the types, including the basic data types (int, real, string, bool) whose instances are static values. A constructor named *moving* is introduced to have the types whose values are dynamic and change over time. For example, the distance between two

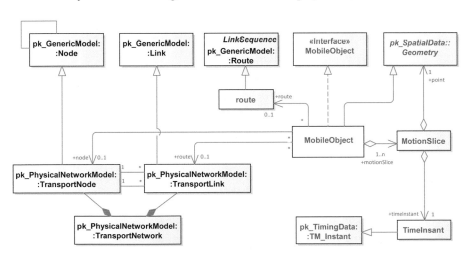

FIGURE 12.3 Moving object data model within multimodal network transport.

airplanes has a value over time and can be expressed by *moving(real)* that is noted *mreal*. Moreover, the location changing point in the Euclidean plan during time is named *moving(point)* and is noted *mpoint*.

A mobile object moving in a transport network is regarded as a mobile point. Based on the 9-intersection model of Egenhofer (Egenhofer 1991) and the definition of the new spatiotemporal data types, new topological relations depending on time can be defined. For example, two objects can be "disjointed" in a certain time as they can be in intersection in another. Therefore, with the spatiotemporal data types, a set of new relations is defined like a temporal version of spatial predicates. This mechanism is called *temporal lifting*. For example, the spatial predicate *inside* is applied to the spatial types {point, region} and it returns a Boolean. By applying the transformation of *temporal lifting*, this predicate will be applicable to the spatiotemporal data types *mpoint* and *mregion* and returns a time-changing Boolean (*mbool*). Some examples of other operations are presented next:

- Trajectory: $mpoint \rightarrow line$
- Distance: $mpoint \times mpoint \rightarrow mreal$

In fact, this model redefines a number of operations proposed in the simple feature specification of the Open GIS Consortium (OGC) by using the formalism presented earlier (OGC 1999). Furthermore, we will also define the spatiotemporal relations in the same way corresponding to the well-known spatial relations. These new relations deal with spatiotemporal data types. Moreover, the spatiotemporal predicates will be added according to the definition given by Erwig (Erwig and Schneider 2002). They are very useful in the data model and help to formulate more complex mobile queries. Following is an example of spatiotemporal query:

— *Find all the taxis that now enter the street "Peace".*

Select t.id, t.location.MPointAt(Now) From Taxicab t, Street s Where s.name= "Peace"

and (t.location.MPointAt(now).Enter(s.type_geo))

12.2.3.2 Further Works on Moving Objects

Finally, the management of spatiotemporal data has become an important axis for research. The extension of this kind of work can relate to the areas with variable surfaces (fires, deserts, tides, changing landscapes). Moreover, the work can also focus on a specific application area where modeling is not linear and probably within a non-Euclidean space.

Furthermore, the uncertainty of the positions, as well as the techniques for indexing the moving objects, still remain important areas of research, especially in a cloud and distributed environment. In the end, there are a few databases supporting moving objects, named moving object databases. We can cite Secondo[1] and Hermes[2] engines as two academic implementations. MobilityDb,[3] which has an industrial purpose, was recently proposed as a moving object database supported by the academic and industrial community. It is a good implementation of all the approaches discussed in recent years.

12.2.4 TOWARD A DISTRIBUTED ENGINE FOR DISTRIBUTED QUERY PROCESSING

In the context of Big Data and the streaming of massive end heterogonous data, there is a real need for new approaches and distributed engines for distributed query processing. After MapReduce and its open-source version Hadoop, many frameworks have been developed—for example, Hadoop-based approaches such as HadoopDB, Hive, Pig, Apache Spark, and Flink and graph processing frameworks such as Pregel or GraphX. They each provide a model of data that can be manipulated and a language to describe distributed processing.

As an extension of a basic engine for moving objects, Parallel Secondo and Distributed Secondo are two already proposed approaches to execute queries in Secondo in a distributed environment (Güting et al. 2021). Parallel Secondo uses Apache Hadoop to distribute tasks over several Secondo installations on a cluster of nodes. Distributed Secondo uses Apache Cassandra as a distributed key-value store for the distributed storage of data and job scheduling. Both implementations use an additional component (Hadoop or Cassandra) to parallelize Secondo. Another approach is also proposed. It is based on a *distributed array* that has fields of some data type of the basic algebra (moving object algebra).

12.2.5 ENRICHED MOVING OBJECT TRAJECTORIES

A trajectory is the path taken by a moving entity through space. The path is not made instantaneously but requires a certain time. So the time remains an inseparable aspect of a trajectory. This is indicated under the term trajectory space-time path as one of the synonyms of "trajectory" (Miller 2005, Hägerstrand 1970). Moreover, most researchers in this field find it necessary to consider not only the trajectories with their spatial and temporal characteristics but also to take into account other aspects that can have a great impact on the behavior of the movement. In particular, the activities and the various phenomena that occur during the movements of the mobile object represent essential data that can help to give answers to many typical questions asked by a set of actors, analysts, and probably the mobile object itself. Spaccapietra et al. define, in addition to the spatiotemporal trajectory, the structured trajectory and the semantic trajectory (Spaccapietra et al. 2008).

The work (Boulmakoul et al. 2015) proposes a unified meta-model to model the different types of trajectories of moving objects, taking into account the structural and geosemantic aspect. The "activity" model is integrated into this modeling; thus, it allows us to capture the activity in a sociogeographical sense. Using the space-time ontology and event-driven aspect, the proposed model uses the object meta-modeling approach to express the different models concerning trajectories. Moreover, this meta-model deals with the trajectories of moving objects taken by different available spatiotemporal protocols and sensors. Also, regions of interest are added in this meta-model. These regions are composite regions that can describe modal networks of transportation, Voronoi spatial networks, etc. (cf. Figure 12.4).

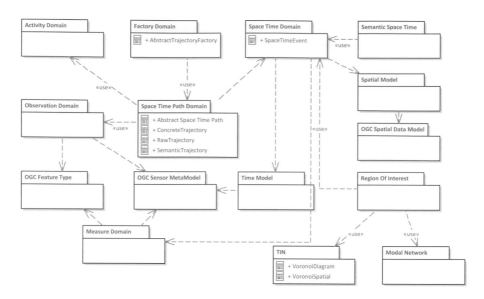

FIGURE 12.4   Meta-model of moving object trajectories.

12.2.6   Simplified Model for a Specific Domain: Model for a Smart City as an Example

An interesting work (Karim et al. 2021) proposes another point of view. A trajectory is considered as an ordered associative container of events that could be fuzzy (cf. Figure 12.5).

An event is characterized by its spatiotemporal attributes and its measurement relative to a given physical observable. The trajectory model can be easily integrated within the OGC CityGML meta-model.

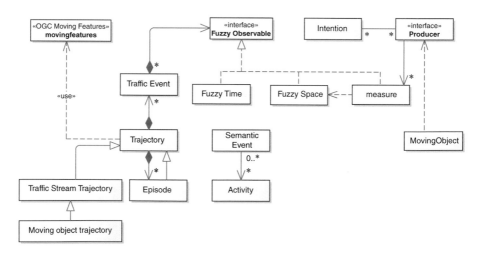

FIGURE 12.5   Simplified model for a specific domain.

12.3 MOBILITY DATA AND SECURED TRANSPORTATION: RISK EVALUATION AND HAZMAT MONITORING

12.3.1 INTRODUCTION

The integration of the SDSS for hazardous materials transportation routing and monitoring is very important to the global system. The adopted technique produces a fuzzy risk graph that models both the transport system and the concept of accident risk (Boulmakoul et al. 2017, Cherradi et al. 2017). This risk fuzzification makes it possible to apply the results concerning fuzzy shortest path search problems in fuzzy graphs (Boulmakoul and Bouziri 2012, Boulmakoul 2006). A new solution is given to the problem of the k-best fuzzy shortest paths. This section presents a short overview of the work that revises the fuzzy shortest path problem and also gives a solution using dioïds for the fuzzy path problem. This algebraic structure is precisely adapted to solve the problem of the k-best fuzzy shortest paths. This work outlines a method for extending Gondran's and Minoux's path algebra results to fuzzy graphs (Gondran and Minoux 1995).

12.3.2 DIOÏDS AND THE SHORTEST PATH PROBLEM

The concept of a dioïd was initially proposed by Kuntzmann (Kuntzmann 1972).

12.3.2.1 Definition of Dioïd

A dioïd is a triplet $(S, \oplus, \otimes)$ made up of the following elements:

- S is a set which has two elements ε and e.
- $\oplus$ is an associative and commutative internal law of composition.
- $\otimes$ is an associative internal law of composition.

Such as:

- $\otimes$ is a *distributive* compared to $\oplus$ on the right and on the left.
- ε is the neutral element for $\oplus$ and absorbing for $\otimes$.
- e is the neutral element for $\otimes$.

This dioïd is known as commutative if the law $\otimes$ is commutative.

12.3.2.2 Generalized Algorithms for the Shortest Path Problem

The general algorithm for a graph without p-absorbing cycle is given hereafter:

Γ is the function successor of the graph.

$$(\alpha)\pi(1) = e, \pi(i) = \alpha_{1i} \text{ for } i \geq 2$$

$$(\beta) \text{ at step k, do (for i = 1 to n):}$$

$$\pi(1) \leftarrow \bigoplus_{j \in \Gamma^{-1}(1)} (\pi(j) \otimes a_{j1}) \oplus e$$

$$\pi(i) \leftarrow \bigoplus_{j \in \Gamma^{-1}(i)} (\pi(j) \otimes a_{ji}) \text{ pour } i \geq 2$$

$$(\chi) \text{ Repeat } (\beta) \text{ until stabilization of } \pi(i).$$

TABLE 12.1
Classical Examples of Dioïds

Type of Problem	S	$\oplus$	$\otimes$	ε	e
Shortest path	$\Re \cup \{+\infty\}$	min	+	$+\infty$	0
Longest path	$\Re \cup \{-\infty\}$	max	+	$-\infty$	0

Where:

- $G = (X, A)$ is a directed graph
- S is a set endowed with a structure of dioïd
- $a_{ij} \in S$ *is* a valuation of arc (i, j) of the set A
- $\pi(j)$ $(j = 1..n)$ is the length of the shortest paths between nodes 1 and j

In the case of the classical shortest path $(S = R^+ \cup \{+\infty\}, \oplus = \min, \otimes = +)$, this generalized algorithm corresponds to a Ford algorithm. For a graph without a cycle, the algorithm corresponds to the optimality equation of dynamic programming (or the generalized algorithm of Bellman). Table 12.1 gives some classical examples of dioïds designed to solve the path finding problem (Gondran and Minoux 1995).

12.3.3 FUZZY GRAPH

Modeling by fuzzy graphs has been applied to various problems. We consider the length of each arc and the length of any path as fuzzy paths. In this case, each arc corresponds to a fuzzy set, which indicates its valuation (cf. Figure 12.6). The elements whose grade of membership is null can be omitted.

12.3.4 NEW DIOÏDS AND THE FUZZY SHORTEST PATH PROBLEM

In the work (Boulmakoul and Bouziri 2012), the first structure is formulated to solve the problem of the k-best fuzzy shortest paths. It is demonstrated that

$$\left(S, \oplus, \otimes, \varepsilon = (+\infty)^k, e = \left(\overbrace{1/0,\ldots,1/0}^{q}, \overbrace{1/+\infty,\ldots,1/+\infty}^{k-q} \right) \right) \text{ is a dioïd and also gives a}$$

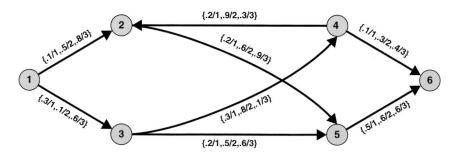

FIGURE 12.6 Fuzzy directed graph without a cycle.

numerical example and interpretation of the operations $\otimes$ and $\oplus$. Since the different routing problems are linked by the concept of the dioïd, a generic library that contains all the algebraic structures supported by the dioïd is developed. The component software of path operators, in ordinary and fuzzy graphs, is integrated into the SDSS for transportation monitoring and planning of hazardous materials. The "fuzzification" of risk has made it possible to highlight the application of fuzzy routing algorithms in a real-time mode. The evaluation of the risk depends on the type of product transported and the consequences in the event of an accident, according to the considered impact (environment, infrastructure, or economy). The risk "fuzzification" process is as follows:

- Let $\phi 1(u, i)$ be the fuzzy subset corresponding to the vulnerability relative to the impact i for the arc u.
- Let $\phi 2(u, i)$ be the fuzzy subset corresponding to the cost associated with impact i.
- Let $\phi 3(u, i) = \phi 1(u, i) \otimes \phi 2(u, i)$ be the fuzzy subset corresponding to the risk associated with impact i, for arc u, where here the operation $\otimes$ is fuzzy multiplication.

The fuzzy subset which corresponds to the fuzzy risk attached to the arc u is given by the max aggregation to obtain the overall risk force. Once all arc valuations are determined, the fuzzy routing can be performed by applying the proposed algorithms. The assignment of the degree of vulnerability and the cost of impact is carried out by the experts of the urban management system.

Finally, an extension of this work is given in (Boulmakoul et al. 2017) that proposes new semirings for shortest path problem with fuzzy stochastic graphs.

12.4 FOG/CLOUD PARADIGMS FOR URBAN COMPUTING

12.4.1 Cloud Computing and the Fog/Edge Solution

With the massive increase in mobile equipment and fast growth in data generated by the IoT, cloud platforms arc widely used and remain the main infrastructure for storage and processing. However, cloud computing cannot handle fast-emerging context-based applications with sensitive requirements for high throughput, latency, and high availability.

To address these concerns, we are shifting from centralized cloud computing to distributed fog/edge computing. Several cloud techniques can be adapted for fog systems, such as resource provisioning and orchestration. But there are many differences. For example, fog computing provides compute and storage resources at the edge of the network to support latency-sensitive services hosted in the mobile client and has location information that can be provided on demand. It can track end users to support mobility and should also provide modules for dynamic service provision (Lan et al. 2019, Yousefpour et al. 2019). In addition to application requirements (latency sensitive, autonomous, privacy, and security), the main characteristics of fog programming are heterogeneity, scalability, quality of service, local context awareness, mobility, and openness.

12.4.2 Challenges and New Directions

To face several challenges when we use cloud computing, in addition to fog computing, other computing paradigms have emerged: mobile cloud computing, mobile ad
hoc cloud computing, multiaccess edge computing, cloudlet, mist computing, etc.
Some paradigms are used interchangeably in some research papers and can give
nuances. There are many issues and research opportunities in this area. We highlight
future challenges:

- Service migration: Due to device mobility, the components and services
  may need to move dynamically at runtime.
- Infrastructure virtualization.
- Dynamic resource-aware software reconfiguration techniques.
- Peer-to-peer (P2P) fog computing.
- Multiobjective fog system design: Many schemes (e.g., offloading, load
  balancing) consider a few objectives and ignore other objectives.

Other challenges can be noted: resource monitoring, green fog computing, bandwidth-
aware system design, resource monitoring, support of high-speed mobile, users, node
security, etc.

12.5 MICROSERVICES-BASED ARCHITECTURE AND HAZMAT TRANSPORT TELEMONITORING: PILOT CASE STUDY

12.5.1 Open Architecture within a Reactive Ecosystem

Among the challenges of this case study is to find a reactive ecosystem for an open
architecture system with broad interoperability in a fog/cloud environment. We
briefly describe in this subsection this architecture with real-time consideration
designed for the transport of hazardous materials and environmental monitoring.
This architecture has known several versions and major evolutions. We recently adopt
a microservices-based architecture to have a real-time environmental sensor system
that has highly scalable applications within a fog/cloud environment. Figure 12.7
presents a global view of the reactive ecosystem integrating the architecture. The
work in (Cherradi et al. 2017) gives more details.

At the cloud level, this architecture consists of a set of loosely coupled services
that can be deployed independently. In addition, the proposed system is driven by an
intelligent data collector deployed at the fog level with a variety of distributed sensors to improve the hazmat transport telemonitoring.

The functions assumed in the ecosystem are Hazmat vehicle authentication,
real-time data collection, real-time monitoring, and incident detection and management. The provided solution is validated and evaluated with a focus on increasing
the safety and efficiency of mobile entities in the transport of hazardous materials.
This type of ecosystem helps to improve decision making and proactively resolve
the problems, in addition to coordinating resources to act effectively and avoid catastrophic accidents.

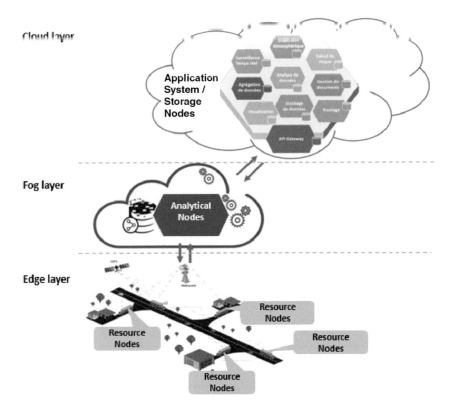

FIGURE 12.7 Global view of a reactive ecosystem.

12.5.2 The Main Components of the Open Architecture

Figure 12.8 describes the different components of the system and their interaction. It will be a generic ecosystem where components (microservices) will be designed to be scalable and efficient. There are several microservices that make up this system.

As previously indicated, the SDSS implements the algorithms and the developed algebraic structures. This architecture efficiently integrates this component into the system prototype, in particular as the hazmat routing container embedded microservice, and illustrates the orchestration with other microservices to improve this system with hazmat routing capabilities using various routing algorithms. As an example, Figure 12.9 illustrates the optimal route for a fuel-carrying vehicle to follow in the city of Mohammedia (the underneath path), as well as the most vulnerable path which must be avoided. In fact, the system also uses the atmospheric dispersion service to calculate the radius of the danger zone of the transported product—and this before the calculation of the optimal path. Finally, this work constitutes an important step to build the hazmat planning aid system that provides valuable data and knowledge to urban planner and decision makers.

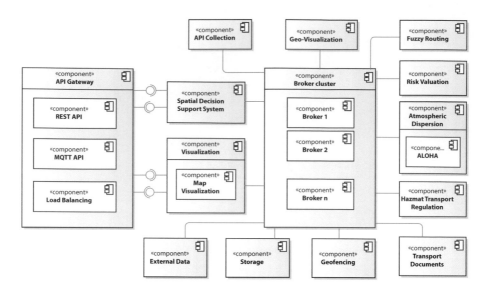

FIGURE 12.8   Component diagram of open microservice-based architecture.

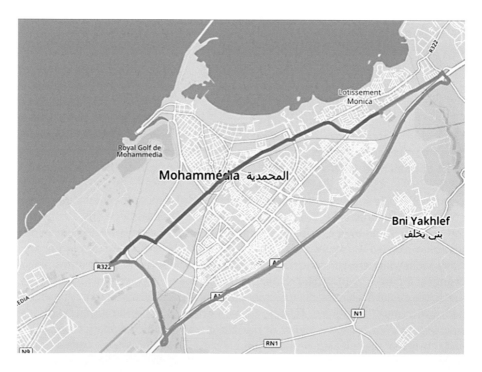

FIGURE 12.9   Optimal path.

12.6 CONCLUSION

All the discussed issues in this chapter are very interesting to promote new architectures and methods for advanced prediction and sophisticated (offline or online) trajectory-based analytics with respect to the constraints and the requirements of the new deployed ecosystems (i.e., stochastic graph model to extract spatiotemporal patterns from spatiotemporal risk trajectories sensitive to the population).

Furthermore, this new fog/cloud-based ecosystem has to deal efficiently with mobility data without ignoring security and privacy issues. Moreover, in particular for critical systems, we must continue to evaluate and propose new methods to minimize the overall risk in urban areas that are constantly evolving. The latter must prepare a smart environment and infrastructure that support the emergence of new technologies and intelligent monitoring in addition to efficient planning.

NOTES

1. https://secondo-database.github.io/
2. http://infolab.cs.unipi.gr/? page_id=1999
3. https://github.com/MobilityDB/MobilityDB

REFERENCES

Boulmakoul, A. (2006). Fuzzy graphs modelling for hazmat telegeomonitoring. European Journal of Operational Research 175(3):1514–1525. doi:10.1016/j.ejor.2005.02.025

Boulmakoul, A. and A. E. Bouziri (2012). Mobile object framework and fuzzy graph modelling to boost hazmat telegeomonitoring. In Garbolino, E., M. Tkiouat, N. Yankevich and D. Lachtar (eds) Transport of Dangerous Goods. NATO Science for Peace and Security Series C: Environmental Security. Springer, Dordrecht. pp 119–149. doi:10.1007/978-94-007-2684-0_5

Boulmakoul, A., A. E. Bouziri and G. Cherradi (2017). New semirings for shortest path problem with fuzzy stochastic graphs. 5ème Congrès International de la Société Marocaine de Mathématiques Appliquées (SM2A), Meknès, Maroc.

Boulmakoul, A., L. Karim, A. E. Bouziri and A. Lbath (2015, 03). A system architecture for heterogeneous moving-object trajectory metamodel using generic sensors: Tracking airport security case study. IEEE Systems Journal 9(1):283–291. doi:10.1109/JSYST.2013.2293837

Cherradi, G., A. E. Bouziri, A. Boulmakoul and K. Zeitouni (2017). Real-time hazmat environmental information system: A micro-service based architecture. Procedia Computer Science 109:982–987. doi:10.1016/j.procs.2017.05.457

Cherradi, G., A. E. Bouziri, A. Boulmakoul and K. Zeitouni (2018). An atmospheric dispersion modeling microservice for hazmat transportation. Procedia Computer Science 130:526–532. doi:10.1016/j.procs.2018.04.075

Egenhofer, M. J. (1991). Point-set topological spatial relations. Geographical Information Systems 5(2):161–174. doi:10.1080/02693799108927841

Erwig, M. and M. Schneider (2002). Spatio-temporal predicates. IEEE Transactions on Knowledge and Data Engineering 14(4): 881–901. doi: 10.1109/TKDE.2002.1019220

Garbolino, E., M. Tkiouat, N. Yankevich and D. Lachtar (2012) (eds). Transport of Dangerous Goods. NATO Science for Peace and Security Series C: Environmental Security. Springer, Dordrecht. doi:10.1007/978-94-007-2684-0

Gondran, M. and M. Minoux (1995). Graphes et algorithmes. Eyrolles, Paris.

Güting, R. H. and M. Schneider (2005). Moving Objects Databases. Morgan Kaufmann Publishers, London. doi:10.1016/B978-0-12-088799-6.X5000-2

Güting, R. H., M. H. Böhlen, M. Erwig et al. (2000). A foundation for representing and querying moving objects. Geoinformatica ACM Trans Databases System 25(1):1–42. doi:10.1145/352958.352963

Güting, R. H., T. Behr and J. K. Nidzwetzki (2021). Distributed arrays: An algebra for generic distributed query processing. Distributed Parallel Databases 39:1009–1064. doi:10.1007/s10619-021-07325-2

Hägerstrand, T. (1970). What about people in regional science?. Papers of the Regional Science Association 24: 6–21. doi:10.1007/BF01936872

Karim, L., A. Boulmakoul, G. Cherradi and A. Lbath (2021). Fuzzy centrality analysis for smart city trajectories. In Kahraman, C., S. Cevik Onar, B. Oztaysi, I. Sari, S. Cebi and A. Tolga (eds). Intelligent and Fuzzy Techniques: Smart and Innovative Solutions. INFUS 2020. Advances in Intelligent Systems and Computing. Vol. 1197,pp. 933–940. Springer. Cham. doi:10.1007/978-3-030-51156-2_108

Kuntzmann, J. (1972). Théorie des réseaux. Dunod, Paris.

Lan, D., A. Taherkordi, F. Eliassen, and G. Horn (2019). A survey on fog programming: Concepts, state-of-the-art, and research challenges. In 2nd International Workshop on Distributed Fog Services Design (DFSD '19), Davis, CA, USA. ACM, New York. doi:10.1145/3366613.3368120

Miller, H. J. (2005). A measurement theory for time geography. Geographical Analysis 37(1):17–45. doi:10.1111/j.1538-4632.2005.00575.x

OGC (1999). OpenGIS Simple Feature Specification for SQL. Open GIS Consortium, Inc., USA.

Spaccapietra, S., C. Parent, M. L. Damiani et al. (2008). A conceptual view on trajectories. Data and Knowledge Engineering 65(1):126–146. doi.org/10.1016/j.datak.2007.10.008

Yousefpour, A., C. Fung, T. Nguyen et al. (2019). All one needs to know about fog computing and related edge computing paradigms: A complete survey. Journal of Systems Architecture 98:289–330. doi:10.1016/j.sysarc.2019.02.009

Zheng, Y. and X. Zhou (2011) (eds). Computing with Spatial Trajectories. Springer, New York, NY. doi:10.1007/978-1-4614-1629-6

13 Utility Assessment of Line-of-Sight Traffic Jam and Queue Detection in Urban Environments for Intelligent Road Vehicles

Zoltan Fazekas, Mohammed Obaid, Azedine Boulmakoul, Lamia Karim, and Péter Gáspár

CONTENTS

13.1 INTRODUCTION

Intelligent and autonomous road vehicles have been foci of research and development efforts throughout the world for over at least two decades. Among these, a small-budget project entitled "Detecting the urban environment and driving a smart road vehicle with a little help from friends" was launched in 2019. One of the research questions posed in this project was as follows: Which is the more important feature for an intelligent road vehicle to have: "connectedness", or "eye sight"? Although the latter notion sounds somewhat anthropomorphic, what is meant by it here is the capability of "seeing" and "recognizing" static and dynamic objects, road vehicles and vehicle queues in a road environment, in particular.

In this chapter, the utility of line-of-sight (LoS) traffic jam and queue detection (TJQD) is assessed via microscopic traffic simulations. In such simulations the individual movement of each vehicle within the target road network is considered, modeled, tracked and recorded. The LoS detection can be accomplished either by the human driver or by one or more LoS sensors (e.g., intelligent camera, LiDAR) on board a road vehicle that is equipped with and managed by advanced driver-assistance systems (ADAS) or else by one or more LoS sensors controlled by an autonomous driving (AD) system. The road traffic simulation is carried out in respect of three urban model environments: two fairly simple synthetic ones and a somewhat more complex, real-world urban subnetwork of roads. Though it is a very interesting LoS traffic detection approach, and in some sense it is also an on-board approach, we do not consider here the cooperation of a connected unmanned aerial vehicle (UAV) equipped with a camera and of a connected and autonomous road vehicle (CAV). The latter vehicle receives and displays the video stream received from the UAV and/or presents the relevant information inferred from the stream, see, e.g., (Nagy et al. 2021).

There are a growing number of urban settlements worldwide that have a highly developed transport and info-communication infrastructure and feature reliable traffic management services. In these cities and towns, traffic information on accidents, congestion, roadworks, etc., reaches connected cars and their drivers fairly soon. With such infrastructure and services at hand, the question arises whether it is worthwhile to invest in the development of an on-board LoS TJQD ADAS subsystem. A similar question arises on the vehicle user's side: Is it worth buying such a subsystem and installing it in a car?

The utility of an on-board LoS TJQD can be evaluated for two transport levels, namely for individual vehicles and for the road network considered, i.e., aggregated for all of the road vehicles traveling there. The actual sensing technology and methodology used on board a road vehicle for TJQD is only of marginal interest to us now. However, we assume that such detection requires an LoS constellation of the ego-vehicle and the forming vehicle queue and some LoS connection between these two (e.g., visible or infrared light perceived by a camera, or the laser beams of a LiDAR device, or the electromagnetic signals of an automotive radar device).

To assess the mentioned utility, the LoS TJQD time, which is measured from the time when a static or slow-moving obstacle appears in the road network and a queue starts to build up behind or because of it until the obstacle/queue is LoS detected, should be compared to the estimated time lag of receiving the same information via various info-communication channels from other transport sensors and information sources. Only a limited utility of the on-board LoS TJQD can be realistically expected in urban environments that have an advanced transport and info-communication infrastructure. However, the low standard of such infrastructure could increase the individual and aggregated utility considerably.

The rest of this chapter is structured as follows. In Section 13.2, we give an overview of the related literature focusing on traffic measurement and queue detection techniques. The image- and video-based methods are applied to images and video data originating from static traffic surveillance cameras and from on-board cameras. Some other LoS detection approaches used in transport networks and on-board smart and autonomous vehicles will be also mentioned. Then, the traffic simulation approaches will be summarized. In Section 13.3, first, the traffic simulation environment and the simulation approach are described. Then, three case studies are presented, and via these, the utility of the on-board LoS TJQD is assessed. Then, in Section 13.4, the results are discussed and evaluated. Finally, in Section 13.5, conclusions are drawn and directions for future work are suggested.

13.2   BACKGROUND

The LoS constellation is a precondition of the effective propagation of light and of microwave radio waves between objects, living beings, sensors, and locations, and therefore, it is essential for various forms of energy transfer, as well as for various forms of LoS detection and of LoS communications (Butcher 2010). Driving a road vehicle is based on LoS detection of the environment. Apart from the everyday forms of interpersonal LoS communications (e.g., gestures used during driving), also various nontrivial LoS communication forms have been in use throughout human history (Huurdeman 2003), and the same applies also for LoS detection (Guth 2004). The flag- and the semaphore-based LoS communication, for example, is still used occasionally in naval and military operations (Sterling 2008). Furthermore, the hand signals of traffic police officers or those of aircraft marshals at airports could be also mentioned as present-day examples of nontrivial LoS communication (Guo et al. 2017). Clearly, traffic signs and traffic lights need to be detected in this manner (Mueller 1970).

The LoS constellation, connection, and detection are essential in road accident prevention for human drivers and ADAS/AD systems alike. In the context of AD in urban environments, a recent paper evaluated the traffic safety risk caused by occlusion due to static elements of the road environment and by the limited sensor range of the on-board optical and radar sensors (Yu et al. 2019). Enabling autonomous vehicles to quantify the risk posed by vehicles and pedestrians turning up within and moving in non-LoS regions allows these vehicles to anticipate and prevent traffic incidents that might occur in the mentioned situations. The algorithm proposed in their paper relies on road layouts known from maps to estimate the mentioned risks. The estimated risks can then be used for planning safer trajectories and driving maneuvers in the vicinity, while risks estimated for a number of junctions could serve the purpose of choosing safer routes. A significantly reduced collision rate compared to that associated with a baseline method was achieved according to the microscopic traffic simulations carried out by the authors of the cited paper. Note that in the traffic simulation terminology "microscopic" means that the movements of the individual vehicles are considered, tracked, and recorded in the simulation.

A similar approach, but making use of LiDAR point clouds, was proposed for the safety assessment of urban intersections considering static occlusions and sight distance experienced by drivers (Kilani et al. 2021). The authors of the cited paper aimed to facilitate risk assessments and traffic safety surveys carried out by transport authorities.

Apart from the LoS constellation, connection and detection between road vehicles, drivers, and road locations considered herein, such constellation, connection, and detection are often studied, evaluated, modeled, and computed in conjunction with aircraft (Bauer et al. 2017), buildings (Feng et al. 2006), and satellites (Matera et al. 2019).

The authors of a recent survey on traffic congestion detection methods and devices opine that cities have been experiencing a profound adverse impact of traffic congestion (Chetouane et al. 2020). They fear that traffic congestion becomes unmanageable, and this leads to unsustainable road transport in major cities. As anyone living in an urban environment could tell, traffic congestion hits cities particularly hard during peak hours. The factors causing unsustainability include waste of time and waste of fuel. Also, the high rate of accidents in urban areas contributes to unsustainability. These factors lead to additional extra costs for the economies. The unsustainability also manifests itself in serious environmental problems for cities and for their surrounding regions. The authors look at and survey research activities in the field of traffic congestion detection and monitoring. Though they name and evaluate various techniques used in the field, they do not consider the on-board LoS TJQD at all. On the one hand, this omission is understandable, as such ADAS/AD subsystems are not widespread, while their static counterparts are installed and used in nearly most of the major crossings and junctions in major cities (Umair et al. 2021). On the other hand, the omission is surprising, as such a real-time system was reported on as early as 2008 (Parisot et al. 2008). In the frame of their road traffic classification project, Parisot and his colleagues mounted an intelligent camera onto the front of a public bus. This camera was then drawn upon to estimate the speed of the vehicles which use the bus lane and of those using the adjacent lanes. In order to speed up the necessary image processing computations and to achieve real-time traffic monitoring capability, the authors employed a 1D approach within each video frame: their method considered only the discrete straight lines corresponding to the lane center-lines.

Some more recent results concerning traffic- and road-related LoS detections were presented in (Levering et al. 2020). The authors of the paper proposed a system for detection and recognition of unsigned physical road incidents. For detection and recognition, either the incident itself, or the damage caused by it, or else the abnormal traffic situation arising needed to be seen in the images. The incident types considered were as follows: animal on road, collapse of the carriageway, car crash, fire, flooding, landslide, snow, and tree fall. The system made use of a popular deep neural network architecture. The authors collected an image dataset of 12,000 images from publicly available sources and fine-tuned the selected convolutional neural network. They reported a recognition accuracy of more than 90% in respect to images on UK roads.

13.3   LoS TRAFFIC SIMULATION STUDIES

In the following, three road networks of increasing complexity will be presented as case studies. These networks have been chosen so that the use and advantage of an early LoS TJQD could be demonstrated through them. Microscopic simulation of the traffic involving vehicles with and without on-board LoS TJQD capabilities is seen as a cost-effective way of investigating this topic. It is a useful step that should precede a comprehensive investigation. PTV Vissim seemed a good choice for supporting such a simulation-based study.

13.3.1 THE SIMULATION ENVIRONMENT AND APPROACH

After decades of research in the field of traffic flow modeling and simulation, Vissim was launched in 1994. It is a microscopic, behavior-based multipurpose traffic flow simulator. Since 1994, it has been used all over the world in the analysis and optimization of traffic flows. It was used by transport and civil engineers, urbanists, traffic safety experts, and economists alike (PTV Vissim 2022). The simulator has been deployed in a wide range of urban and highway applications, and in these, has helped its users in considering and managing different geographical scales. It has been employed in integrated modeling of public and private modes of transportation. The traffic models created in Vissim make use of the simulator's own traffic flow models. These describe the vehicle movements in longitudinal and lateral directions on roads and handle roads with different lane structures. Vissim has traffic conflict resolution models that are applicable for the areas of potential vehicle-vehicle and vehicle-pedestrian conflicts. Vissim has its own dynamic assignment and social force model for handling pedestrian movements in the simulations.

Several authors have emphasized the importance of and proposed methods for validating the microscopic traffic flow simulation models both on vehicle-level and traffic flow-level measurement data (Fang and Tettamanti 2021). To illustrate this need, let us consider the traffic flow models created in Vissim that rely on its car-following algorithm. The algorithm adheres to a psycho-physical car-following model that can realistically mimic a range of different driver behaviors. This model, on the one hand, offers and, on the other, requires a number of model parameters; these should be calibrated using real measurement data from the considered region.

Given the fact that visibility analysis, which includes also LoS detection, connection, and filtering, has not been incorporated into conventional traffic flow modeling, it is not surprising that common traffic simulation software versions do not support such analysis. Even so, LoS detections, connections (e.g., the LoS detection of queues), and situations can still be tentatively modeled in an indirect manner by using proxy parameters (e.g., layout-dependent time delays). These parameters can be estimated through the use of a simple model, or by using Monte Carlo estimation methods (Yan et al. 2013), or through learned choice of credible delays.

In the case studies considered in the present section, the spatiotemporal process of LoS queue detection within the road network, as well as the reactions of drivers and of autonomous vehicles to such detections, are modeled through three proxy parameters. Only one of these parameters is actually varied in our investigation, while the other two are used with fixed real values. The varied proxy parameter is real-valued and falls in the range of 0.0 and 1.0, and for convenience, it is given as a percentage. This parameter is assigned different values in the consecutive simulation runs; e.g., in case of the first case study, the considered values are 0%, 5%, …, 45%, and 50%, and represent the portion of vehicles diverted in the simulation from one or more blocked paths to other paths, or to a single path. It models the portion of drivers or of the intelligent vehicles that decide to take some other routes to their respective travel destinations upon LoS detection of the forming vehicle queues.

The two proxy parameters used herein as fixed values quantify specific delays introduced in the simulation. The first one specifies the delay before a driver or an intelligent car can carry out a certain driving action, while the second specifies the delay before a driver or an intelligent car can take a routing decision. One could argue that the first of these fixed parameters is related to the geometry of the road block (e.g., its size with respect to the road or lane width) and to the seriousness of the cause (e.g., a fender-bender or some more serious car crash), while the second one is more descriptive of the delay due to the need of LoS connection required for the detection of the queues. Clearly, also the proxy parameters used as fixed values herein could and should be varied in a more comprehensive investigation.

13.3.2 THREE CASE STUDIES

Three relatively small subnetworks of urban road networks will be considered here. These will be simulated assuming steady flows of cars through the undisturbed subnetworks. Then, in each case study, something happens in the subnetwork that blocks/disturbs the steady traffic flow. Such a disturbance could arise, for instance, due to a large garbage truck that turns up in the area and keeps stopping to empty the garbage containers, thereby in effect blocking a whole lane. As another example, the disturbance could be due to a road accident in which two or more cars have crashed into each other in a junction, thereby blocking one or more lanes. Further possibilities are mentioned and studied in (Levering et al. 2020); moreover, a taxonomy was proposed therein for unsigned physical incidents.

When does the information on the road block and/or the queue building up due to it reach a car approaching the obstacle location based solely on a LoS TJQD carried out on-board? This is a question we intend to answer here, albeit in an indirect and approximate way, as it was explained and outlined in Subsection 13.3.1.

In our simulation-based study, we look at the cars that would be affected, i.e., delayed, by the queue(s) that have formed but are still in the position to choose some alternative routes to their respective destinations. We investigate whether the information on a road block or on a traffic jam reaches the driver/car sooner if on-board LoS TJQD is used on the ego-car or if this information is gained through some public info-communication channel. Examples for info-communication relevant to the topic include, e.g., crowdsourced warning messages sent through such a channel or warning messages broadcast by the vehicle-to-infrastructure network in regard to some detrimental traffic situation within the area.

If one intends to answer these questions in their full depth, then one needs to consider in the given urban neighborhood the visibility of the road block from other vehicles, the visibility of the individual vehicles queuing up due to the road block from other vehicles, and the visibility of the last vehicle standing in the queue from other vehicles. At the same time, one needs to estimate the propagation times of the traffic information through the various info-communication channels and means mentioned earlier. In this case, not only the physical aspects need to be considered but also the network security and integrity, e.g., the traffic data and information gained from the public should be checked and moderated. As it has been stated before, we seek herein only indirect and tentative answers to these questions.

We expect that the on-board LoS TJQD has a rather limited region of utility within the road network for avoiding traffic jams and in escaping queues. This applies particularly to urban areas with a highly developed transport and info-communication infrastructure.

13.3.2.1 First Case Study: A Simple Synthetic Urban Road Network

A simple synthetic urban network was devised for our first case study and was analyzed via microscopic traffic simulations using the simulation software described briefly in Subsection 13.3.1. More concretely, the effect of a static or slow-moving obstacle that appears in the road network and that slows down or even blocks the normal road traffic is looked at in some detail. With this small synthetic road network and its three associated scenarios, we intend to demonstrate the advantage of early/timely LoS TJQD in route selection, i.e., in route replanning, for the cars moving within, or through the network. Such an advantage may manifest itself at two transport levels, namely in respect of individual cars with their drivers and passengers and on the road network level, which includes all the drivers and passengers concerned, as well as the vehicle owners and other transportation stakeholders relying on the particular road network within the analyzed time period.

In the first scenario associated with the network, a light, unsaturated, and undisturbed traffic is considered. Then, in the second and third scenarios, one and two obstacles appear in the road network causing a disturbance in its traffic. The second and third scenarios without any intervention implemented in them will be referred to as the second and the third base-scenarios, respectively.

For the purpose of this analysis, in line with the approach outlined in Subsection 13.3.1, each of the second and third scenarios is split up into a number of parameterized subscenarios. These subscenarios serve the purpose of modeling the reaction of drivers and/or of ADAS/AD systems to the LoS detection of the forming queues via the use of a proxy parameter. Then, the traffic indicators gained through the simulation of the subscenarios are compared to those derived for the respective base-scenarios.

The base-scenarios can be thought of as scenarios with all the drivers being patient. We call those drivers patient who do not search for and do not use alternative routes to their respective travel destinations, even if they must wait in long queues with their cars. This notion is applied to autonomous vehicles as well. In the implemented simulations, the fact that drivers are patient means that no interventions in the traffic flows are carried out.

In the traffic simulations presented herein, we analyze the traffic for one-hour time intervals. In the second and third scenarios and also in their subscenarios, the simulation starts from the appearance of the obstacle(s). The selected one-hour interval is long enough for such simple networks to settle and reach either a steady traffic flow or a complete lockdown. Also, the queue(s) due to the obstacle(s) can form, grow, and stabilize. However, this duration is too long to consider if we want to assess the utility of the on-board LoS TJQD and compare it to that of other traffic information sources and channels that could also be relied on in route replanning.

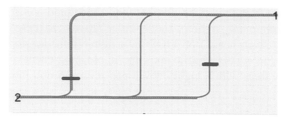

FIGURE 13.1 A simple synthetic urban road network used in the first case study. There is one-way traffic on its roads. In the first scenario of the case study, there are no obstacles. Obstacles turning up in the second and third scenarios are marked with short bars.

This aspect will be taken into account when discussing the utility of the on-board LoS TJQD in Section 13.4 in light of the simulation results obtained.

Indeed, on-board LoS TJQD is relevant in route replanning only for a few minutes, or for a quarter of an hour at most. After this period, the majority of the drivers in the vicinity will have been informed through some info-communication channel (e.g., through the Internet, through live-traffic map services, via V2X communication, or through traditional radio broadcast) of the traffic hindrance within the road network (Meneguette et al. 2016). This is particularly true within busy urban areas equipped with proper traffic surveillance facilities (e.g., traffic surveillance cameras, inductive loops, vehicle-to-infrastructure road-side units) and reliable traffic management services.

A simple synthetic road network is shown in Figure 13.1. This network is considered in the first case study. It is assumed that there are buildings along the roads in the area and these buildings block the view. So, for instance, Location 1 is not visible from Location 2, and vice versa. In other words, there is no LoS connection between these two locations, and therefore, no LoS detection is possible between cars or drivers located there. The traffic flows from the right to the left and from the top to the bottom in Figure 13.1. As a consequence, Location 2 can be reached from Location 1 within this network, but not vice versa.

The obstacle locations for the second and third scenarios, as well as the traffic demand, have been chosen in such a way that before the appearance of the obstacle(s), the traffic flows without hindrance; after their appearance, however, the traffic situation considerably worsens, and one or more queues start to build up.

In the second and third scenarios, the buildings initially constrain the LoS detection of obstacle(s) and of the forming queue(s). However, after a little while the LoS detection of the queue(s) that have formed and grown long enough becomes possible for more and more drivers. Some of these drivers opt out of queueing and choose alternative routes to their travel destinations.

Let us now specify the traffic demand chosen for the simulations associated with the first case study. The road network shown in Figure 13.1 moves 1000 vehicles from Location 1 to Location 2 via three left-reverse-bend paths. The rightmost path is Path 1, the leftmost is Path 2, and the middle one is Path 3. *The lengths of these paths are given in Table 13.1. The traffic demand is distributed*

TABLE 13.1
Lengths of the Three Paths within the Simple Synthetic Urban Road Network

Path Number	Path Length (m)
1	967
2	976
3	971

TABLE 13.2
Traffic Volumes Used in the Simulation for the First Scenario

Path Number	Traffic Volume (cars/hour)
1	314
2	346
3	340

over the three paths nearly equally resulting in the traffic volumes given in Table 13.2.

The simulation models a one-hour interval of the undisturbed traffic through the road network. Within this interval, vehicles emerge in the network during the first 45 minutes, while in the last 15 minutes, the network is left on its own to discharge.

The most relevant traffic indicators describing the undisturbed road traffic of the first scenario are summarized in Table 13.3. The table shows, e.g., that the average speed of cars was about 47 km/h, which is quite acceptable within an urban network. Furthermore, about every second car had to stop while completing its journey through the network, and the stop lasted about 3.2 seconds on average. Most of these stops must have occurred in the two convergence conflict areas located on the "lower horizontal" straight patch of road. Note that the specified traffic demand is fully satisfied by the undisturbed network.

The second and third scenarios associated with the road network shown in Figure 13.1 feature one obstacle, namely Obstacle 1, which appears in Path 1, and two obstacles, namely, Obstacles 1 and 2. Obstacle 1 is as before, while Obstacle 2 appears in Path 2.

TABLE 13.3
Simulation Results for the First Scenario

Average Delay (s)	Average Delay from Stops (s)	Average Speed (km/h)	Average Number of Stops	Total Travel Time (s)	Number of Arrived Vehicles
7.9	1.6	47	0.5	74,965	1000

TABLE 13.4

Realized Traffic Volumes over the
Paths for the Second Base-Scenario

Path Number	Traffic Volume (cars/hour)
1	141
2	161
3	164

In the Vissim implementation of the second scenario, Obstacle 1 is modeled to stop each vehicle passing the obstacle location for 60 seconds. The two-obstacle case of the third scenario is modeled similarly for each of the two obstacles.

The traffic indicators for the second base-scenario are summarized in Table 13.4 and in the first data row of Table 13.5. These reveal that the obstacle has caused a considerable slowdown of the traffic; e.g., the average delay grew from 7.9 s to nearly 10 minutes, while the average speed dropped from 47 km/h to walking speed. A vehicle had to stop about 20 times on average in the second base-scenario, which is to be compared to half a stop per vehicle on average in the first scenario. The total travel time grew to about four times of its undisturbed value, while the number of arrived vehicles decreased severely. The diminished traffic volumes are shown in Table 13.4. It indicates that only less than half of the vehicles could now enter the network in the analyzed period. Only about 75% of these vehicles have actually arrived at Location 2.

TABLE 13.5

Simulation Results for the Parameterized Subscenarios of the Second Scenario

Ratio of Vehicle Diversions (%)	Av. Delay (s)	Av. Delay from Stops (s)	Av. Number of Stops	Av. Speed (km/h)	Total Travel Time (s)	No. of Arrived Vehicles	Av. Queue Length (m)
0	577	457	20	5	270,953	348	331
5	466	370	16	6	273,142	440	327
10	384	303	13	7	273,829	531	320
15	354	276	12	7	276,987	576	314
20	316	249	11	8	276,870	633	308
25	278	217	10	9	273,652	701	298
30	252	195	9	9	273,109	758	288
35	247	185	9	9	281,316	788	276
40	231	170	9	9	279,004	822	258
45	227	158	10	10 km/h	293,996	880	234
50	247	160	13	9	319,702	895	211

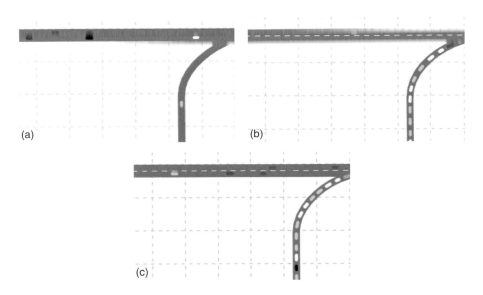

FIGURE 13.2   Road traffic at the first road fork near Location 1 in the first scenario (a), in the second base-scenario (b), and in a subscenario of the second scenario (c).

To model the reaction of drivers and/or of ADAS/AD systems to the LoS detection of the forming vehicle queue in case of the second base-scenario, a proxy parameter was introduced into the Vissim simulation. The scenario is split up into a number of parameterized subscenarios. In these subscenarios, 0%, 5%, 10%, ..., 45%, and 50% of vehicles are removed from the blocked Path 1 to the unblocked Paths 2 and 3 with a 60-s delay for each vehicle concerned. The vehicle numbers were rounded to the nearest integer numbers.

The traffic indicator values for the parameterized subscenarios of the second scenario are summarized in Table 13.5. According to the results, the obstacle that appeared in the network caused considerable slowdown of the traffic, but the growing diversions of the traffic from the blocked path eased the traffic situation gradually. *The effect of such a diversion can be verified in the Vissim simulation snapshots shown in Figures 13.2 and 13.3.*

The snapshots in Figure 13.2a to c present the road traffic in the vicinity of the first road fork near Location 1 for the first scenario, for the second base-scenario, and for a subscenario of the second scenario, namely for the one associated with the diversion ratio of 10% at a particular point in time. Light traffic without any hindrance can be observed in Figure 13.2a. In Figure 13.2b, an arc of the queue can be seen that has built up on Path 1 because of the appearance of Obstacle 1, while in Figure 13.2c, a similar queue can be seen despite the diversion. Note that in case of a real LoS simulation, this queue would not disappear either, as this arc and the following straight patch of road in Path 1 are not visible from Location 1.

The snapshots in Figure 13.3a to c present the road traffic along the middle section of Path 3 in the respective scenarios/subscenarios at a particular moment. In Figure 13.3a, the traffic is very light there indeed: a single car has just turned on

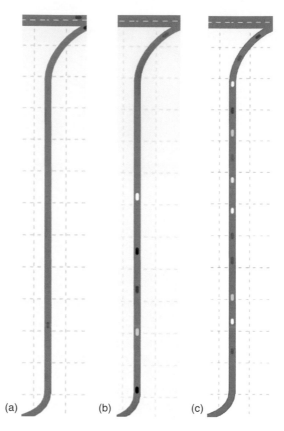

(a)   (b)   (c)

FIGURE 13.3   Road traffic along the straight middle section of Path 3 in the first scenario (a), in the second base-scenario (b), and in a subscenario of the second scenario (c).

this road section. Figure 13.3b pictures light traffic on this road section, while in Figure 13.3c, somewhat more intense traffic can be perceived. It is due to the 10% vehicle diversion from Path 1 to this path and to Path 2.

Comparing the corresponding traffic data from Tables 13.2 to 13.5, one can see, for instance, that the average delay for a vehicle was 7.9 s in the undisturbed network of the first scenario. This delay grew to nearly 10 minutes in the second base-scenario. Then, this delay dropped to less than 4 minutes after having diverted 45% of the vehicles intending to use Path 1 to Paths 2 and 3 in the simulation. The realized traffic volumes over the paths after the gradual diversions are not given herein, but the sum of these traffic volume shows an increasing trend that eventually reaches the original traffic volume.

In the Vissim implementation of the third scenario's parameterized subscenarios, 0%, 5%, 10%, …, 45%, and 50% of vehicles intending to use Paths 1 and 2 are diverted from these blocked paths to Path 3 with a 60-s delay. In this base-scenario, the traffic is very light on Path 3, as some of the vehicles intending to use this path cannot do so because of the two long queues that have built up due to the obstacles.

TABLE 13.6

Simulation Results for the Parameterized Subscenarios of the Third Scenario

Ratio of Diversions (%)	Av. Delay (s)	Av. Delay from Stops (s)	Av. Speed (km/h)	Av. Number of Stops	Total Travel Time (s)	No. of Arr'd Vehicles	Av. Length of Queue 1 (m)	Av. Length of Queue 2 (m)
0	872	658	3	36	560,645	460	260	309
5	773	600	4	28	506,859	498	260	257
10	722	551	4	28	535,932	555	251	286
15	660	509	4	25	534,643	610	240	291
20	568	440	5	21	511,287	676	234	271
25	504	387	5	19	465,516	708	224	232
30	412	321	6	14	397,810	740	206	184
35	349	273	6	12	379,626	818	198	169
40	343	268	7	12	372,663	824	198	160
45	311	244	7	10	353,361	850	187	138
50	261	207	8	8	318,769	886	170	122

In the consecutive subscenarios, however, traffic on Path 3 gets gradually more intense as the vehicles/drivers originally intending to use Paths 1 and 2 are gradually diverted to the unblocked Path 3. This diversion mimics the reactions of the drivers, who decide to take another route on the LoS detection of the queues.

The traffic indicators for the third base-scenario are summarized in the first data row of Table 13.6. These data reveal that the two obstacles caused an even worse slowdown than just a single one. The traffic indicator values for the consecutive subscenarios are summarized in the second to the eleventh data rows of Table 13.6.

Comparing the corresponding traffic data from Tables 13.3 and 13.6, one can see, for instance, that the average delay for a vehicle was 7.9 s in the undisturbed network. This delay grew to nearly a quarter of an hour for the third base-scenario. Then, it dropped to somewhat less than 5 minutes after having diverted 50% of the cars from Paths 1 and 2 to Path 3.

If one compares the respective traffic indicator values given in Tables 13.5 and 13.6, it is clear that the traffic according to the third scenario is more severely hindered than that according to the second scenario. The former recovers slower than the latter. It is by no means surprising, and one expects a similar relation between the respective traffic indicator values in case of a real LoS-based simulation and, of course, in real traffic making use of LoS detections.

13.3.2.2   Second Case Study: A Somewhat More Complex Urban Road Network

In this case study, a somewhat more complex urban road network is considered and analyzed. *The network is shown in Figure 13.4.* In the first scenario, the road network moves 3000 vehicles in an hour between the six numbered locations. The lengths of the shortest paths between these locations varies between 259 m and 507 m.

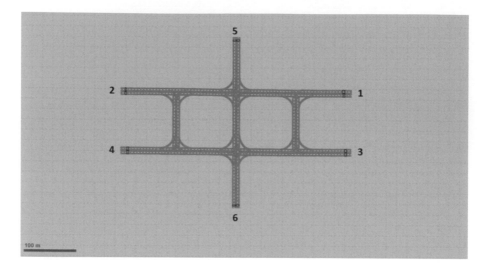

FIGURE 13.4 A somewhat more complex, ladder-like urban road network with six numbered locations. These locations generate and resorb traffic.

There are slight differences between lengths of the shortest path from Location i to Location j and that of the return journey.

The simulation models again a one-hour interval. Within this interval, vehicles are generated in the first 45 minutes, while in the last 15 minutes the network is left on its own to discharge. In the first 45 minutes, each numbered location generates 100 vehicles that head for the other five numbered locations, i.e., the mentioned total of 3000 vehicles is generated for the whole network in an hour. *The most relevant traffic indicators describing the undisturbed road traffic within the road network are summarized in Table 13.7.* The table shows, for instance, that the average vehicle speed is about 35 km/h, which is quite acceptable for an urban network. Furthermore, about every fourth to fifth car had to stop, on average, while completing its journey within the network, and the stopping lasted about 3 s on average.

The second base-scenario associated with the road network shown in Figure 13.4 features one obstacle. It turns up in the network and hinders the traffic flow. The obstacle location is close to the middle of the road section that corresponds to the left spoke of the ladder-like road network. A queue counter is associated with the queue building up in each of the two traffic directions. The modeled traffic

TABLE 13.7

Simulation Results for the Network Shown in Figure 13.5 According to the First Scenario

Average Delay (s)	Average Delay from Stops (s)	Average Speed (km/h)	Average Number of Stops	Total Travel Time (s)	No. of Arrived Vehicles
7.1	0.6	35	0.23	126,347	3000

TABLE 13.8

Simulation Results for the Road Network Shown in Figure 13.4 According to the Parameterized Subscenarios of the Second Scenario

Ratio of Vehicle Diversions (%)	Av. Speed (km/h)	Av. Delay (s)	Total Travel Time (s)	No. of Arrived Vehicles	Av. Queue Length at Counter 1 (m)	Av. Queue Length at Counter 2 (m)
0	6	217	620,307	2236	251	188
5	7	162	466,544	2504	151	155
10	10 km/h	106	336,467	2664	105	95
15	14 km/h	67	256,301	2977	48	23
20	13 km/h	74	275,632	2905	61	50
25	12 km/h	95	301,452	2766	65	78

is keep-to-the-right. In the Vissim implementation of this scenario, the obstacle is modeled to stop each vehicle passing the obstacle location for 60 seconds (in both traffic directions). *The traffic indicators for the second base-scenario are summarized in the first data row of Table 13.8.* It reveals that the appearance of the obstacle slowed down the traffic considerably. For instance, the average delay for a vehicle grew from 7.1 seconds to more than 3 minutes, while the average speed of the vehicles dropped from 35 km/h to walking speed.

In the Vissim implementation of the second scenario's parameterized subscenarios, 0%, 5%, 10%, 15%, 20%, and 25% of vehicles that would use the blocked path are diverted to alternative compatible paths with a 60-s delay. Obviously, the nearest integer numbers are used when calculating the diversions. The traffic indicator values for the parameterized subscenarios of the second scenario are summarized in the further rows of Table 13.8.

Comparing the corresponding traffic indicators from Tables 13.7 and 13.8, one can see, for instance, that the average delay for a vehicle was 7.1 s in the undisturbed network. This delay grew to more than 3 minutes in the second base-scenario. Then, this delay dropped to somewhat more than 1 minute after having diverted 15% of the vehicles to alternative paths. According to the results, the obstacle that appeared in the road network caused considerable slowdown of the traffic, but the gradual diversions of the traffic from the blocked path somewhat eased the traffic situation. This is also confirmed by the gained traffic data in respect to origin and destination pairs. These data are not fully presented herein—we just provide the relevant data between Location 2 and Location 4. For this origin and destination pair, the average travel time was 30 s for the first scenario, it increased to 419 s for the second base-scenario, and that decreased to 245 s after having diverted 5% of the hindered traffic flow to compatible alternative paths.

13.3.2.3 Third Case Study: A Busy Subnetwork of Roads In Újbuda

A busy subnetwork of roads within the Újbuda district of Budapest, Hungary, was chosen as a real-world example for assessing the utility of an on-board LoS road

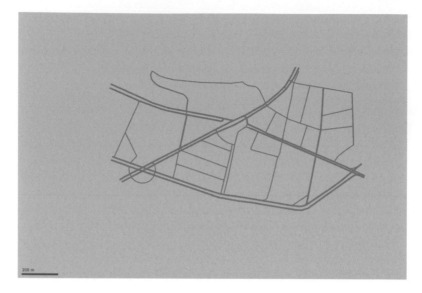

FIGURE 13.5   The subnetwork of roads within the Újbuda district of Budapest, Hungary. The subnetwork is centered around the Móricz Zsigmond circus.

TJQD system. The subnetwork is located on the west bank of the River Danube. The Móricz Zsigmond circus is situated in the center of the chosen subnetwork. The circus is within walking distance from the River Danube. The circus is situated at the convergence of major boulevards and at the intersection of significant transport corridors that cross the Danube. *The chosen subnetwork of roads is presented as a line drawing in Figure 13.5.* In this graph, the Móricz Zsigmond circus is at the intersection of the mentioned major boulevards, which form a distinct X-shape within the subnetwork. In the simulation model, there are four numbered locations that are specifically referred to in the subnetwork. Location 1 is at the top left end of the X-shape; Location 2 is at its bottom right end. Location 3 appears at the top right end of the X, while Location 4 is at its bottom left end. The lengths of the shortest routes between these numbered locations fall between 1291 m and 2459 m.

The Móricz Zsigmond circus was known by the local drivers for its long vehicle queues in peak hours for many decades. The vehicle queues are not that frequent nowadays, as some of the through traffic has been redirected to other routes, and several public bus terminal stations that used to be there have been relocated to more peripheral transport centers of the capital city. Nevertheless, there still exist some road transport vulnerabilities in the area that can cause queues in the peak hours. One of these vulnerable road locations is investigated and discussed here; it was used for instigating a virtual vehicle queue in our traffic simulations.

The photograph shown in Figure 13.6 gives an idea of the place and also a feel for the mentioned road transport vulnerability. In the figure, a patch of road near the circus is shown. This patch is located on the road that leads from Location 2 to the circus. It is about 50 m from the arc of the circus. This patch of road is next to a busy tram terminal station, and with that in place, the road traffic towards the

FIGURE 13.6   The modeled road near and viewed from the Móricz Zsigmond circus in the Újbuda district of Budapest, Hungary. The obstacle associated with the second scenario was "inserted" onto this patch of road in the simulation.

circus has to make do with a single lane for cars and a separate, very narrow lane for cyclists. In the second scenario associated with this case study, an obstacle (e.g., a refrigerator van delivering products to a shop) was "inserted" for the purpose of traffic modeling onto this patch of road. We just note here that the simulation modeled a much busier period of the day than shown in the figure.

In our simulation experiments, we assumed that the road network shown in Figure 13.5 moves 3600 vehicles between the four numbered locations in an hour. The simulation models a one-hour interval, and within this interval the vehicles are "generated" in the first 45 minutes, while in the last 15 minutes the network is left on its own to discharge. In the simulated period, each numbered location generates 300 vehicles heading for each of the other three numbered locations. The major junctions within the modeled subnetwork were implemented as signaled junctions in the traffic simulation.

This traffic demand is not that strong, and the subnetwork can handle it fairly well. *In Table 13.9, the average travel times between the numbered locations are shown for the first scenario.* The traffic flow that is expected to be hit hardest by the obstacle inserted in the second scenario is from Location 2 to Location 1.

The second base-scenario associated with the subnetwork features one obstacle; it is located on that half of the road which leads from Location 2 to the circus, and it is about 50 m from the perimeter of the circus. A single queue counter was used

TABLE 13.9

Average Travel Times between the Numbered Locations for the First Scenario Associated with the Újbuda Subnetwork

From Location Numbered	To Location Numbered	Average Travel Time (s)	Arrived Vehicles
1	2	221	300
1	3	356	300
1	4	258	300
2	1	391	300
2	3	242	300
2	4	427	300
3	1	128	300
3	2	212	300
3	4	160	300
4	1	515	210
4	2	505	228
4	3	552	226

in the traffic simulation. The obstacle was modeled to stop each vehicle passing the obstacle towards the circus for 60 seconds.

Again, the base-scenario has been split up into a number of parameterized subscenarios. In this case, only the 0.0 to 0.3 range is presented for the proxy parameter. The corresponding seven discrete values were used in separate simulation runs. In these, 0%, 5%, 10%, …, 25%, and 30% of vehicles were diverted with a 60-s delay from the blocked path to alternative compatible paths. *Table 13.10 presents the average travel times and the number of arrived vehicles heading from Location 2 to Location 1 for the subscenarios of the second scenario.* It shows that the diversion

TABLE 13.10

Average Travel Times and the Number of Arrived Vehicles from Location 2 to Location 1 for the Subscenarios of the Second Scenario

Ratio of Diversions (%)	Average Travel Time (s)	Arrived Vehicles
0	468	271
5	535	299
10	503	299
15	447	300
20	461	300
25	430	300
30	448	300

of 25% of the traffic from the blocked path somewhat improved the traffic situation on this hard-hit path.

13.4 DISCUSSION

In the previous section, we presented three case studies: we considered and modeled the traffic of three increasingly complex road networks relying on a widely used microscopic traffic simulator. Two out of these three road networks were synthetic, while one was a real road network from Budapest. These networks were devised and chosen so that the advantage of early LoS TJQD in respect to route selection and replanning could be demonstrated. Furthermore, the cases, i.e., the networks and the associated scenarios/subscenarios, serve the purpose of comparing and time-wise restricting the obtained results with the corresponding traffic indicators gained through other detection and communication means and methods. However, due to the lack of preprogrammed support for checking whether a constellation of locations, cars, drivers, and/or pedestrians is LoS or not, we opted for indirect and tentative modeling of the LoS TJQD process through three proxy parameters.

In conjunction with individual vehicles modeled in our simulations, we can state that on average several minutes of delays due to stops can be spared per vehicle through diversions in the severe traffic situations that we analyzed. We expect similar reductions in delays due to stops in the case of real LoS TJQD as well.

Clearly, in many transport applications, a few-minute delay is not a problem. In these applications, many vehicle drivers and autonomous vehicles are prepared to stand in the forming queues for a while. However, there are cases, trips, deliveries, and services when the drivers, passengers, vehicle owners, and vehicle users would be prepared to pay for a reasonably priced on-board LoS TJQD ADAS/AD subsystem, even if the same traffic information would reach them soon anyway.

On a road network level, the utility of an on-board LoS TJQD ADAS/AD subsystem depends on its prevalence. For example, in the presented case studies, the total travel time grew considerably with the appearance of obstacle(s). The traffic situation, however, eased gradually with increasing the percentages of the diversions applied. We expect that the traffic situation would also ease with a real LoS TJQD function being available to growing portions of vehicles. As in our experiments for the gradually increasing diversion ratios, there could be a limit to this easing phase also for the increasing prevalence of real LoS TJQD subsystems.

13.5 CONCLUSIONS AND FURTHER WORK

In this chapter, we have endeavored to assess the utility of on-board LoS TJQD capability, especially in conjunction with road vehicles equipped with ADAS/AD systems. Such a capability requires and heavily relies upon LoS sensors. The utility was considered on two levels, namely in respect to individual vehicles and their users and in respect of all the current users of the road network. Road traffic simulations considering the movements of individual vehicles were carried out to this end in respect to simple model road networks. These were studied under different scenarios: first without and then with the disturbance. The disturbance was caused by one or more obstacle appearing in the modeled road network. Upon their appearance, one or more vehicle queues started to build up and the traffic slowed down perceivably.

As the traffic simulator version that was available for us in this experimental research does not support LoS filtering of locations/vehicles/people, we made do with a rough model of the spatiotemporal LoS TJQD process.

Though each of our simulation runs covered one-hour duration of road traffic, the results obtained via these simulations need to be restricted and recalculated for the first few minutes, at most for the first quarter of an hour. This restriction is necessary so that we can realistically assess the utility of the studied subsystems; as after the mentioned period, the relevant traffic information originating from traditional and/or wireless static vehicle/traffic sensors, and/or based on crowdsourced data will have most likely reached the connected cars in the area through some info-communication channels. This latter process, however, requires a highly developed urban transport infrastructure and an info-communication environment of a high standard.

With a view on the mentioned alternative ways of exploring the current road traffic situation and locating traffic problems, we see only a limited utility of on-board LoS TJQD subsystems, if these are to be used in developed urban environments. Still, certain segments of car users may value even the few-minute travel time shortening that we have measured in our simulations concerning the presented simple road networks. Furthermore, the poor state of the local road and info-communication infrastructure raises the perceived utility of the proposed and studied subsystems.

We plan to explore in the near future the effect of varying also the two proxy parameters that we have used as fixed delays herein.

ACKNOWLEDGMENTS

The research was supported by the Ministry of Innovation and Technology National Research Development and Innovation (NRDI) Office within the framework of the Autonomous Systems National Laboratory Program. It was also supported through the 2018-2.1.10-TÉTMC-2018-00009 research contract by the NRDI Office.

REFERENCES

Bauer, P., A. Hiba, B. Z Daróczy et al. (2017). Real flight demonstration of monocular image-based aircraft sense and avoid. *ERCIM News*, 110, pp. 42–43.

Butcher, G. (2010). *Tour of the Electromagnetic Spectrum*. National Aeronautics and Space Administration. Washington, D.C.

Chetouane, A., S. Mabrouk, and M. Mosbah (2020). Traffic congestion detection: Solutions, open issues and challenges. In *International Workshop on Distributed Computing for Emerging Smart Networks* (pp. 3–22). Springer, Cham, doi:10.1007/978-3-030-65810-6_1.

Fang, Xuan, and Tettamanti, T. (2021). Change in microscopic traffic simulation practice with respect to the emerging automated driving technology. *Period. Polytech-Civ.*, 66(1), pp. 86–95, doi:10.3311/PPci.17411.

Feng, Qixing, E. K. Tameh, A. R. Nix, and J. McGeehan (2006). WLCp2-06: Modelling the likelihood of line-of-sight for air-to-ground radio propagation in urban environments. IEEE Globecom, pp. 1–5, San Francisco, CA, USA. doi:10.1109/GLOCOM.2006.917.

Guo, Fan, Jin Tang, and Xile Wang (2017). Gesture recognition of traffic police based on static and dynamic descriptor fusion. *Multimed. Tools. Appl.*, 76(6), pp. 8915–8936, doi: 10.1007/s11042-016-3491-9.

Guth, P. L. (2004). The geometry of line-of-sight and weapons fan algorithms. In *Studies in Military Geography and Geology* (pp. 271–285). Springer, Dordrecht.

Huurdeman, A. A. (2003). *The Worldwide History of Telecommunications*. John Wiley & Sons, p. 638 Hoboken, NJ, doi:10.1002/0471722243.

Kilani, O., M. Gouda, J. Weiss, and K. El-Basyouny (2021). Safety assessment of urban intersection sight distance using mobile LiDAR data. *Sustainability*, 13, Paper-id 9259, doi: 10.3390/su13169259.

Levering, A., M. Tomko, D. Tuia, and K. Khoshelham (2020). Detecting unsigned physical road incidents from driver-view images. *IEEE Trans. Intell. Veh.*, 6(1), 24–33, doi:10.1109/TIV.2020.2991963.

Matera, E. R., A. Garcia-Pena, O. Julien et al. (2019). Characterization of line-of-sight and non-line-of-sight pseudorange multipath errors in urban environment for GPS and Galileo. Proceedings of the 2019 International Technical Meeting of The Institute of Navigation, pp. 177–196, doi:10.33012/2019.16687.

Meneguette, R. I., Fillho, G. P. R., Bittencourt, L. F., Ueyama, J., & Villas, L. A. (2016). A solution for detection and control for congested roads using vehicular networks. IEEE Latin America Transactions, 14(4), 1849–1855. DOI: 10.1109/TLA.2016.7483525

Mueller, E. A. (1970). Aspects of the history of traffic signals. *IEEE Trans. Veh. Technol.*, 19(1), pp. 6–17, doi: 10.1109/T-VT.1970.23426.

Nagy, M., P. Bauer, A. Hiba et al. (2021). The forerunner UAV concept for the increased safety of first responders. 7th International Conference on Vehicle Technology and Intelligent Transport Systems (VEHITS), pp. 362–369, doi:10.5220/0010408203620369.

Parisot, C., J. Meessen, C. Carincotte, and X. Desurmont (2008). Real-time road traffic classification using on-board bus video camera. 11th International IEEE Conference on Intelligent Transportation Systems, pp. 189–196, doi:10.1109/ITSC.2008.4732628.

PTV VISSIM. (2022). Online: https://www.ptvgroup.com/en/solutions/products/ptv-vissim/ (last accessed: 19 January 2022).

Sterling, C. H. (ed.). (2008). *Military Communications: From Ancient Times to the 21st Century*. ABC-CLIO, Santa Barbara, CA. 565.

Umair, M., M. U. Farooq, R. Raza et al. (2021). Efficient video-based vehicle queue length estimation using computer vision and deep learning for an urban traffic scenario. *Processes*, 9(10), Paper-id 1786, doi:10.3390/pr9101786.

Yan, X., F. Gu, X. Hu, and C. Engstrom (2013). Dynamic data driven event reconstruction for traffic simulation using sequential Monte Carlo methods. Winter Simulations Conference, pp. 2042–2053, IEEE, doi:10.1109/WSC.2013.6721582.

Yu, Ming-Yuan, R. Vasudevan, and M. Johnson-Roberson (2019). Occlusion-aware risk assessment for autonomous driving in urban environments. *IEEE Robot. Autom. Lett.*, 4(2), pp. 2235–2241, doi:10.1109/LRA.2019.2900453.

14 Risky Trajectory Prediction for Safe Walkability under an Intuitionistic Fuzzy Environment

Ghyzlane Cherradi, Azedine Boulmakoul, Lamia Karim, Zoltan Fazekas, and Péter Gáspár

CONTENTS

14.1 INTRODUCTION

Every transport journey, no matter how big or small, begins and ends with walking. Walking is often the only way that many people can access daily activities. Unlike other modes of transport, walking is not linked to a vehicle on a road, and the surrounding infrastructure is highly heterogeneous. In addition, walking is directly affected by environmental factors (traffic lights, public equipment, advertisements, etc.), as well as the total wait time for pedestrians, the mean space between vehicles, and atmospheric conditions (wind, rain, etc.). Moreover, pedestrians move with a high

DOI: 10.1201/9781003255635-14

229

degree of stochasticity, so multiple plausible and distinct future behaviors can exist. In this context, modeling pedestrian safety while respecting the randomness and uncertainty of influencing factors is an important research goal.

Predicting pedestrian trajectories is an important but difficult process, as pedestrian directions and behaviors can change due to sudden events and depending on vehicles, human interactions, other moving objects, etc. Pedestrian position sensing and tracking are key components of today's road safety applications. The technological development and the availability of geolocation-based solutions are leading to the formation of new paradigms, such as walking as a service (WaaS), which aims to provide a mainstream, comprehensive, and simple service. Path prediction technologies enable smart environments to predict future pedestrian movements to increase road safety and reduce road fatalities.

In literature, methods for modeling the problem of predicting the path of pedestrians have evolved, moving from physics-based models to data-based models using deep neural networks (DNNs). Rather than physics-based models that generate and visualize pedestrian conditions based on what should be, data-based models are specifically designed to carry the full features of available observations by exploring the statistical correlation between pertinent variables and discovered processes. Recently, DNNs represent one of the most popular data-based models, thanks to their interesting property of being able to learn complex characteristics from data, and they have demonstrated that they exceed the prediction accuracy of traditional models in many scientific fields.

In this chapter, we present a new model featuring a fusion between pedestrian trajectory prediction and pedestrian risk assessment. Specifically, we develop a fuzzy intuitionist model-based approach to extract spatiotemporally risk models from pedestrian risk trajectories. Then, following the essential fulfillment of the neural network in the area of modeling nonlinear problems, we build a new network among risk patterns for risky trajectory prediction using approaches based on DNNs, in particular, stochastic techniques, to deal with the large uncertainties on future trajectories and the nonrigid aspect of the pedestrian.

The remainder of this chapter is organized as follows. Section 14.2 addresses pedestrian risk models, deep learning models, and tracking technologies used in the field of pedestrian trajectory prediction. Section 14.3 discusses the fuzzy intuitionistic model used to extract spatiotemporal risk patterns from pedestrian risk trajectories, as well as the construction of the new network among risk models for risk trajectory prediction using network-based approaches with deep neurons, in particular, stochastic techniques. Section 14.4 presents our microservice development approach, with a special emphasis on the advantages of this pattern on the basis of a real infrastructure under development. We present an integral system to determine, gather, and offer a variety of pedestrian-related control functions to indicate how it is possible to develop a scalable and integrated mobility service for safe walkability. Finally, Section 14.5 concludes the chapter.

14.2 PEDESTRIAN SAFETY

Efforts to make pedestrian walking safer are driven by many research interests, including pedestrian risk modeling, predicting trajectories, and urban navigation. In this section, we review these areas of research and their relationship to our study.

14.2.1 PEDESTRIAN RISK MODELING

Typically, the risk to pedestrians has been assessed using accident frequency models, using historical data (Brüde and Larsson 1993), (Cameron, 1982). These models were developed using geographic data at intersections. However, the frequency of accidents depends on many factors such as traffic volume, speed, geometry, environment, etc. Several studies have been developed to understand the association of risk factors with the estimation of pedestrian accidents. Research is increasingly interested in screening risk factors to the pedestrian accident occurrence and gravity, using data augmentation methods to make statistical deductions about the parameters related to risk exposure (Ayala et al., 2012). A problem regarding a lack of data will lead to a deceptive assessment of risk factors. Recognizing the risk factors for pedestrian crashes is critical to developing a planning strategy and determining appropriate countermeasures to reduce pedestrian crashes. Some other studies have focused on modeling pedestrian behaviors. For instance, Yanfeng et al. (Yanfeng et al., 2010) proposed a model using the statistical approach for pedestrian violation crossing behaviors at signalized intersections. The work presented in (Hamed, 2001) mentions low pedestrian waiting times as a factor of risky behavior. Earlier studies on pedestrian crossing behavior patterns focus on the evaluation of traffic control characteristics and road design. Given the uncertain nature of influencing factors, the authors of (Boulmakoul and Mandar, 2011) proposed a new method for evaluating a virtual indicator of mutual pedestrian-vehicle accident risk. In the work given by (Mandar et al., 2017), pedestrian dynamics are modeled using the basic fuzzy ant model, into which artificial capacitance is incorporated. From these examples, pedestrian risk models are developed to study potential risky behavior and its influencing indicators. Data allied to risk behavior and influencing factors are usually obtained via detection and observation systems. Using a model, it is possible to understand incidents, explain the most pertinent factors, and predict the occurrence of an event.

14.2.2 TRAJECTORY PREDICTION METHODS

Trajectory prediction remains an important problem in road safety. Although spatiotemporal data are scarce, pedestrian monitoring data do exist and have increased in quantity and quality over the past decades. Data-driven modeling is based on the reasoning that influencing relationships can often be distinguished between relevant variables and observed outcomes. Data-driven modeling provides the fundamental foundation for machine learning (ML) and is especially beneficial when the full physics of a problem are fuzzy or unknown, when elementary physics are not easily modeled numerically, or when huge amounts of relevant data are available.

The artificial neural network (ANN) is one of the popular data-driven models—a biologically inspired nonlinear regression model whose notable power is its theoretical capacity to fit any function. Neural networks have demonstrated promising potential in a variety of ways (Lippmann, 1987), (Lau, 1991), (Payeur et al., 1995). More recently, DNNs have been applied to the task of route prediction, as they use a data-driven approach to elicit relationships and influences that may not be obvious. These DNN-based approaches have revealed impressive results.

Practically all of these approaches are based on recurrent neural networks (RNNs) given that a trajectory is a temporal series (Mandic and Chambers, 2001). Since RNNs share parameters over time, they are able to condition the model on all previous positions of a trajectory.

14.2.3 Urban Navigation and Sensing Technologies

In designing pedestrian navigation systems, care should be taken to provide various path choices to reduce potential hazards and road qualities (including safe crossings, traffic volume, speeds, etc.). Recently, systems with novel navigation goals that are not limited to shortest path determination, but mainly adapted to the individual interests and spatiotemporal constraints of users have appeared in the literature (Gonzalez et al., 2007), (Kanoulas et al., 2006). Absolutely, navigation and pedestrian guidance systems must be developed in a user-friendly way to boost road safety and provide an optimal user experience. Road safety management includes geolocation and real-time tracking. In fact, certain crucial elements are considered for safe walking: detection, information management, intelligent alerting, and decision making. Sensing involves collecting data from connected objects and transmitting them to a database or cloud platform. Sensing technologies can be divided into two classes. The first, seen as internal, makes it possible to detect the state of pedestrians, such as behavior, speed, and position. The second informs about the surrounding environment state, such as gap detection, environmental conditions, traffic volume, etc. In fact, portable positioning technologies using systems such as radio frequency identification (RFID) and global positioning system (GPS) have developed a lot (Djuknic and Richton, 2001), (Jaselskis et al., 1995). Moreover, the power of radio detection and ranging technology (RADAR), together with the integrated smart cameras, allow the recognition of distances, directions, paths, and moving objects. All of these technologies provide useful information that allows a person to make decisions and act automatically and immediately.

14.3 METHODOLOGY

14.3.1 Problem Formulation

Intuitionist fuzzy sets have been proposed by Atanassov (Atanassov, 1986). Which take into account the degree of membership, non-membership and hesitancy. It can characterize the more delicate uncertainty of the objective world. Therefore, this theory is commonly used to model real life issues such as sales analysis, financial services, negotiation processes, psychological research, etc. Unexpected pedestrian behavior, as well as the uncertainty in signal and traffic control, are the primary motivation behind the fuzzy logic approach to dealing with this ambiguity and uncertainty. Fuzzy logic aims to provide a mathematical context that helps define the uncertainty associated with pedestrian and traffic data. In fuzzy set theory, precise reasoning is seen as a restricted case of approximate inference. Everything is always a matter of the membership degree knowledge is thus interpreted as a set of uncertain constraints on a set of fuzzy variables (Zadeh, 1968), (Zadeh, 1999). Researchers have considered decision making to be one of the most attractive appli cation areas of fuzzy set theory. A fuzzy set A of the universe U is defined by its membership function $\mu_{\tilde{A}}(x)$ over a universe U and then there is a set of ordered pairs: $A(u) = \{x, \mu_{\tilde{A}}(x) | x \in U, \mu_{\tilde{A}}(x) \in [0,1]\}$. The membership function $\mu_{\tilde{A}}(x)$ is a single value between 0 and 1 that combines the proof for $x \in U$ and the proof against $x \in U$, without mentioning the quantity of each. The intuitionistic fuzzy sets (IFS) are quite helpful to express ambiguity and uncertainty more precisely as compared to fuzzy sets (Atanassov, 2016). In intuitionistic fuzzy logic, we study values with two types of degrees, namely, membership and nonmembership. Thus, the uncertainty and variety of pedestrian risk can be well covered and modeled using IFS (Atanassov, 2016), (Kahraman et al., 2017).

14.3.2 Basic Concepts

Definition 1: IFS (Atanassov, 2016). An IFS A for a given nonempty set X, X = $(x_1, x_2,..., x_n)$ is represented by a pair $\langle \mu_A(x), v_A(x) \rangle$ of functions, A = $\{[x, \mu_A(x), v_A(x)] | x \in X\}$. $\mu_A(x)$ is membership of x to A, and $v_A(x)$ is nonmembership of x to A, where $\mu_A(x) \in [0, 1]$, $v_A(x) \in [0, 1]$ and $0 \le \mu_A(x) + v_A(x) \le 1$. The value $\pi_A(x) = 1 - \mu_A(x) - v_A(x)$ describe the uncertainty of x to A.

Definition 2: The triangular intuitionistic fuzzy number (TIFN) (Atanassov, 2016), TIFN is defined as an intuitionistic fuzzy number that applies traditional triangu lar fuzzy numbers to represent the membership $\mu_{\tilde{A}}(x)$ function and nonmembership function $v_{\tilde{A}}(x)$. We introduce some fundamental concepts about TIFN, per the following:

Let $\tilde{A}$ be a TIFN, where the membership $\mu_{\tilde{A}}(x)$ and nonmembership $v_{\tilde{A}}(x)$ degrees are represented as follows:

$$\mu_{\tilde{A}}(x) = \begin{cases} \dfrac{x - a_1}{a_2 - a_1} & \text{if } a_1 \le x < a_2 \\[2ex] \dfrac{a_3 - x}{a_3 - a_2} & \text{if } a_2 \le x < a_3 \\[2ex] 0, & \text{otherwise} \end{cases} \tag{14.1}$$

$$v_{\tilde{A}}(x) = \begin{cases} \dfrac{a_2 - x}{a_2 - \acute{a}_1} & \text{if } \acute{a}_1 \le x < a_2 \\[2ex] \dfrac{x - a_2}{\acute{a}_3 - a_2} & \text{if } a_2 \le x < \acute{a}_3 \\[2ex] 1, & \text{otherwise} \end{cases} \tag{14.2}$$

where $\acute{a}_1 \le a_1 \le a_2 \le a_3 \le \acute{a}_3$, $0 \le \mu_{\tilde{A}}(x) + v_{\tilde{A}}(x) \le 1$ and TIFN is denoted by $\tilde{A}_{\text{TIFN}} = (a_1, a_2, a_3; \acute{a}_1, a_2, \acute{a}_3)$ as illustrated in Figure 14.1.

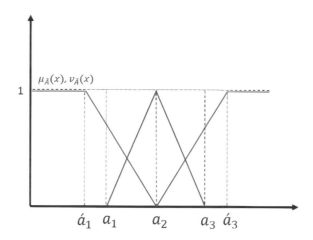

FIGURE 14.1 Membership and nonmembership function of triangular intuitionistic fuzzy number.

We use the TIFN proof given by Kahraman et. al. (Kahraman et al., 2017). A new illustration of the membership and nonmembership functions of the TIFN is given in Figure 14.2.

Definition 3. (Addition of two TIFNs) The addition of two TIFNs is as follows (Kahraman et al., 2017):

$A = ([a_1, b_1, c_1]; \mu_A, v_A)$ and $B = ([a_2, b_2, c_2]; \mu_B, v_B)$, with $\mu_A \neq \mu_B$ *and* $v_A \neq v_B$,

$A + B = \langle [a_1 + a_2, b_1 + b_2, c_1 + c_2]; \text{ MIN}(\mu_A, \mu_B), \text{ MAX }(v_A, v_B) \rangle$.

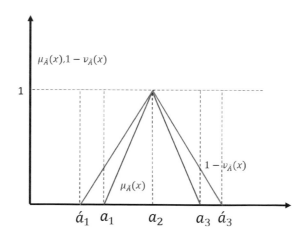

FIGURE 14.2 Membership and nonmembership functions of TIFN.

14.3.3 TRIANGULAR INTUITIONISTIC RISK EXPOSURE MODEL

The risk modelling of the urban road network is essentially the assignment of a risk score R to each road section, which is proportional to the probability of an accident occurring on this part of the road with the risk factors to which pedestrians are exposed. In this work, we estimate that the duration of exposure to risk is linked to the travel time on the portion of the road traveled. During this time, pedestrians are exposed to risks according to various risk indicators. This walking time is one of the various sources of uncertainty in pedestrian behavior that cannot be exactly known or represented. We represented this uncertainty by an IFS. For simplicity, we will focus on the TIFN. We propose a construction of an intuitionistic fuzzification of the pedestrian's travel time T_s by means of TIFNs. The severity of pedestrian accidents associated with each factor is multiplied by the probability of the accident occurring and the travel time to obtain an estimation of the risk exposure for each segment of the road. The formula used in this work is as follows:

$$R_s = \sum_j P_{sj} * G_j * T_s \qquad (14.3)$$

where:

R_s: the risk on a section road s

P_{sj}: the probability of accident occurrence on a road section s with respect to a risk factor j

G_j: the severity of accidental pedestrian injury on a road section s with respect to a risk factor j

T_s: the walking time associated with a road section s

In formula (14.3) some parameters are TIFN and some are invariable; consequently, the obtained risk is also in the form of a TIFN ([a, b, c]; μ_α, v_α). The full risk of a road, denoted by $R(x) = ([a, b, c]; \mu_r, v_r)$, is the sum of the risk score of segments s_k. where s_k is the kth segment for a road x. The risk $R(x)$ is calculated for each segment in the network N using (14.3). There are a variety of risk values for each segment s, depending on θ and $\acute{\theta}$ (predefined confidence levels) to be determined according to the minimum risk not to be exceeded. $\mu_R(x < \lambda) = \theta$ and $(1 - v_R(x < \alpha)) = \acute{\theta}$; where $\lambda \in [0, 1]$ and $\alpha \in [0, 1]$. We need to search values of λ and α for the evaluation of an edge according to a θ and $\acute{\theta}$ risk level. There are two triangles for TIFN: one for the membership function and one for nonmembership function, as shown in Figure 14.3.

Therefore, the values of λ and α are calculated as follows. We know that $P(x < \theta) = 1 - p(x > \theta)$, $P(x < \acute{\theta}) = 1 - p(x > \acute{\theta})$. Using the surface of triangles $a_2B\,a_3$ and $a_2B\,\acute{a}_3$, we get the following results:

$$\text{Surface } (a_2B\,a_3) = 1 * (a_3 - a_2)/2$$
$$\text{Surface } (a_2B\,\acute{a}_3) = 1 * (\acute{a}_3 - a_2)/2$$

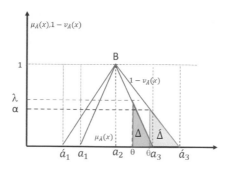

FIGURE 14.3 Acceptable risk value λ, α according to θ and $\acute{\theta}$ risk level.

Using the Thales theorem, we get the following results:

$$1-\lambda = \frac{\theta-a_2}{a_3-a_2} ; \lambda = 1 - \frac{\theta-a_2}{a_3-a_2}$$

$$1-\alpha = \frac{\acute{\theta}-a_2}{\acute{a}_3-a_2} ; \alpha = 1 - \frac{\acute{\theta}-a_2}{\acute{a}_3-a_2}$$

Now for every segment s, we have the values of λ and α corresponding to the θ and $\acute{\theta}$ level of risk, respectively. When $\lambda \ll 1$ the risk grows significantly, and hence it is the fuzzy intuitionistic indicator of risk that we will use. Ultimately, the problem simply becomes a matter of determining the safest path in a time-dependent network.

14.3.4 RISKY TRAJECTORY PREDICTION

Practically, given a case where pedestrians are present, their positions are recorded for a determined period of time, called the space-time trajectory denoted by T = $\{p_1, p_2, ..., p_n\}$ as illustrated in Figure 14.4. A risk trajectory (RT) is a spatiotemporal trajectory weighted by a risk score calculated using the model proposed earlier. A trajectory at risk is described as $RT = \{(p_1, r_1),(p_2, r_2),....,(p_n, r_n)\}$ where p_n indicates the position at the nth point of the trajectory RT, and r is the associated risk score. A risk trajectory contains not only the spatiotemporal location data of a pedestrian but also the potential risk at each position.

The goal is to predict the future trajectory that is dangerous for pedestrians. Our prediction approach takes the calculated risks for each path using the model presented in the previous section, as input to an RNN, due to the fact that the trajectories are continuous in character, have no complex "state", and have a strong spatiotemporal correlation, which can be exploited by computationally efficient convolutions. The architecture of RNNs aims to effectively model temporal dynamic behavior by summarizing inputs into hidden internal states, used as a memory mechanism for capturing dependencies during a temporal sequence. RNNs have achieved remarkable results in various research fields.

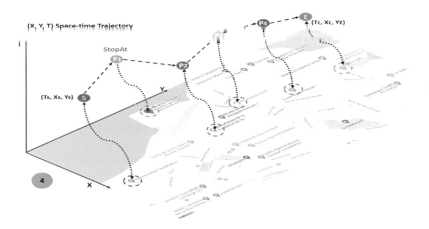

FIGURE 14.4 A space-time trajectory.

14.4 SYSTEM ARCHITECTURE

The proposed system is an integrated solution that provides a variety of services such as real-time traceability, geofencing, and pedestrian navigability. We are devoted to microservices architecture and aim to develop and run some microservices (Cherradi et al., 2017a). These microservices are developed, tested, deployed, and run independently of other microservices. In addition, this implementation reveals the benefits of the microservices architecture such as extensibility, scalability, availability, and flexibility. Figure 14.5 briefly shows the key components and the proposed system architecture and their connections. The fog-cloud paradigm uses a node-oriented approach that is especially useful for integrating Internet of Things (IoT) devices. It includes three elements that are considered the principal criteria to take into account to decide whether a fog-cloud architecture is required by an IoT application.

14.4.1 DATA COLLECTION SERVICE

Improving the quality of analytics and decision making requires powerful, scalable, and interoperable data collection services. Various paradigms are proposed to overcome the lack of resources on connected objects while allowing scaling and low-latency calculations. The fog-cloud paradigm uses a node-oriented approach that is especially useful for integrating IoT devices. It includes three elements that are considered the principal criteria to take into account to decide whether a fog-cloud architecture is required by an IoT application (Cherradi et al., 2017b) (Cherradi et al. 2022b).

- **Data resource nodes:** Include the organization of distributed compute nodes that represent message broker, data connection, IoT tool connector, and data processing. Geographically, adjoining computing nodes deployed at the edge, fog, and cloud are usually linked via a big number of communication networks.

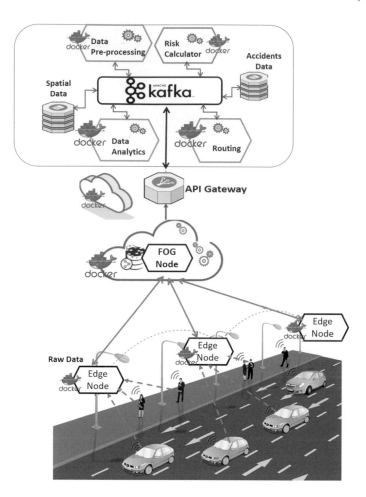

FIGURE 14.5 An overview of the system architecture.

- **Analytical nodes:** Include maximum splendid practical solutions for performing analytical functions that are vital to meeting the demand of IoT applications.
- **Storage nodes:** Include efficient distributed storage solutions deployed to control the vast amount of data generated via means of huge IoT devices.

To enhance pedestrian protection, the surrounding environment must be monitored in space (route, crosswalk, etc.) and time (status, speed, etc.). The data collection system was designed to be part of a bigger IoT infrastructure—this means that it needs to provide data to a bigger IoT platform. In this way, the data collected can be used in different contexts, for instance, associated with hazmat transportation or fleet management. This has become inspired by the fact that sensors and platforms operate in various places and are delivered through diverse entities.

With the leverage of distributed architectures and the fog computing paradigm, the system is designed as event-driven microservices. This design offers the system nice flexibility in scalability and adaptability to varied information applications. The system must integrate dynamic data streams to form predefined tasks in close to real time. Many streams of spatiotemporal data are sent frequently from the smartphone via the Message Queuing Telemetry Transport (MQTT) protocol.

14.4.2 PEDESTRIAN NAVIGATION

When walking from a source node to one or more destinations, the route taken is an important factor to ensure a safe walk. Determining an appropriate walking route, known as the routing problem, involves choosing a route for each pedestrian from among their alternative routes between origin-destination (OD) pairs to minimize risk. In this work, we formulate the pedestrian routing problem as follows: we consider a pedestrian network $G = (V, E)$, where V is the set of nodes containing the origin and destination locations, and E is the set of arcs that connect the nodes. Each arc is identified by its risk cost. To define the problem, the following notations are first defined:

- $O \{o_1, o_2, ..., o_n\} \in V$: set of original nodes
- $D \{d_1, d_2, ..., d_n\} \in V$: set of destination nodes
- R_{ij}: cost of the risk from node i to node j

Each arc is assigned an expected risk cost based on one of the approaches proposed in the previous sections. With a known cost per arc, the pedestrian's navigation problem can be reduced to a classical shorter path problem. Therefore, we have proposed an optimal route service based on the pgRouting library to provide a set of functions to calculate the optimal route and also to provide the best alternative route to use when the usual route is obstructed by an accident. In general, two types of data are required as input for the proposed service, namely spatial data, which represents the transport network as a collection of arcs with nodes loaded from OSM (OpenStreetMap) into the PostgreSQL/PostGIS database, and nonspatial data such as name, category, length, risk cost, etc. The realization of this service is premised on combining a geographic information system (GIS) with spatial information (PostgreSQL) and a geospatial routing library (pgRouting) (Cherradi et al. 2022a). The pgRouting functions are mainly based on dynamic costs and therefore are often applied to different types of real emergency situations on the road network.

14.5 CHAPTER SUMMARY

This chapter addresses the issue of safety associated with smart walking, which is a crucial element to support the daily life of pedestrians and provide decision makers with comprehensive and complete information on the extent of the risks associated with each trajectory. To achieve this goal, we introduced a real-time reactive distributed system powered by an intuitionistic fuzzy risk model. The proposed model is an acceptable classification methodology for intuitionistic fuzzy numbers and will be applicable to decision-making issues in intuitionistic fuzzy environments. In addition, we offered an optimal route service to choose the safest routes for the pedestrian.

This work also presents the design and general description of a georeferenced data collection microservice based on IoT protocols for a smart walk. This has general advantages that would benefit almost any IoT application. Thus, we used the edge-fog-cloud architecture to offer a high level of isolation, autonomy, and responsiveness to applications in sensitive areas such as road safety. In addition, the proposed ecosystem can be reused and easily readapted for other application fields.

ACKNOWLEDGMENTS

This study was funded by Ministry of Equipment, Transport, Logistics and Water, Kingdom of Morocco, the National Road Safety Agency (NARSA), and National Center for Scientific and Technical Research (CNRST). Road Safety Research Program# An intelligent reactive abductive system and intuitionist fuzzy logical reasoning for dangerousness of driver-pedestrians interactions analysis.

REFERENCES

Atanassov K. (1986) Intuitionistic fuzzy sets, Fuzzy Sets and Systems, 20(1986) 87–96.
Atanassov, K. (2016). Intuitionistic fuzzy sets. International Journal of Bioautomation, 20, 1.
Ayala, I., Mandow, L., Amor, M., & Fuentes, L. (2012, December). An evaluation of multiobjective urban tourist route planning with mobile devices. In International Conference on Ubiquitous Computing and Ambient Intelligence (pp. 387–394). Springer, Berlin, Heidelberg.
Boulmakoul, A., & Mandar, M. (2011). Fuzzy ant colony paradigm for virtual pedestrian simulation. The Open Operational Research Journal, 5(1).
Brüde, U., & Larsson, J. (1993). Models for predicting accidents at junctions where pedestrians and cyclists are involved. How well do they fit? Accident Analysis & Prevention, 25(5), 499–509.
Cameron, M. H. (1982). A method of measuring exposure to pedestrian accident risk. Accident Analysis & Prevention, 14(5), 397–405.
Cherradi, G., Boulmakoul, A., Karim, L., & Mandar, M. (2022a). Toward a safe pedestrian walkability: A real-time reactive microservice oriented ecosystem. In Networking, Intelligent Systems and Security (pp. 439–451). Springer, Singapore.
Cherradi, G., Boulmakoul, A., Karim, L., & Mandar, M. (2022b). A geo-referenced data collection microservice based on IoT protocols for smart HazMat transportation. In Internet of Things (pp. 85–100). CRC Press. Boca Raton, https://doi.org/10.1201/9781003219620.
Cherradi, G., Bouziri, A. E., Boulmakoul, A., & Zeitouni, K. (2017a). Real-time microservices based environmental sensors system for Hazmat transportation networks monitoring. Transportation Research Procedia, 27, 873–880.
Cherradi, G., El Bouziri, A., Boulmakoul, A., & Zeitouni, K. (2017b). Real-time hazmat environmental information system: A micro-service based architecture. Procedia Computer Science, 109, 982–987.
Djuknic, G. M., & Richton, R. E. (2001). Geolocation and assisted GPS. Computer, 34(2), 123–125.
Gonzalez, H., Han, J., Li, X., Myslinska, M., & Sondag, J. P. (2007). Adaptive fastest path computation on a road network: A traffic mining approach. In 33rd International Conference on Very Large Data Bases (VLDB), Vienna (pp. 794–805). Association for Computing Machinery.

Hamed, M. M. (2001). Analysis of pedestrians' behavior at pedestrian crossings. Safety Science, 38(1), 63–82.

Jaselskis, E. J., Anderson, M. R., Jahren, C. T., Rodriguez, Y., & Njos, S. (1995). Radio-frequency identification applications in construction industry. Journal of Construction Engineering and Management, 121(2), 189–196.

Kahraman, C., Onar, S. C., Cebi, S., & Oztaysi, B. (2017). Extension of information axiom from ordinary to intuitionistic fuzzy sets: an application to search algorithm selection. Computers & Industrial Engineering, 105, 348–361.

Kanoulas, E., Du, Y., Xia, T., & Zhang, D. (2006, April). Finding fastest paths on a road network with speed patterns. In 22nd International Conference on Data Engineering (ICDE'06), Atlanta, GA, USA (pp. 10–10). IEEE.

Lau, C. (1991). Neural Networks: Theoretical Foundations and Analysis. IEEE Press.

Lippmann, R. (1987). An introduction to computing with neural nets. IEEE ASSP Magazine, 4(2), 4–22.

Mandar, M., Boulmakoul, A., & Lbath, A. (2017). Pedestrian fuzzy risk exposure indicator. Transportation Research Procedia, 22, 124–133.

Mandic, D., & Chambers, J. (2001). Recurrent Neural Networks for Prediction: Learning Algorithms, Architectures and Stability. Chichester: Wiley.

Payeur, P., Le-Huy, H., & Gosselin, C. M. (1995). Trajectory prediction for moving objects using artificial neural networks. IEEE Transactions on Industrial Electronics, 42(2), 147–158.

Yanfeng, W., Shunying, Z., Hong, W., Bing, L., & Mei, L. (2010). Characteristic analysis of pedestrian violation crossing behavior based on logistics model. In 2010 International Conference on Intelligent Computation Technology and Automation, Changsha, China (vol. 1, pp. 926–928). IEEE.

Zadeh, L. A. (1968). Fuzzy sets as a basis for a theory of possibility. Fuzzy Sets and Systems, 1(1), 3–28.

Zadeh, L. A. (1999). Fuzzy sets as a basis for a theory of possibility. Fuzzy Sets and Systems, 100, 9–34.

15 A Real-Time Reactive Service-Oriented Architecture for Safe Urban Walkability

Kaoutar Bella and Azedine Boulmakoul

CONTENTS

15.1 INTRODUCTION

During the early days of technology, computers were designed to calculate artillery tables, tides, crack codes, and other precise, difficult, yet rote mathematical applications. These custom-built devices later become programmable machines. With time-sharing operating systems, they could execute many applications on a single computer, but they remained independent. Client-server architectures came about as more and more computers were connected to each other (Wu 2020). This allowed a computer in one room or building to use the power of a mainframe in

DOI: 10.1201/9781003255635-15

another room or building. The growth of the internet and large-scale data centers with thousands of low-cost commodity computers networked together fueled the widespread development of distributed systems in the early 2000s. Unlike client-server architectures, distributed system applications consist of numerous apps operating on various computers, or many replicas running on different workstations, all interacting to build a system like a web search platform or a retail sales platform. Distributed systems are naturally more dependable when organized appropriately. They may also result in more scalable organizational structures for the software developers who created them. Sadly, these advantages come at a cost. Distributed systems are more challenging to design, implement, and troubleshoot. Making a dependable distributed system requires more technical ability than making single-machine applications like mobile or web frontends (Banzai and Kanbayachi 2010). Regardless, trustworthy distributed systems are in high demand. As a consequence, the tools, techniques, and processes used to generate them are in high demand because even a few seconds of system or service downtime may cost millions of dollars, if not more.

15.1.1 REACTIVE PRINCIPLES

15.1.1.1 Reactive Manifesto

Fortunately, technological advancements have made it easier to design distributed systems. A document called the reactive manifesto was created in response to companies attempting to adapt to these changes in the software landscape. Multiple groups of engineers developed similar patterns for solving similar problems independently. The reactive manifesto's authors, Jonas Boner, Dave Farley, Roland Kuhn, and Martin Thompson (Boner and Thompson. 2014), got together and looked at what was going on in the industry before coming up with the document. The aspects of a reactive system were previously recognized individually by the various groups that were all working towards the manifesto's goals. And the manifesto attempts to condense all of these common ideas into a single unified set of principles that anyone can look at and follow. A reactive system operates on four fundamental principles. First and foremost, the system must be responsive. According to the reactive manifesto, a system must always respond promptly.

All of the arrows in Figure 15.1 point to responsiveness. Following the arrows around the circle, everything eventually returns to responsiveness. However, for a system to be responsive, it must also be resilient. This means that a reactive system must remain responsive even when there is a failure. That is significant because if we have a system that is responsive except when a failure occurs, it is not a responsive system, especially if failures happen frequently. When a failure occurs, it must be resilient. It must be able to respond to that failure while continuing to operate to the best of its abilities.

A reactive system will remain responsive even if the load changes. It's called elasticity. The reactive manifesto originally used the word scalability, but the concept was modified to elasticity since scalable typically connotes growth. And this is critical: we must be able to scale up to meet demand, but also down. But we don't want to be in a situation where we expand up to our limit and never scale back down

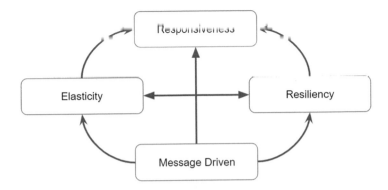

FIGURE 15.1   Reactive manifesto key concepts.

because that's costly. Elasticity means we can scale up as required but also down to save resources. The foundation is asynchronous nonblocking messaging. Earlier versions of the manifesto called it event-driven. However, the word "event" has certain negative connotations. Not that a message-driven or reactive system requires simple events. There is nothing wrong with having events; however, there are other forms of communication that are not the events of a traditional event-driven system. Those messages are also completely acceptable. Eventually, they switched from an event-driven system to a message-driven system. The prominent factor of a message-driven system is that the messages are asynchronous and nonblocking as the primary way to communicate between the components forming the system.

15.1.1.2   Performance vs Elasticity

If we dive deeply into elasticity, it is about the number of requests the system can handle at a time; however, how much it takes to respond is not the issue here. How much the system or the request takes to respond is called performance.

Figure 15.2 plots response time against the number of requests or loads. Response time measures the time it takes to respond to a single request. While the number of

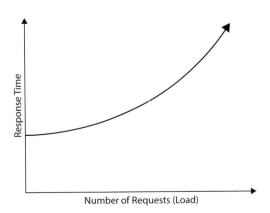

FIGURE 15.2   Performance graph.

requests or loads indicates how many queries we can handle at the same time, the performance aims to improve response time. It aims to improve the system's capacity to process single responses quickly. Elasticity, on the other hand, aims to increase the system's load-management capabilities. A metric like requests per second measures both. It's essential because we can use it to check whether we've increased performance or scalability, but it's also limiting since we can see if we've improved either, but not which one. Our requests per second will be one if our system processes one request per second and can only handle one at a time. One request per second. If we lower our response time from one second to half a second, we can now handle two requests instead of one. We will witness an increase in requests per second. If we take the same system, one request per second, and enable ourselves to process two requests simultaneously, we double our requests per second. We went from one request per second to two requests per second. So if we alter our code and notice that our requests per second move from one to two, we don't know whether we've improved performance or scalability. It's a useful measure since it demonstrates progress, but it doesn't tell us where it originated from. On the graph in Figure 15.3, if we want to improve performance, we improve our response time. That pushes the graph down, but it has a limit and we can only push it so far.

The graph shown demonstrates zero reaction time. We will never achieve zero due to physics and other factors. And going farther will cost more. Sooner or later, striving for improved performance will be obsolete. It may be physically or financially—it depends on what we're trying to change, but the number of requests we handle per second may not vary despite the faster response time. The axis has not changed. A move ahead. Elasticity boosts our load-carrying capacity, pushing the graph down the x-axis. It moves the graph along with the number of requests. Again, the performance of each request is potentially unlimited. On the x-axis, we could theoretically keep going. In reality, this is rarely the case. We are generally limited by some constraints. It might be too costly to keep pushing it, or something else. The program may have a design constraint that prohibits us from going any farther, but in principle, we could go on indefinitely, which isn't often the case when attempting to

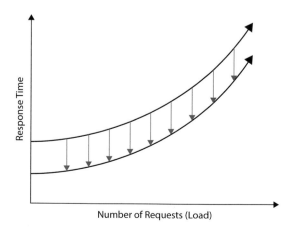

FIGURE 15.3 Performance improvement graph.

enhance response time. Because we know that scalability has no theoretical limit, we prefer to concentrate on enhancing elasticity while building reactive microservices. That doesn't imply we don't examine performance, but we tend to concentrate on it less.

15.2 SAFE URBAN WALKABILITY SYSTEM

15.2.1 PREVIOUS WORK

Our system is primarily concerned with pedestrian safety and navigation in the context of urban walkability (David and Flach 2010). According to the World Health Organization (WHO), 1.35 million people die annually as a result of traffic accidents and more are injured. Road accidents were and still are one of the concerns of all urban cities. To improve pedestrians' walkability, it is important to identify a suitable route for walking (Boulmakoul and Lbathet. 2021). Many pieces of research were conducted to improve pedestrian safety (Valencia 2019). The authors (Kreps and Raoet. 2010) proposed a composite indicator of pedestrian exposure that considers pedestrian attributes, road and traffic conditions, and pedestrian adherence to traffic rules. The authors (Escoffier and Finnigan 2022) introduced an Android mobile application, WalkSafe, that aids people who talk while walking, enhancing pedestrian smartphone users' safety (Boulmakoul and Mandar 2014). It identifies vehicles approaching the user using the mobile phone's back camera, alerting the user to a risky situation. This application is primarily intended for pedestrians who walk and chat. Because it involves the use of the camera, it may be battery draining. The system is primarily concerned with recognizing automobiles in the proximity of the phone holder, which requires a trained dataset; at the moment, the algorithms are not yet trained on motorbikes trucks, or buses. In the United States of America, 47% of accidents (National Highway Traffic Safety Administration 2010) occur between the hours of 6 and 12 p.m., which is why the system must be effectively trained on automobile identification at night, which will also need at least medium-quality camera and eventually increase the phone's battery consumption.

15.2.2 SYSTEM DESIGN

Our system includes a web application for safe routing, in which users may provide a starting and a finishing point. The application will then display the safest route on the map. Additionally, the user may utilize the mobile application to navigate securely, since the program identifies cohesions in real-time (Baljak and Salawayct. 2018). The application is on minimal battery usage.

We used Docker as a container and Kubernetes as an orchestrator to boost our system's efficiency and responsiveness. Our deployment technique will be explored in Section 15.4 along with alternative approaches. We used Valhala for advanced routing and Kafka for stream messaging (Bella and Boulmakoul 2020). Managing enormous amounts of data becomes difficult. We have a lot of data to interact with this study, and it's all interconnected. Previous research suggests Cypher might be a standard graph query language. This supports our choice of the graph database.

A graph database stores data as nodes connected by edges (association). It uses Cypher as a query language for the graph database. The Cypher syntax matches node patterns and relationships in a graph visually and logically. We can also use the sink connection with Kafka to move data from Kafka topics to Neo4j using Cypher templates (Bella and Boulmakoul 2021). Rather than the system's functionality, this chapter focuses on the patterns and models used to create it. We will also learn and debate essential topics along the way. Performance criteria requirements must be made early in the system design process. Other performance difficulties originate from poor design choices. Changing an architecture before building a system is cheaper than developing a system from scratch. Our system must be scalable and handle enormous numbers of requests per second and also add features without slowing down the system. This will assure system uptime, facilitate system maintenance, and allow future application growth.

15.2.3 DIVING INTO MICROSERVICES

Microservices break software up into distinct components that each does one thing. Due to the smaller scope, a single team can build and manage each service. A smaller team also reduces the overhead associated with keeping a team motivated and on track. Explicit application programming interfaces (APIs) across microservices decouple teams and provide a reliable contract between services. The team producing the API understands the surface area it must maintain stability, and the team consuming the API can rely on a reliable service without worrying about its intricacies. This decoupling enables teams to independently manage their code and release dates, allowing them to innovate and enhance their code. Scalability is improved by decoupling microservices. Because each component is its service, it may be grown independently. In a larger application, it is rare for all services to grow in the same way. Some systems can be scaled horizontally while others need sharding or other ways to scale. Because one service is distinct from the others, it may grow independently. This is impossible if all services are monolithic. This separation will help our system to be more elastic by allowing each component to be scaled up and down independently depending on demand. Isolation also allows faults and malfunctions to not spread widely throughout the entire application, thereby increasing our system's resiliency, which will ultimately improve our system's responsiveness.

However, the use of microservices in system architecture has several drawbacks. Debugging the system when faults occur is substantially more difficult due to the system being loosely connected, which is one of the biggest downsides. We can no longer just load a single application into a debugger and debug it. Many systems, typically on multiple computers, are the source of any faults. In a debugger, it is difficult to replicate this scenario. Similarly, microservices-based systems are challenging to design and construct. In a microservices-based system, there are numerous ways of communication between services, as well as multiple patterns of coordination and control among the services that as, synchronous, asynchronous, message-passing, etc.

Distributed patterns are a response to these challenges. An architecture that is built out of patterns is simpler to build since many design principles are already defined by the patterns, making it easier to implement (U.S. Department of Transportation 2017)

(Wang and Comphellet 2012). Another benefit is that using patterns makes it simpler to troubleshoot problems in many systems that share common patterns, like the birth of reactive manifesto as explained.

15.3 SINGLE-NODE DISTRIBUTED PATTERNS

15.3.1 Deployment and Management of a Distributed System

The next section elaborates on single-node distributed patterns for long-running serving systems like our web applications. Message-driven architectures and replicating and scaling patterns are examined.

How can we best deploy and maintain our reactive system on the cloud? Containers and Kubernetes solve this issue. Containerization is used to bundle microservices and their dependencies into lightweight packages that can operate anywhere. Kubernetes has evolved as the de facto open-source standard for managing containerized applications in production and development. Kubernetes is a cluster orchestration solution that brings "reactive systems" to container management. A Kubernetes cluster is a group of Kubernetes nodes (virtual machines [VMs] or physical computers). A pod is a Kubernetes deployment unit. It contains one or more app containers, including network and storage resources. Controllers are Kubernetes services that monitor the cluster and resolve errors when the intended state is not met. Kubernetes' cluster management capability provides both robustness and elasticity at the infrastructure layer. Elasticity is provided by Kubernetes auto-scaling, which may alter the number of nodes in our cluster as well as scale the number of pods operating in our cluster depending on workload. Automated pod recovery capabilities improve resilience by restarting failing pods or rebuilding deleted pods. A reactive architecturally designed application will extend, shrink, and redistribute itself in response to changes in the underlying infrastructure. Running our reactive microservices application on a Kubernetes-managed infrastructure may give a high degree of robustness and elasticity.

15.3.2 The System Firewall

The majority of us do not build websites with the expectation of being targeted by a denial-of-service attack. A denial-of-service attack can be caused by a developer incorrectly configuring a client, a site-reliability engineer accidentally running a load test against a production installation, or it can be caused by the actions of a malicious cyber threat actor targeting our site. Users will be unable to access the system, which will harm our responsiveness. Our system is protected by a firewall supplied by Nginx; we prevent all cross-site scripting (XSS), XML, and SQL injections, as well as known malicious activity, and we offer rate limitations. The firewall secures our system from the vast majority of external threats that arrive from the internet.

15.4 REPLICATED LOAD-BALANCED SERVICES

A replicated load-balanced service is the most basic distributed pattern, and it is also the one that most people are acquainted with. In such a service, every server is similar to every other server, and every server is capable of serving the amount of traffic

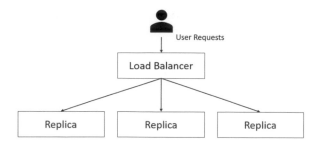

FIGURE 15.4   Load balancer.

that is required of it. Essentially, the pattern is made up of a scalable large number of servers with a load balancer in between them. The load balancer is normally either totally round-robin or employs some type of session stickiness to ensure that no one is left behind. Our stateless servers, client application, and the global positioning system (GPS) server interface are all served by a load balancer, which is the first and most important step. Stateless services are those that do not rely on the previously stored state to function properly. If a stateless application is simple enough, even individual requests may be routed to several instances of the service to fulfill them, as presented in Figure 15.4.

15.4.1   Scalability and Redundancy of Stateless Systems

Regardless of size, a "highly available" service level agreement (SLA) needs two copies. Consider this: Our target is a 39% performance (99.9% availability) and daily downtime of a three-nine service (24 60 0.001). To meet our SLA, we must be able to update software in less than 1.4 minutes. Using daily software rollouts, to meet our 99.9% uptime SLA, the team must be able to deploy software in 3.6 seconds or less. We lose 0.01% of our data if the procedure takes more than 3.6 seconds. Sure, we could build two copies of our service and place a load balancer between them. The second copy of the service will serve our users while we deploy or if our software fails. Services are copied to accommodate additional users as they grow in size. Figure 15.5 illustrates how horizontally scalable systems can support an increasing number of users by adding additional replicas. This is possible via the use of the load-balanced replicated serving pattern.

15.4.2   Readiness Probes for Load Balancing

Naturally, duplicating our service and adding a load balancer is just a portion of a comprehensive architecture for stateless replicated serving. When developing a

FIGURE 15.5   Horizontal scaling of a replicated stateless application.

replicated service, it is critical to develop and deploy a readiness probe that notifies the load balancer. We examined how a container orchestration system may utilize health probes to identify whether an application requires restarting. In comparison, a readiness probe identifies when an application is prepared to handle user requests. The distinction is necessary because many programs need time to start before they are ready to serve. They may be required to connect to databases, load plugins, or obtain network-based serving files. In each of these instances, the containers are alive but unprepared. To address this issue, we've developed a uniform resource locator (URL) that implements the readiness check of our GPS server interface.

15.4.3   CACHING LAYER

Occasionally, despite being stateless, the code in our stateless service is still costly. It may do database queries or perform extensive rendering or data mixing to satisfy the request. A caching layer makes a lot of sense in such a world. Between our stateless program and the end-user request, there is a cache. A caching web proxy is the simplest kind of caching for online applications. The caching proxy is nothing more than an HTTP server that caches user requests in memory. If two users make the identical web page request, only one will be sent to our backend; the other will be handled in memory by the cache, as shown in Figure 15.6.

Now, how do we deploy our cache? The simplest way to deploy the web cache is alongside each instance of our web server using the sidecar pattern, as shown in Figure 15.7.

Although this solution is straightforward, it has certain drawbacks, including the fact that we will have to grow our cache at the same rate as our web servers. This is often not the method we choose. We want as few replicas as possible with plenty of resources for each copy (for example, instead of 10 replicas with 1 GB of RAM each, we would want two replicas with 5 GB of RAM each). Consider how each page will

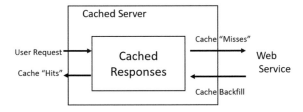

FIGURE 15.6   Cache server workflow.

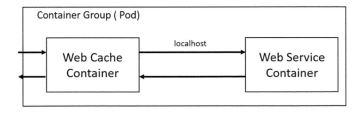

FIGURE 15.7   Deploying the cache layer as a sidecar.

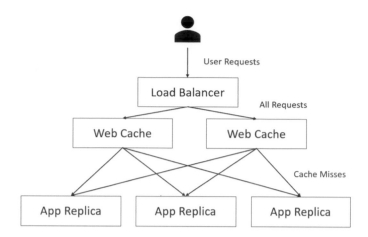

FIGURE 15.8   Final replicated service with the caching layer.

be kept in each replica to see why this is better. With 10 replicas, we will store each page 10 times, lowering the total number of pages that may be kept in memory in the cache. This reduces the hit rate or the proportion of the time that a request may be serviced out of the cache, lowering the cache's utility. While we may desire a few large caches, we may also want a huge number of tiny duplicates of our web servers. Numerous languages can only use a single core; thus we want many replicas to be able to use many cores, even if they are on the same system. As a result, configuring our cache layer as a second stateless replicated serving tier atop our web-serving tier makes the most sense, as illustrated in Figure 15.8.

15.4.4   Message-Driven Services

Message-oriented microservices in a reactive system collaborate to achieve responsiveness, resilience, and elasticity, which are mostly accomplished through being message-driven. We will now examine two degrees of "message driven-ness": intraservice, or how service components interact, and interservice, or messaging between services. Creating many instances is one method of attaining scalability and consistency within a particular service (Boulmakoul and Lbathet. 2021). We should be able to grow our microservice in a manner that is transparent to the rest of the system. For example, the microservice would manage load balancing and instance routing internally, and more instances would be spun up or down as required in response to load and other external variables. None of this is easy, but it is made achievable via messaging between the microservice collective's components.

There are open-source libraries available that can assist us. Akka, for example, offers libraries that enable us to manage the components of our microservice, in this instance, actor systems, as a "system of systems". Vert.x is another open-source framework that employs an event bus to connect our microservices (verticals) and supports both point-to-point and publish/subscribe (pub/sub) communications. Both Akka and Vert.x allow the best effort or, at most, once message delivery by default. Apache Kafka is a well-known framework for implementing long-term pub/sub communications. We've used Apache Kafka to handle the messaging layer. Kafka has provided a solution for streaming data because of its versatility and data type independence. It is a powerful messaging platform that can handle massive volumes of data without losing any of it.

Kafka is set up with a single node of two brokers (replication-factor = 2), allowing it to be designed to be scalable. When one broker fails to supply the topic's data, another must step in. More producers may be supported by this design. However, since Kafka is distributed, a cluster often comprises numerous nodes with various brokers (Bella and Boulmakoul 2021). We adopted a two-broker design to disclose the first step in distributing the system. While Zookeeper manages the load across the nodes, using Kafka is another kind of abstraction that provides us developers with a standard for controlling data flow across application resources, allowing us to focus on the core functionality. As recommended, we use two brokers, one for each subject with specialized data and one producer for each topic. Once the data (e.g., pedestrian locations) is uploaded to the cluster, our consumers have generated it automatically. The data in question are processed by the consumer.

This implementation gives us a graceful way to handle unbounded streams of data across asynchronous boundaries with backpressure. This makes it perfect to integrate our reactive microservices with external systems and to implement messaging between services within our reactive system

15.4.5 Netty Network Pipeline

Many frameworks that provide reactive application functionality have emerged in recent years. Their objective is to make it easier to develop reactive apps. They accomplish this by providing higher-level primitives and APIs for dealing with events and abstracting nonblocking input/output (I/O). Employing nonblocking I/O is not as straightforward as it seems. Combining this with a reactor pattern (or a version) is complicated. Fortunately, libraries and toolkits are working alongside frameworks to accomplish the hard lifting. Netty is an asynchronous event-driven network application framework that uses nonblocking I/O to enable the development of highly concurrent applications. It is the most widely used nonblocking I/O library in the Java world.

Netty, on the other hand, may be challenging. Our GPS server, Traccar, which is the source of our streaming data, is the essential component of our system. And the Netty network framework serves as the foundation for this open-source GPS server. Traccar generates a pipeline of event handlers for each network channel or connection. Incoming GPS messages are received as binary buffers, divided into frames, decoded into an internal position model, and subsequently saved in the database.

15.5 SUMMARY

The first section explored the reactive manifesto and its concepts. Section 15.2 examined performance and scalability. Reactive systems change the way we create distributed systems. They are required for today's computer programs' reliability, agility, and scalability. The third section summarized our system's features and dispersion. The following section examined single node distributed patterns. Using containers and container orchestration as a basis, we may create patterns and components. These patterns and components help us develop more dependable and efficient systems. The last section proposed a stateless replication concept. Then it elaborated on how to add two additional replicated load-balanced layers for performance caching. The layers are connected with three deployment and service load balancers in Kubernetes. The Netty framework is the final ingredient that encourages reactive applications to use nonblocking I/O and prohibits the establishment of too many OS threads.

15.6 CONCLUSION

Nowadays, road accidents are one of the most serious problems confronting the transportation sector (Bella and Boulmakoul 2021). Pedestrian safety is a critical problem that must be addressed to transition to a smart, safe city ecosystem. Our study's goal was to showcase our reactive service-oriented architecture for safe pedestrian navigation. It creates a secure path between two locations specified by the user in the web application. We used Docker and Kubernetes to optimize our system's resource utilization and scalability along with patterns. We utilized the Traccar GPS server to collect real-time location data, Kafka to manage stream messaging, and Valhala for routing. Our data is stored in a graph database Neo4j. After the stress test, we determined that the system is sufficiently scalable and robust, but some enhancements can be addressed. From a performance standpoint, we want to focus on low-level performance, such as functional performance. We aim to eliminate wasteful interactions and enhance database searches. And we discovered that blurred regions and collected coordinates are not precise or are not captured at all, posing a risk to the user's safety. Therefore, we attempt to stabilize the GPS server to decrease location frequency and faulty locations. In this chapter, we propose a microservice-based architecture to calculate the safest path for pedestrians' walkability and cohesion prediction in real-time (Maguerra et al. 2020). Furthermore, several critical technical concerns were addressed to create strong and scalable software components.

Building distributed microservices-based systems was previously inconceivable (Marz and Warren 2015). Today's businesses demand solutions that can adapt to the company while also accommodating users' whims. We can now elastically scale systems to handle big datasets and large numbers of users. It is now possible to safeguard systems so well that downtime is measured in seconds rather than hours. But as technology evolves, it is difficult to predict. We can only watch for trends and prepare for change. Here are some trends to keep an eye on. HTTP/3 improves flow control and parallel request delivery, boosting system communication. Message brokers are becoming more popular. Pulsar, NATS, and KubeMQ are emerging brokers. NATS and KubeMQ are both designed for Kubernetes. Several groundbreaking initiatives are revolutionizing message and event processing. Querying data from event streams like Apache Kafka is one. Machine learning and artificial intelligence (AI) are influencing reactive system design. Machine learning can manage system faults or demand spikes. They are collecting system data and adapting them to the current workload.

REFERENCES

Baljok, V., Ljubovin, A., Michol, I., Montgomery, M., Calanay, R. (2010). Volumes 9–10.

Banzai, T., Koizumi, H. and Kanbayashi, R. (2010). D-Cloud: Design of a software testing environment for reliable distributed systems using cloud computing technology, 10th IEEE/ACM International Conference on Cluster, Cloud and Grid Computing.

Bella, K., Boulmakoul, A. (2020). Real-Time Messaging System in Kafka-Neo4j Pipeline Architecture. INTIS.

Bella, K., Boulmakoul, A. (2021). Containerised real-time architecture for safe urban navigation, Dec.

Bella, K., Boulmakoul, A. (2021). Real-Time Distributed Pipeline Architecture for Pedestrians' Trajectories Networking, Intelligent Systems and Security, 2022, Volume 237. ISBN: 978-981-16-3636-3.

Boner, J., Farley, D., Kuhn, R., Thompson, M. (2014). The Reactive Manifesto.

Boulmakoul, A., Bella, K., Lbath, A. (2022). Real-Time Distributed System for Pedestrians' Fuzzy Safe Navigation in Urban Environment INFUS. In: Kahraman, C., Cebi, S., Cevik Onar, S., Oztaysi, B., Tolga, A.C., Sari, I.U. (eds) Intelligent and Fuzzy Techniques for Emerging Conditions and Digital Transformation. INFUS 2021. Lecture Notes in Networks and Systems. Volume 308. Springer, Cham. https://doi.org/10.1007/978-3-030-85577-2_77

Boulmakoul, A., Mandar, M. (2014). Virtual pedestrian risk modeling. International Journal of Civil Engineering and Technology 5(10), 32–42.

David, K., Flach, A. (2010). Car-2-x and pedestrian safety. Vehicular Technology Magazine 5(1), 70–76.

Escoffier, C., Finnigan, K. (2022). Reactive Systems in Java Resilient, Event-Driven Architecture with Quarkus. ISBN: 1492091723.

Kreps, J., Narkhede, N., J. Rao. (2010). Kafka: A Distributed Messaging System for Log Processing. In Proceedings of the NetDB (June), 11, 1–7, 2011.

Maguerra, S., Boulmakoul, A., Karim, L. al. (2020). Towards a reactive system for managing big trajectory data. Journal of Ambient Intelligence and Humanized Computing 11, 3895–3906. https://doi.org/10.1007/s12652-019-01625-3.

Marz, N. and Warren, J. (2015). Big Data: Principles and Best Practices of a Scalable Real-Time Data System. Manning Publications. ISBN: 9781617290343.

McKee, H. (2017). Designing Reactive Systems. Sebastopol, CA, O'Reilly Media.

National Highway Traffic Safety Administration. (2010). Traffic Safety Facts 2009. Washington, D.C.

U.S. Department of Transportation. (2017). Synthesis of Methods for Estimating Pedestrian and Bicyclist Exposure to Risk at Areawide Levels and on Specific Transportation Facilities. Publication No. FHWA-SA-17-041.

Valencia, J. (2019). Inattention is the leading cause of deadly pedestrian accidents in El Paso.

Wang, T., Cardone, G., Corradi, A., Torresani, L., Campbell, A.T. (2012). WalkSafe: A pedestrian safety app for mobile phone users who walk and talk while crossing roads. In: Proceedings of the Twelfth Workshop on Mobile Computing Systems & Applications, 1–6.

Wu, J. (2020). Distributed System Design, Computer Science (General). CRC Press. https://doi.org/10.1201/9781315141411.

16 Safest Trajectories for Pedestrians Using Distributed Architecture Based on Spatial Risk Analysis and Voronoï Spatial Accessibility

Aziz Mabrouk and Azedine Boulmakoul

CONTENTS

DOI: 10.1201/9781003255635-16

16.1 INTRODUCTION

For a long time, several research studies have focused on the subject of the safety and security of pedestrian individuals while traveling in urban areas. Indeed, several authors have proposed algorithmic and technical solutions allowing individuals to take safe paths away from dangerous areas. This latter, which represent threats to their security, can be criminal areas, contaminated area (virus, pollution, etc.), or even installations containing dangerous materials which can have thermal, mechanical, toxic, or even radioactive effects.

In addition, intelligent management of pedestrian safety, and particularly the recommendation of safe paths, requires the implementation of an exhaustive set of functions, from data collection to the recommendation of safe paths for pedestrians and also the implementation of a dashboard for the authority (security, emergency services, etc.) and experts concerned with the safety of individual pedestrians. This diversity of actors involved in the geospatial ecosystem requires the implementation of internal technological mechanisms ensuring the coordinated sharing and distributed processing capacity of a large mass of geospatial data. In addition, these actors must have secure access—in some cases instant access—to the right information at the right time and in the right place. That said, the ecosystem must have enough resources and use the power of advanced technologies. In this chapter, we propose a distributed architecture based on a new approach to spatial risk analysis. In fact, this approach, based on microservices architecture, allows calculating the safest pedestrian paths and this, by using the Voronoï spatial diagrams and also based on the evaluation of the zones at risk. This architecture is designed to provide individual pedestrians with a mobile platform allowing them to take the safest paths. On the one hand, this platform is a real-time alert tool for the dangerous areas near pedestrians. On the other hand, it makes it possible to monitor minor pedestrians or incapable adults by their guardians.

16.2 RELATED WORK

Among several studies, we find the research work of (Lee 2014), which focuses on the identification of impacted areas and on the resident populations who can potentially also be affected by hazardous materials released following an accident. By exploiting the properties of hyperbolas, the authors (Aljubayrin et al. 2015) aim to find paths that minimize the distance traveled outside the security zones by proposing a transformation of the continuous data space with zone security in a graph. The research work of (Mandar and Boulmakoul 2018) and (Karim et al. 2020) investigate factors related to pedestrian and driver behavior traits that contribute to pedestrian crashes. These authors develop a set of approaches to measure the pedestrians' exposure to accident risk based on fuzzy modeling and also on intuitionist fuzzy logic. However, modeling of urban spaces used by pedestrians, which is based on Voronoï spatial diagrams, is used to provide a set of spatial relationships (spatial association, adjacencies, etc.) between all space objects (current position pedestrian, paths, danger zones, etc.). In this context, several works have been carried out to find short and safe paths. (Erwig 2000) proposes, to avoid pollution in the event of an accident, to keep a minimum distance from a set of obstacle sites in the Euclidean plane. (Mabrouk and Boulmakoul 2017b) propose an approach by adapting the approach of

(Erwig 2000) to a spatial environment where people move on a real spatial network. (Mabrouk et al. 2017a) by using the spatial properties of the Voronoï spatial diagrams division points, they propose a computational process for the search for short and safe routes by traversing as much as possible the most distant nodes compared to the vulnerable sites. In addition, there are research studies based on a microservices architecture that address the pedestrian movement issues in an urban environment. (Melis et al. 2016) present a microservice architecture use case that aims to create a scalable and integrated mobility service for people with disabilities. Based on an intuitionistic modeling of the fuzzy pedestrian risk, (Cherradi et al. 2022) propose a reactive microservice-oriented ecosystem that aims to provide the safest route in real time. (Bella and Boulmakoul 2021) implement a containerized real-time system offering pedestrians safe urban displacement.

16.3 RISK AREAS AND SPATIAL MODELING OF PEDESTRIAN MOVEMENTS IN A RISKY URBAN SPACE

Spatial risk analysis, for pedestrian individuals, is mainly based on the use of an undirected graph. This spatially models the paths potentially taken by pedestrians who move freely and spontaneously. This assumes, in addition to the urban road network, the availability of linear spatial objects which are topologically connected and which represent the paths of pedestrians within neighborhoods. In previous work, we proposed the generation of linear geometric objects describing pedestrian paths from geometric objects representing obstacles (construction, fence walls, lakes, etc.). The approach we have followed is the computation of the Voronoï diagram of the obstacles. In addition, the risk analysis, based on the physical and spatial proximity to dangerous areas and also on the concept of Voronoï spatial accessibility (Mabrouk and Boulmakoul 2012), makes it possible to have, at each point of the spatial network of pedestrians, a set of useful information on the position of the pedestrian and their level of risk and also on the nearest danger zone, namely the geographical position and the distance to reach this zone by taking the shortest route. In fact, individual pedestrians move on the paths in a free and spontaneous way without being limited exactly to the marked routes, which are described in (Mabrouk and Boulmakoul 2020). In addition, the Voronoï planar diagram (VPD) generated by the spatial accessibility points is a solution to associate the risk levels and all the information on the nearest danger zone (the geographical position and the distance of the shortest path) to the space surrounding the footpaths. The Voronoï Planar Diagram, which is generated and associated with the risk levels, comprises a set of polygons in which each group, with a determined risk level, represents a spatial risk zone. The latter includes information on the level of risk and all the information on the nearest danger zone, namely its geographical position and the distance of the shortest path.

16.4 CHARACTERISTICS OF THE PROPOSED ECOSYSTEM

The proposed ecosystem is based on a microservices architecture. That said, it consists of a set of independent processes, which run as a service. These processes communicate through a well-defined interface and using lightweight application programming interfaces (APIs). These processes are developed for business capabilities, and each process performs a single function, and it can be updated, deployed, or scaled to meet the demands of the specific functions determined. The performance of a distributed system is a key factor in its success. Indeed, the recommendation of safe trajectories in real time, the early warning of the approach of dangerous areas, and the instantaneous and continuous monitoring of miners require good performance. In fact, the proposed geospatial ecosystem is designed based on a microservices architecture that favors having a result of geospatial data processing and execution of functions adequate to the available technological resources deployed and to the fixed objectives. Moreover, in order to allow the processes to be carried out in real time on a large mass of geospatial data concerning the management of the safety of pedestrian individuals, these processes need greater scalability. Indeed, the distributed architecture, based on microservices, offers the proposed ecosystem high scalability since each microservice can be scaled independently of other services. Then, the ecosystem allows its processes to scale up autonomously. Consequently, this ecosystem is able to provide an increasing number of its microservices to simultaneously execute its functionalities and to manage an increasing number of users and ultimately be susceptible to expansion. In addition, ecosystem functionalities, to manage the pedestrians, are distributed among several services. If the execution of functionality is interrupted, this functionality is offered by other microservices. In addition, the ecosystem uses asynchronous communication, between microservices, to increase decoupling and manage communication by subscriptions, and therefore it continues to function and be available. In fact, requests to access available nodes are sent by a load balancing server. Moreover, the containerization of the microservices deployed to ensure the objectives of the proposed ecosystem makes it portable with the ability to be moved and executed consistently in all environments and on all infrastructures, regardless of their operating system.

16.5 DISTRIBUTED ECOSYSTEM ARCHITECTURE BASED ON MICROSERVICES AND SPATIAL RISK ANALYSIS

The distributed system proposed in this chapter is hosted in a cloud backend. It is based on a microservices architecture.

This system communicates, on the one hand, with the geographic information system (GIS) platforms of geospatial data providers (urban network and hazardous area) and those of the authority (security, emergency services, etc.); on the other hand, it communicates in real time with the applications hosted in the mobile devices (smartphone, smart watch, other custom mobile device) of pedestrians via a structure based on an edge-fog-cloud architecture, as shown in Figure 16.1.

16.5.1 Geospatial Data Management in the Cloud Backend

Using a relational database service (RDS) helps install, manage, and scale a relational database in the cloud. However, the data processed within the backend are geospatial data; hence, the need to use an extension of a geospatial database management system (DBMS). This provides extensive geographic functionality and allows

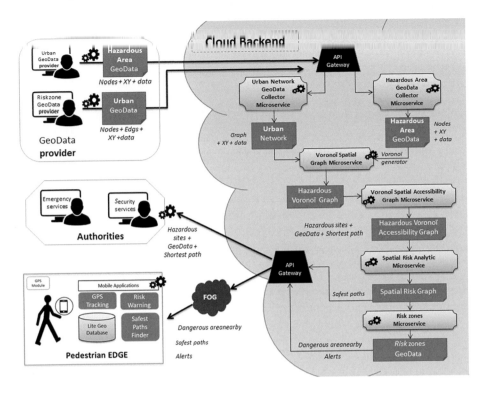

FIGURE 16.1   Global ecosystem architecture.

the system to support geographic objects and perform location queries in SQL. In addition, the system supports multiple storage models for different data types, formats, and uses. In fact, geospatial data are represented as points, lines, or polygons. These vector objects are associated with semantic alphanumeric data relating to urban geospatial data and geospatial data of hazardous sites.

In fact, the urban data providers provide geospatial data that concern urban spaces: urban networks (roads, pedestrian paths, etc.), infrastructure facilities (schools, hospitals, hotels, etc.), buildings, etc. In addition, the hazardous site data providers provide geospatial data. In fact, these data concern spaces that represent threats to the individuals' safety. They can be criminal areas, contaminated areas (virus, pollution, etc.), or facilities containing hazardous materials that can have thermal, mechanical, toxic, or radioactive effects.

16.5.2   Upload Geospatial Data to the Cloud Backend

Geospatial data providers, who may be public or private institutions, are often GIS experts and have similar experiences. These establishments may also be specialized in the acquisition and preprocessing of data. In order to collect large amounts of raw input data, data providers use methods such as photogrammetry and aerial or ground laser scanning. These input data are then cleaned, processed, and converted into

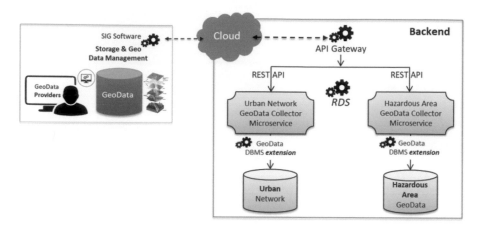

FIGURE 16.2   Uploading geospatial data via a REST API microservice.

standardized file formats to make it available for the storage process and geospatial data management. In our context, these providers are classified into two categories: urban geospatial data providers and providers of geospatial data on hazardous sites. Geospatial data providers, concerning the safety management, have user accounts to connect to the instance of the geospatial database created on the backend, using a geospatial DBMS. Additionally, using local GIS software, as shown in Figure 16.2, geospatial data providers have the ability to connect to the geospatial database instance created on the cloud backend.

These providers are now able to upload and update geospatial data in this remote geospatial database instance using the database management tools of the local GIS software. The geospatial data providers (urban network and hazardous area) have the necessary permissions to interact with other parts of the ecosystem in the cloud backend, in particular, the API gateway and the extension of a geospatial DBMS using RDS. Indeed, they can also successfully query and expose geospatial data through a RESTful API in microservices. Indeed, the "Urban Network GeoData Collector" microservice and "Hazardous Area GeoData Collector" microservice, respectively, take care of the collection of geospatial data from the urban network and hazardous area and then use the extension of the geospatial DBMS to store this data in their own "Urban Network" and "Hazardous Area" geospatial databases, as shown in Figure 16.2.

16.5.3   VORONOÏ SPATIAL GRAPH MICROSERVICE

With each creation or update carried out by the providers of geospatial data (urban network and hazardous area), the "Voronoï Spatial Graph Microservice" proceeds to create or update the "Hazardous Voronoï Graph", as shown in Figure 16.3. Indeed, the "Voronoï Spatial Graph Microservice" divides this spatial network into several Voronoï subnetworks, while associating the nodes and arcs of each of these subnetworks to the closest "Hazardous Area".

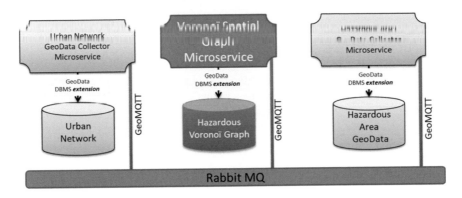

FIGURE 16.3   Voronoï spatial graph microservice.

This operation makes it possible to extract geospatial data describing the different shortest paths and the distances between each hazardous area and all the nodes and arcs of each subnetwork. Using the DBMS geospatial extension, the "Voronoï Spatial Graph" microservice stores or updates these data in its own geospatial database named "Hazardous Voronoï Graph".

16.5.4   VORONOÏ SPATIAL ACCESSIBILITY GRAPH MICROSERVICE

Based on geospatial data provided by the Voronoï Spatial Graph microservice, the Voronoï Spatial Accessibility Graph microservice calculates the Hazardous Voronoï Accessibility Graph in real time, as shown in Figure 16.4. In fact, the latter constitutes

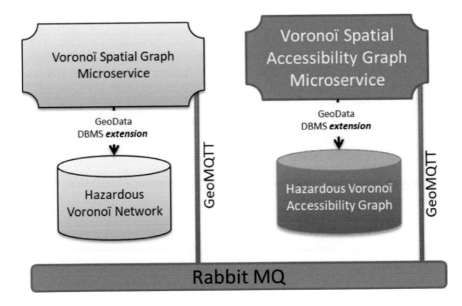

FIGURE 16.4   Voronoï spatial accessibility graph microservice.

a detailed Hazardous Voronoï Graph. In the Hazardous Voronoï Graph, the points of a given arc have the same information (nearest hazardous area and shortest path distance) associated with that arc. In order to minimize this inaccuracy, the "Voronoï Spatial Accessibility Graph" microservice receives as input the list of nodes, the list of arcs, the hazardous area, and the cutting distance, and then it proceeds to cut these arcs by what we have called " accessibility points" (Mabrouk and Boulmakoul 2012). These are points created on these arcs and which are separated by intervals of small equal distances, with each of these points containing information on the hazardous area associated with the cut arc and the distance of the shortest path between this point and this hazardous area.

16.5.5   SPATIAL RISK ANALYTICS MICROSERVICE

When the Voronoï Spatial Accessibility Graph is created or updated by the "Voronoï Spatial Accessibility Graph" microservice, the Spatial Risk Analytic microservice performs the spatial risk analysis and determines the degree of risk in each point of the spatial network, as shown in Figure 16.5. The Voronoï Spatial Accessibility Graph received by this microservice is a valued spatial graph that represents urban streets and pedestrian paths. It is a structure denoted $G = (N, A)$, composed of a set of nodes N and a set of links A to represent the segments of the spatial network, which will be associated with a risk value R. For each link $(i, j) \in A$, there are two attributes, the probability of damage pij (viral infection, criminal assault,

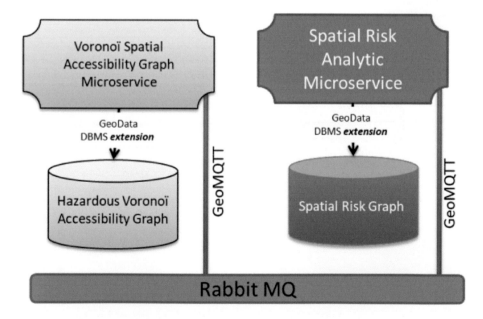

FIGURE 16.5   Spatial risk analytics microservice.

explosion of hazardous material, etc.) and the consequence of danger C_{ij}. Suppose a path l consists of an ordered set of links.

$$A^l = \left\{ (i,j)^k \,\middle|\, i,\ j \in N, k = 1,2,\ldots,m^l \right\} \qquad (16.1)$$

Indeed, in a path l which belongs to the set of paths available for an expedition s, the risk R^l is measured by Jin and Batta (1997) by the combination of the probability of occurrence of a hazard and the consequences that can result in pedestrian individuals, according to the following formula:

$$R^l = \sum_{(i,j) \in A}^{l} p_{ij} C_{ij} \qquad (16.2)$$

Based on the distance traveled by the pedestrian between two nodes i and n, and on the risks measured on the paths A^l and also on the direction of the pedestrian, the process analyzes the spatial risk associated with the path of the pedestrian. In contrast, when the individual pedestrian changes direction heading towards the secure zone, this says that the pedestrian is moving away from the dangerous zone. Indeed, the probability of danger decreases and consequently the risk decreases in parallel with the distance of the individual pedestrian from the dangerous zone. Following this analysis, the Spatial Risk Analytic microservice associates the risk values to each arc of the Voronoï Spatial Accessibility Graph and then it stores this new spatial graph in its geospatial database, "Spatial Risk Graph".

16.5.6 RISK AREA MICROSERVICES

The Spatial Risk Analytic Microservice bases risk analysis on physical and spatial proximity to danger zones and also on the concept of spatial points of accessibility (Mabrouk and Boulmakoul 2012). Indeed, it makes it possible to have, at the level of each point of the pedestrian spatial network, a set of useful information on the position of the pedestrian and their level of risk, and also on the nearest danger zone, namely the position and the minimum distance to reach this area. As shown in Figure 16.6, with each creation or update of these behind geospatial data, the "Risk Zones" microservice associates these data on the set of points surrounding each node of the "Spatial Risk Graph" and this, by calculating the Voronoï Planar Diagram of the set of these nodes. Indeed, for each node, a risk zone (polygon) is created and includes all the information generated by the Spatial Risk Analytic microservice.

16.6 PEDESTRIAN MOBILE SYSTEM

In an ecosystem based on an edge-fog-cloud architecture, pedestrians use mobile devices (smartphone, smart watch, other custom mobile device) to communicate with the Spatial Risk Analytic microservice and Risk Zones microservice hosted in the cloud backend and this, using the GeoMQTT API, as shown in Figure 16.7.

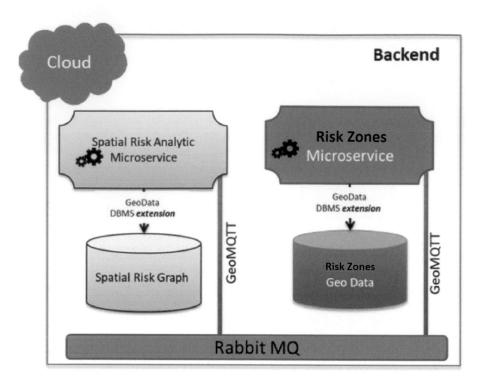

FIGURE 16.6   Risk area microservices.

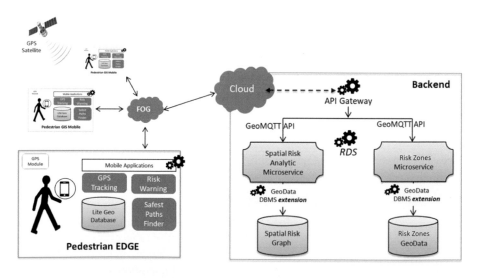

FIGURE 16.7   Edge-fog-cloud architecture to alert and recommend safe trajectories.

Each mobile device hosts three mobile applications allowing, in real time, these pedestrians to find the safest trajectories, to have very useful spatial information for the safety of movements, and finally, to allow the tutors to monitor minors or severely incompetent pedestrians.

16.6.1 SAFEST PATHS FINDER

This process, hosted in a mobile application, uses a GeoMQTT API to communicate with the "Risk Zones" microservice to retrieve geospatial data stored in its database and then calculates, in real time, the safest paths that are potentially taken by pedestrians (pedestrian paths and urban roads), as shown in Figure 16.8.

Indeed, this process allows the pedestrian moving in an urban environment to obtain a set of safe trajectories, since each arc of this graph is valued by a weight which represents the level of risk of the current position. Indeed, this process chooses the safest path that weighs a minimum value of the risk.

16.6.2 RISK WARNING AND GPS TRACKING

This process, hosted in a mobile application, uses the GeoMQTT API to communicate with the "Spatial Risk Analytic" microservice to send real-time geolocation data provided by the global positioning system (GPS) module. Then, it retrieves the geospatial data relating to the risk areas, which are stored by this microservice in its database.

The risk warning process then allows mobile pedestrian individuals to have, in real time, very useful spatial and semantic information for their safety in traveling,

FIGURE 16.8   Safest paths.

FIGURE 16.9   Risk warning.

namely dangerous zones, which are close to their current geographical position; the minimum time and distance to reach these dangerous areas; and also the distance and time to reach their destination, as shown in Figure 16.9.

In addition, by using geolocation data provided by the GPS module and associating them with geospatial data relating to risk areas, this process makes it possible to monitor underage pedestrians or a severely incapacitated person by their guardians. In fact, this process makes it possible to create geographical barriers to know if the pedestrians have left the zones defined by the tutors. Indeed, this process sends these guardians' alerts based on the location of the monitored pedestrians.

16.7   AUTHORITY MONITORING PLATFORM

The risk analysis must include the influence of the emergency services. In some cases, the real consequences of a criminal assault or an accident are mitigated by an effective intervention of the emergency services in a reduced time. Indeed, in order to take preventive decisions and provide citizens with security agents and vehicles at the right time and in the right place, the authority (security, emergency services, etc.) uses the monitoring platform to have useful information concerning the safety of people, namely their geographical positions, the number of people who are close to a risky zone, the secure zones, etc. The emergency response time is therefore a parameter to be taken into consideration.

16.8 SECURITY ASPECT OF THE PROPOSED ECOSYSTEM BASED ON MICROSERVICES

This granularity and the heterogeneity which characterize the proposed system considerably extend its surface of attack for malicious users between the multiplication of the services which communicate between them and a more complex architecture. This requires real security thinking based on flow control and correct encryption of connections and on effective event management solutions adapted to the needs of microservices. In fact, a profound aspect of security goes beyond the scope of this work. Nevertheless, we give an overview of the most important security-related points that should be taken into account. In this section, we only deal with the security aspect that concerns microservices, since our contribution in this chapter mainly concerns the microservices architecture. In fact, the distributed architecture requires a different approach in the implementation of security practices. The works (Hofmann, M. 2017), (Daya, S. 2015), (Lewis, J. 2014) and (Nishanil. 2018) suggest setting up authentication mechanisms. They also advise controlling and minimizing reach and access. Data must be protected during service-to-service communication. These last authors also recommend using the HTTPS protocol instead of HTTP to ensure data encryption between the two services and to promote additional protection against external traffic penetration. Also, it is recommended to secure data that are not currently in use by making nonpublic all end points where the data are stored. In addition, microservices are inseparable from containerization technologies. Bélair, Laniepce & Menaud (2019) and Sultan, Ahmad & Dimitriou (2019) report that container security is still in its infancy and faces unresolved challenges. They also recommend improving vulnerability management and digital forensics.

16.9 CONCLUSION

While traveling in urban locations, pedestrians need to move safely away from dangerous areas. This means providing them with intelligent tools that guide them to take the safest paths, and also allowing their guardians, in the event that these pedestrians are minors or incapable, to monitor their movement in complete safety.

Indeed, in this chapter, we have proposed an ecosystem ensuring intelligent management of pedestrian safety and, in particular, the recommendation of safe trajectories. Well-resourced and using the power of advanced technologies, this ecosystem encompasses a set of functionalities, ranging from data collection to the recommendation of safe paths for pedestrians, through the establishment of a dashboard for the authorities (security, emergency, etc.) and for experts concerned about the safety of pedestrians. In fact, in order to achieve the objectives set, the distributed system that we have proposed in this chapter communicates with the GIS platforms of the geospatial data providers and with those of the authority (security, emergency, etc.); it also communicates in real time with the applications hosted in the mobile devices of the pedestrians via a structure based on an edge-fog-cloud architecture. The distributed system which we propose is mainly based on a microservices architecture. The latter offers this system a high performance, scalability, and availability to favor a geospatial data processing result and to perform functions appropriate to the available technological resources deployed and the objectives set. Furthermore, our system is based on a novel approach to spatial risk analysis that takes into consideration spatial proximity to hazardous areas and assessment of Voronoï spatial accessibility. This provides information on the level of risk and all information relating to the nearest danger zones. Finally, although the deep aspect of security is beyond the scope of this work, we have discussed the most important points that are related to the security of the proposed system, since it is based on microservices and it is characterized by its granularity and heterogeneity.

ACKNOWLEDGMENTS

This work was partially funded by the Ministry of Equipment, Transport, Logistics and Water-Kingdom of Morocco, the National Road Safety Agency (NARSA), and National Center for Scientific and Technical Research (CNRST). Road Safety Research Program# An intelligent reactive abductive system and intuitionist fuzzy logical reasoning for dangerousness of driver-pedestrians interactions analysis.

REFERENCES

Aljubayrin, S., Qi, J., Christian S., Zhang, R., He, Z. and Wen, Z. 2015. The safest path via safe zones, IEEE 31st International Conference on Data Engineering, DOI: 10.1109/ICDE.2015.7113312 pp. 13–17. ISBN: 978-1-4799-7964, April 13–17 2015. IEEE: Seoul, South Korea

Anil, N., Parente, J. and Wenzel M., 2018. Microservices Architecture.. NET Microservices and Web Applications, published by Microsoft Developer Division,. NET and Visual Studio product teams, Retrieved March 12, 2022, from https://docs.microsoft.com/en-us/dotnet/architecture/microservices/architect-microservice-container-applications/microservices-architecture

Bélair, M., Laniepce, S. and Menaud, J. M. 2019. Tirer parti des mécanismes de sécurité du noyau pour améliorer la sécurité des conteneurs: une enquête, Actes de la 14e Conférence internationale sur la disponibilité, la fiabilité et la sécurité. Association for Computing Machinery, New York, NY United States pp. 1–6, August 26–29, 2019 Canterbury CA United Kingdom

Bella, K. and Boulmakoul, A. 2021. Containerised real-time architecture for safe urban navigation, International Conference on Decision Aid Sciences and Application (DASA), DOI: 10.1109/DASA53625.2021.9682354, ISBN:978-1-6654-1634-4, pp. 7–8 Dec. 2021 Sakheer, Bahrain.

Cherradi, G., Boulmakoul, A., Karim, L., Mandar, M. and Lbath, A. 2022. Towards a safe pedestrian walkability under intuitionistic fuzzy environment: A real-time reactive microservice-oriented ecosystem, Intelligent and Fuzzy Techniques for Emerging Conditions and Digital Transformation, Springer International Publishing, ISBN 978-3-030-85577-2, pp. 40–47.

Daya, S. 2015. Microservices from theory to practice: Creating applications in IBM Bluemix using the microservices approach. ISBN-13: 978-0738440811 Vervante, 2015

Erwig, M. 2000. The graph Voronoi diagram with applications, Networks,Edited By: Dr. B. L. Golden and Dr. D.R. Shier, https://doi.org/10.1002/1097-0037(200010)36:3%3C156::AID-NET2%3E3.0.CO;2-L, 156–163, September 2000.

Hofmann, M., Schnabel, E. and Stanley, K. 2017. Microservices Best Practices for Java. IBM Redbooks, ISBN 0738442275, March 2017

Jin, H., R. Batta. 1997. Objectives derived from viewing hazmat shipments as a sequence of independent Bernoulli trials. Transportation Science 31(3) 252–261.

Karim, L., Boulmakoul, A., Mandar, M. and Nahri, M. 2020. A new pedestrians intuition-istic fuzzy risk exposure indicator and big data trajectories analytics on Spark-Hadoop ecosystem, Procedia Computer Science 170:137–144, DOI: 10.1016/j.procs.2020.03.018.

Lee, M. 2014. "GIS-Based Route Risk Assessment of Hazardous Material Transport". Civil Engineering Theses, Dissertations, and Student Research.

Lewis, J. 2014. Episode 213: James Lewis on Microservices. Retrieved March 12, 2022, from https://www.se-radio.net/2014/10/episode-213-james-lewis-on-microservices

Mabrouk, A. and Boulmakoul A. 2012. Modèle spatial objet base sur les diagrammes spatiaux de voronoï pour la geo-gouvernance des espaces urbains, Workshop International sur l'Innovation et Nouvelles Tendances dans les Systèmes d'Information (INTIS 2012), ISBN: 2168/2008 978-9981-1-3000-1, pp. 105–116, Nov 23–24, 2012 Mohammadia, Morocco.

Mabrouk, A. and Boulmakoul A. 2020. Pedestrians' safest path using Voronoï spatial dia-grams based on risk zones, the Ninth international Conference on Innovation and New Trends in Information Systems (INTIS 2020), ISBN: 2168/2008 978-9981-1-3000-1. pp. 94–102, Dec 18–19, 2020, Tangier, Morocco

Mabrouk, A. and Boulmakoul A. 2022. Smart COVID-19 GeoStrategies using Spatial Network Voronoï Diagrams, Machine Learning and Deep Learning in Medical Data Analytics and Healthcare Applications Book, Edited By Om Prakash Jena, Bharat Bhushan, Utku Kose, Chapter 14, eBook ISBN 9781003226147. DOI: https://doi.org/10.1201/9781003226147 February 2022

Mabrouk, A. and Boulmakoul, A. 2017b. Nouvelle approche basée sur le calcul des itinérai-res courts et sûrs pour le transport des matières dangereuses favorisant l'accès rapide aux secours, The Sixth International Conference on Innovation and New Trends in Information Systems, (INTIS 2017), ISBN: 978-9954-34-378-4, ISSN: 2351-9215 pub-lished by EMSI-Casablanca, pp 61–73 November 2017, Casablanca.

Mabrouk, A., Boulmakoul, A., Karim, L. and Lbath, A. 2017a. Safest and shortest itiner-aries for transporting hazardous materials using split points of Voronoï spatial dia-grams based on spatial modeling of vulnerable zones, Procedia Computer Science 109C:156–163.

Mandar, M. and Boulmakoul, A. 2018, Indicateurs de risque d'accidents piétons: Vers une décision floue intuitionniste, Conference: ASD 2018: Big data & Applications 12th edition of the Conference on Advances of Decisional Systems At: Marrakech Morocco, ISBN: 978-9920-35-679-4

Melis, A., Mirri, S., Prandi, C., Prandini, M., Salomoni, P. and Callegati, F. 2016. A Microservice Architecture Use Case for Persons with Disabilities, the 2nd EAI International Conference on Smart Objects and Technologies for Social Good. DOI: 10.1007/978-3-319-61949-1_5.

Sultan S., Ahmad I. and Dimitriou T. 2019. Sécurité des conteneurs: enjeux, défis et chemin à parcourir. Accès IEEE. 7:52976–52996. DOI: 10.1109/ACCESS.2019.2911732.

17 Toward a Predictive Simulation Framework of Accident Risks for Pedestrians Based on Distributed Artificial Intelligence and Intuitionist Fuzzy Modeling

Meriem Mandar and Azedine Boulmakoul

CONTENTS

17.1 INTRODUCTION AND MOTIVATION

Every day we are all pedestrians, even if only for a short time, since each trip begins and ends with a walk even to reach a means of transport. But this natural and health-promoting travel mode is not as safe as it should be. Unprotected pedestrians are vulnerable users of the road system. They are either alone or driving a baby carriage, sick carriage, or any other small vehicle without a motor; either pushing a bicycle or moped by hand; or either disabled people in wheelchairs driven by themselves or traveling in a walking space.

DOI: 10.1201/9781003255635-17

There is a close association between the walking environment and pedestrian safety. Walking in an environment that lacks a dedicated pedestrian infrastructure and allows the use of high-speed vehicles increases the risk of pedestrian accidents. In addition to the general case, this risk is all the more important in children, the elderly, and pedestrians in a state of intoxication. The first is because they cannot yet adapt to traffic and the second is because their reflexes and the quality of hearing decrease, the field of vision narrows, and visual acuity weakens with age.

Pedestrian accidents mainly occur in urban areas due to the greater risk exposure in towns. The severity rate expressed by the ratio of the number of people killed and the number of those injured varies according to the place of occurrence of the accidents: This rate is higher outside built-up areas on motorways and national roads. This is partly because of the importance of traffic speeds and densities and because drivers expect less to find pedestrians. In addition, many less serious accidents occur at pedestrian crossings where traffic speeds are lower and drivers are warned of the presence of pedestrians. Drivers are more likely to stop when pedestrians seem determined to cross, unlike those who passively wait for their turn to cross.

However, there are differences depending on the categories of pedestrians: children are more exposed to accident risk at the start of the crossing since they rush off without looking, while elderly pedestrians are at the end of the crossing since they do not have time to cross when traffic has already resumed. Changing modes of transport (getting in or out of a car, public transport, etc.) also leads to accidents. Child pedestrians are particularly concerning since they do not realize that they become very vulnerable when leaving the first means of transport. Pedestrian-vehicle collisions usually involve only one vehicle. Heavyweights generate a higher rate of gravity than other types. And the drivers relate the accidents to the problems of visibility that they encounter vis-à-vis pedestrians. The ability to address pedestrian safety is an important component of efforts to prevent road traffic accidents. Reducing or eliminating risks to pedestrians is an important and achievable goal.

After a short introduction, Section 17.2 presents the fuzzy ant pedestrian model using artificial potential fields. Section 17.3 presents a route crossing process. Section 17.4 presents the developed intuitionistic fuzzy risk model. Next Section 17.6 presents the proposed prediction framework of accident risk for pedestrians. Finally, we present a conclusion.

17.2 FUZZY ANT PEDESTRIAN MODEL USING ARTIFICIAL POTENTIAL FIELDS

Collective pedestrian movements are governed by collective self-organization processes. These result from a series of actions and local interactions between pedestrians, as well as with the components of their environment. Pedestrians begin their walking processes (operational level) with self-motivation to reach a given destination at an appropriate speed (strategic level)—and this while using their abilities to explore the surrounding space and plan paths (tactical level). Thus, their movements can be subdivided into stages, each of which requires a particular level of analysis. The modeling of all these levels requires the consideration of several cognitive mechanisms, such as intellectual capacities, personal motivations, and the motor control of pedestrians, whose objectives are different from ours. Indeed, we are particularly interested in the behavior of pedestrians emanating from the operational level, since they are responsible for the emergence of collective structures.

Furthermore, it should be noted that the collective self-organization of pedestrians shows a striking resemblance to those of natural swarms, and more specifically ants. This remark does not take into consideration the cognitive abilities of pedestrians, however, since generally we are guided by the imitation behavior of others, in a systematic approach, especially in the event of panic. We are particularly interested in the small-scale approach to model the movement of pedestrians, which we have named "virtual pedestrians", in a two-dimensional space. Like real pedestrians, these virtual entities do not have a complete perception of their environment and can only perceive the neighborhood in which they move. Their movements are then based on information collected locally, allowing them to find a way to their destinations while avoiding obstacles. However, this information is often presented in an imprecise manner, hence the need to adopt a fuzzy approach.

In this perspective, and inspired by swarms of ants, we have based our model on the metaheuristic of optimization by ant colony (ACO) in its simple version, without focusing on an aspect of optimization, which diverges from our goal, which is to develop a model of the movement of virtual pedestrians, without providing them with any form of personal intelligence. The bioinspired character of the swarm intelligence technique will treat virtual pedestrians as a swarm, being able to act and interact with the components of their environment, while having a distributed artificial collective intelligence, allowing them to produce global self-organized structures which are not even considered at the local level. Furthermore, the model will use the paradigm of two-dimensional cellular automata, due to their fundamental characteristic of temporal and spatial discretization. The basic idea of the model is to draw inspiration from the pheromone deposit behavior of ants to create a virtual trace that will play the same role for pedestrians. By the same principle, as real pedestrians, we feel more confident to take a path already taken. We projected this chemical trace into a virtual trace that we generalized to two soil fields. The first is dynamically created from the traces of pedestrian movements, and the second is static, representing the static obstacles of the environment. This virtual trace is reinforced with the passage of pedestrians without being permanent and weakens with a given rate of evaporation. Indeed, for real pedestrians, unpaved roads or tracks created in natural environments fade over time. To better respond to the desirability of undertaking a path, we have modified the visibility term of ants in the ACO metaheuristic. We used a walking direction preference matrix, two soil fields to represent the virtual trace of pedestrians, and their land occupations. The knowledge of the first three elements remains imprecise, and they are therefore represented by fuzzy numbers. However, the land occupation remains precise, since each zone can only be occupied by one pedestrian at a time (Boulmakoul and Mandar 2011). We have extended the developed model using the basic mechanisms of artificial potential fields. The notion of preference matrix has been replaced by artificial potential fields, which better represent the spatial topology of the environment and provide guided navigation to pedestrians. The artificial potential field method considers each pedestrian as a particle moving through an artificial potential field to achieve a given goal.

The latter acts as an attractive force on pedestrians, while obstacles play the role of a repulsive force. The superposition of all the forces considerably influences the movement of pedestrians. This approach is characterized by its simplicity and speed of implementation. In addition, it allows continuous or discrete modeling of the pedestrian space, as well as local navigation with a reduced computational cost (Boulmakoul and Mandar 2016).

17.3 ROUTE CROSSING PROCESS

The crossing process requires, on the one hand, an ability to perceive all the moving elements close to the environment of the pedestrians and, on the other hand, the understanding and projection of these elements shortly. However, previous experience and knowledge can serve as filters to select relevant information. This process can be divided into basic steps starting with the selection of the place and time of passage, after visual exploration of the near environment, and the selection of the relevant information to better estimate a potential collision and ending with a walking activity (see Figure 17.1). The estimation of a probable collision relies heavily on the estimation of the vehicle's stopping distance at the front of the flow compared to the route width.

Pedestrians cross the road safely if the stopping distance of the vehicle D_v with which a collision can occur is greater than the distance D to cross the road (see Figure 17.2). The choice of the crossing trajectory is generally based on a compromise between the pedestrian's perception of risk and their ability to cross with the greatest possible comfort while adjusting their speed according to their situations. On the crossing area, the path usually takes the form of a line perpendicular to the road. And outside the crossing zone, pedestrians tend to round the corners and choose oblique lines.

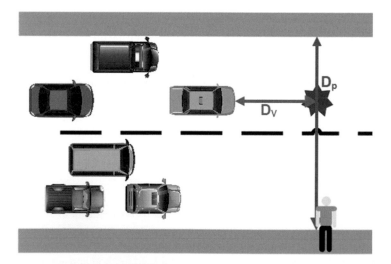

FIGURE 17.1   Accident risk factors for pedestrians.

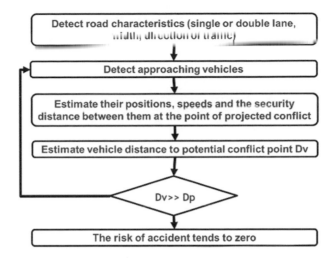

FIGURE 17.2 Pedestrian crossing process.

17.4 INTUITIONISTIC FUZZY RISK MODEL

The state of the art of road safety is based on two approaches. First, a reactive approach based on statistical studies of data collected from accidents. In this approach, accidents are already happening without any prevention. The second approach, called proactive, is based on indicators measuring the temporal and/or spatial proximity of road users (pedestrians and their vehicles). Work presented in (Mandar et al. 2020) gives a survey on pedestrians' risk indicators. We proposed three approaches to measure the exposure of pedestrians to accident risks. There is no better measure for pedestrian exposure. However, depending on the specific needs and objectives, some measures are better suited than others. Figure 17.8 schematizes our approaches to modeling risk indicators and their positioning in the literature.

The first approach is classic. It presents a measure of the risk indicator of mutual accidents of virtual vehicles and pedestrians, according to their flows and their densities (Boulmakoul and Mandar 2014). The second approach presents a new formulation of a pedestrian accident risk indicator. This indicator is based on pedestrian crossing time, vehicle safety stopping time, vehicle density, and speed. Pedestrian crossing time and vehicle safety stopping time are modeled as fuzzy numbers due to their imprecision. In the third approach, we aligned the theory of intuitionistic fuzziness with the objective of modeling the risk of accidents for pedestrians. Indeed, in modeling the risk of pedestrians interacting with vehicles, only the kinematic parameters of the vehicle and the pedestrian and certain geometric criteria of the road are considered. The integration of behavioral factors related to the perception of space and the decision of the two road users remains absent. To overcome this problem, we use the intuitionist approach, which allows us to link the two realities perceived by pedestrians and drivers (Mandar and Boulmakoul 2016). The approach is based on both objective and subjective information from the driver and the pedestrian. On the one hand, we integrated the two antagonistic perceptions as intuitionistic

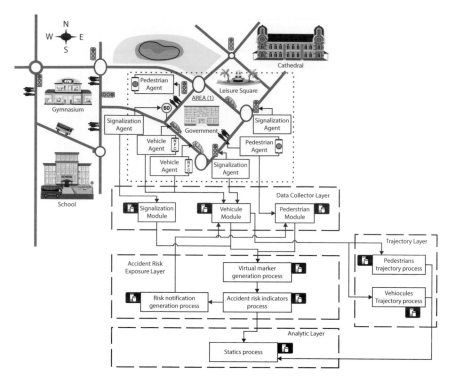

FIGURE 17.3   Global architecture system of the accident risk prediction framework

fuzzy numbers and, on the other hand, we developed a relative ranking method to derive risk exposure indicators (Mandar and Boulmakoul et al. 2017) (Mandar and Karim et al. 2017) (Mandar et al. 2018).

17.5   TOWARD A PREDICTIVE SIMULATION FRAMEWORK OF ACCIDENTS RISKS FOR PEDESTRIANS

Nowadays, all users of the urban network have physical terminals, "objects", integrating sensors, system software, and other technologies allowing them to connect to the internet. These objects can share and collect data with minimal human intervention. And digital systems can examine and adjust any interaction between these connected objects. This allows the physical and digital worlds to cooperate. In the field of pedestrian safety, these systems compensate for the behavior of road users from perception to reaction.

The basic idea of this framework is to equip road users who find themselves in a situation of risk of a "pedestrian-vehicle" type accident with risk notifications on their physical terminals. In the case of the vehicle agent, this notification indicates the presence of a pedestrian with whom a potential accident is likely to occur if he continues to advance at the same speed and therefore avoidance maneuvers are necessary such as deceleration is required (see Figure 17.6). And in the case of the

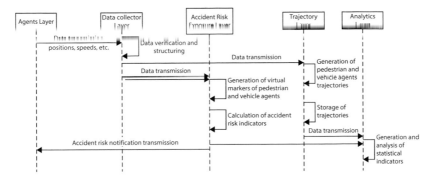

FIGURE 17.4   Nominal flow of the framework of accident risk prediction for pedestrians.

pedestrian, this notification allows him to take the necessary actions to avoid the potential accident, such as changing the time of crossing if he has not yet started, changing the crossing trajectory, adaptation of walking speed, etc. This will partially circumvent the problem of mutual visibility of road users. In pedestrian-vehicle inter-actions, if such prescanning of the conflict environment can be performed, the effi-ciency and reliability of pedestrian protection systems will be significantly improved.

Based on this principle, we propose a framework for predictive simulation of acci-dent risks for pedestrians based on distributed artificial intelligence and intuitionistic fuzzy modeling. This framework consists of four layers (Figure 17.3):

1. **Data collector layer:** This layer is responsible for collecting data from pedestrians and vehicles, particularly in terms of positions and speeds.

   In addition it includes the types and positions of signals deemed useful in our context

2. **Accident risk exposure layer:** This layer is responsible for generating vir-tual markers for pedestrians and vehicles in the studied area of the urban network. These markers are used to calculate the risk of accidents for pedes-trians. Then notifications are sent to the physical terminals of pedestrians and vehicles affected by possible collisions to prevent them.

3. **Trajectory layer:** This layer uses the data collected by the data collection layer to build the trajectories of pedestrians and vehicles, on the one hand and, on the other hand, the storage of these trajectories for analytical purposes.

4. **Analytics layer:** This layer is responsible for collecting and analyzing all the statistical indicators and representing them graphically.

Figure 17.4 shows the nominal flow of the proposed framework. Agent data (pedes-trians, vehicles, and infrastructure) are collected by the data collector layer, which checks their integrity and structures. These data are transmitted, on the one hand, to the accident risk exposure layer to calculate this exposure in real time. Agents at risk of an accident are then notified. And on the other hand, the trajectory layer builds the trajectories of pedestrians and vehicles and takes care of their storage. The analytical layer is responsible for generating and analyzing the statistical indicators developed.

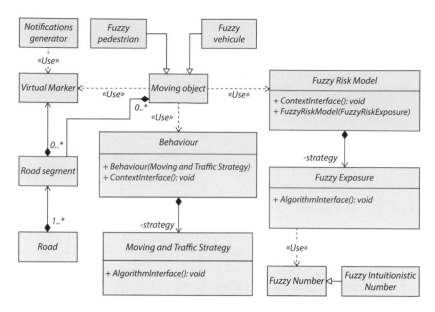

FIGURE 17.5   General class diagram of the accident risk prediction framework.

Figure 17.6 schematizes the overall component diagram of the accident prediction framework. This diagram includes several modules interacting with each other:

- **The pedestrian module** grouping the pedestrian agents present and connected in the studied area of the urban network.
- **The vehicle module** grouping the vehicle agents present and connected in the studied area of the urban network.
- **The urban network configuration module** is responsible for setting up the general layout of the studied area of the urban network in terms of roads, the direction of traffic signs deemed relevant, etc.

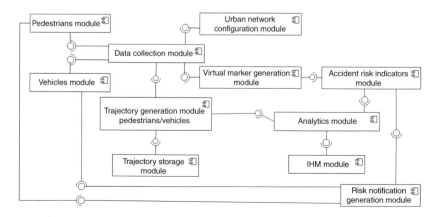

FIGURE 17.6   General component diagram of the accident risk prediction framework.

- **The data collection module** is responsible for collecting data from the pedestrian module, vehicle module, and urban network configuration module.
- **The virtual marker generation module** relies on the data collection module to generate virtual markers for pedestrian and vehicle agents.
- **The trajectory generation module** relies on the data collection module to generate the trajectories of pedestrian and vehicle agents.
- **The trajectory storage module** is responsible for storing the generated trajectories of pedestrian and vehicle agents.
- **The accident risk indicators module** is responsible for calculating these indicators based on the virtual markers generated by pedestrian and vehicle agents.
- **The accident risk notification generation module** is based on the calculated indicators to send notifications to the pedestrian agents and vehicles concerned.
- **The analytical module** is based on the indicators calculated at the level of the accident risk indicators module and the trajectories generated by pedestrian and vehicle agents to identify the various statistical indicators.

Our pedestrian accident risk prediction framework is based on the basic class diagram shown in Figure 17.5 where pedestrians and vehicles are considered as objects moving according to the chosen behavior model, which is fuzzy and provided.

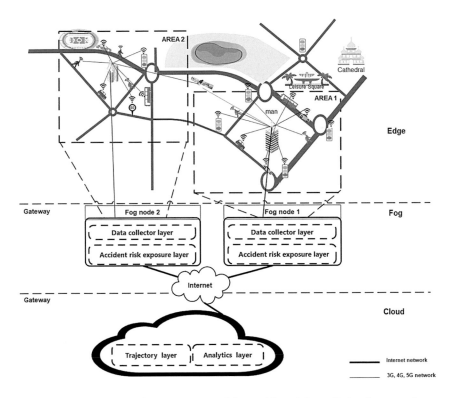

FIGURE 17.7   Global architecture system of the accident risk prediction framework.

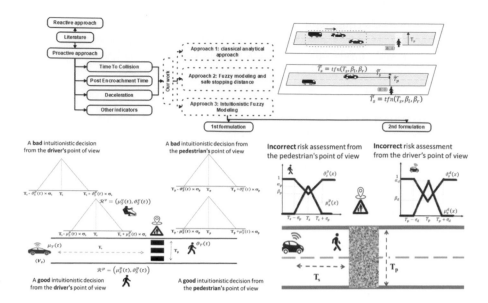

FIGURE 17.8   Schematization of the approaches developed to measure the risk of accidents for pedestrians.

The risk of pedestrian accidents is assessed for each road. Each agent (pedestrian or vehicle) has a virtual marker, which is used by the notification generator for the transmission of the latter in case of risk.

Thus, as shown in Figure 17.7, the overall system architecture of the accident risk prediction framework consists of three essential layers, which are as follows:

- The Edge layer is composed of the telecommunications infrastructure and connected objects such as connected vehicles, connected pedestrians, and connected infrastructure. This layer includes the agents: pedestrians, vehicles, and urban traffic signs. The data of these agents are transmitted via a wireless network.
- The Fog layer is responsible for the direct interaction with the agents of the Edge layer and includes the data collection layer and the accident risk exposure layer. This layer manages the real-time processing of massive data generated by connected agents.
- The Cloud layer allows global visualization of traffic situations and in-depth data analytics. It includes the trajectory layer and the analytical layer.

17.6   CONCLUSION

In this chapter, we have presented a framework for predicting accident risks for pedestrians. This framework is based on the foundations of multiagent systems, the Internet of Things, and cloud computing. The fundamental idea of the framework is to collect data from road users to measure in real time the exposure of pedestrians to the risk of accidents and to notify pedestrians and vehicles of a potential accident risk situation to be able to take the necessary actions. This will significantly reduce the risk of pedestrian-vehicle accidents. Accident risk measurements are based on intuitionistic fuzzy modeling. This approach can model the behavioral psychology of pedestrians and drivers with intuitive methods based on the theory of intuitionistic fuzzy sets. Admittedly, the idea is interesting, allowing the integration of both abductive reasoning and uncertainty, thanks to the theory of intuitionistic fuzzy sets, since it makes it possible to model uncertain hypothetical decision making for drivers and pedestrians. Other psychological factors must be included in our approach. A software system is being developed for this purpose which reuses pedestrian simulation models developed in our previous work. A detailed study of the proposed framework and the development of a prototype are necessary and are part of our ongoing work. Further future developments will consider the multiattribute preference problem and the intuitionistic fuzzy number ranking process to improve the intuitionistic fuzzy model of accident risk measurement.

REFERENCES

Boulmakoul, A., and Mandar, M. (2011). Fuzzy ant colony paradigm for virtual pedestrian simulation. The Open Operational Research Journal, 19–29, ISSN: 18742432, DOI: 10.2174/1874243201105010019.

Boulmakoul, A., and Mandar, M. (2014). Virtual pedestrians' risk modeling. International Journal of Civil Engineering and Technology (IJCIET) 5 (10 Oct. 2014), 32–42, ISSN: 0976 – 6308.

Boulmakoul, A., and Mandar, M. (2016). Fuzzy pheromone potential fields for virtual pedestrian simulation. Advances in Fuzzy Systems, Volume 2016, Article ID 4027687. https://doi.org/10.1155/2016/4027687

Cardona, O.D., van Aalst, M.K., Birkmann, J., Fordham, M., McGregor, G., Perez, R., Pulwarty, R.S., Schipper, E.L.F., and Sinh, B.T. (2012). Determinants of risk: exposure and vulnerability. In: Managing the Risks of Extreme Events and Disasters to Advance Climate Change Adaptation [Field, C.B., V. Barros, T.F. Stocker, D. Qin, D.J. Dokken, K.L. Ebi, M.D. Mastrandrea, K.J. Mach, G.-K. Plattner, S.K. Allen, M. Tignor, and P.M. Midgley (eds.)]. A Special Report of Working Groups I and II of the Intergovernmental Panel on Climate Change (IPCC). Cambridge University Press, Cambridge, UK, and New York, NY, USA, pp. 65–108.

Hatfield, J., Fernandes, R., Job, R.S., and Smith, K., (2007). Misunderstanding of right-of-way rules at various pedestrian crossing types: observational study and survey. Accident Analysis and Prevention 39 (4), 833–842.

Mandar, M., Karim, L. Boulmakoul, A., and Lbath, A. (2017). Triangular intuitionistic fuzzy number theory for driver pedestrians interactions and risk exposure modeling. Procedia Computer Science 109 (2017), 148–155, ISSN: 1877-0509.

Mandar, M., and Boulmakoul, A., (2016). Pedestrians' fuzzy intuitionistic risk exposure model: foundation and first development. 5ème Edition du Workshop International sur l'Innovation et Nouvelles Tendances dans les Systèmes d'Information INTIS 2016, Fès. ISSN: 2351-9215.

Mandar, A., Boulmakoul, A., and Karim, L. (2018). Indicateurs de risque d'accidents piétons: Vers une décision floue intuitionniste. ASD 2018: Big data et Applications 12th edition of the Conference on Advances of Decisional Systems, ISBN: 978-9920-35-679-4.

Mandar, M., Boulmakoul, A., Karim, L., and Cherradi, G. (2020). Survey on pedestrians' accident risk indicators. The Ninth International Conference (INTIS'2020).

Mandar, M., Boulmakoul, A., and Lbath, A. (2017). Pedestrian fuzzy risk exposure indicator. Transportation Research Procedia 22, 124–133, ISSN: 2352-1465.

18 Trajectory to a New Shape of Organizational Structure, Enterprise Architect, and Organizational Audit for Governance of Information Systems Processes

Zineb Besri and Azedine Boulmakoul

CONTENTS

DOI: 10.1201/9781003255635-18

18.1   INTRODUCTION

Aligning technologies with business strategy has been discussed for quite a while now. If the subject of the information system's governance has become so pressing, it is quite simply because the stakes are now crucial. The information system has become a strategic resource which determines the success of the development of the company. However, faced with the increase in IT and technological costs and the gap between expectations and the results obtained from projects, decision makers have become very selective.

The contribution of the information system to the overall performance of the company therefore becomes a priority in a context of improvement in economic profitability. Aware of this problem, most decision makers have set their main objectives: aligning IT investments with the company's strategy, controlling, and reducing costs, and earning the benefits of investments already made.

For several years, the hospital organization has been facing major issues and challenges, from risk control and the quality of its services, to reducing operating costs and the pressing demands of politico-socio-economic partners. The hospital organization is required more than ever to register in the logic of organizational innovation. In terms of scientific research, the issue of hospital organization has occupied a major place in the various works relating to the management and engineering of hospital systems. Currently, domains gradually digitize, as is the case of healthcare.

The healthcare workforce is being augmented by technology to help meet the needs of a changing patient population. A growing number of healthcare facilities have expanded their use of telemedicine to deliver services to patients in hospitals as well as in remote locations. In a previous work, we proposed a new approach to drive the transformation of the organization regarding digital transformation. Indeed, we proposed a fuzzy intuitionist approach to discover organizational structures that support digital transformation (Deiser 2018, Laloux 2016, Sommerfeld and Moise-cheung 2016).

In this research work, we focus on the processes and on the organizational units resulting from an analytical approach to show that the development of digital strategies reveals new processes and new organizational structures resulting from emerging strategies.

The purpose is to understand, through this case study, how digital transformation on a organization's structure operates. In this chapter, we deal with a case study of an intuitionistic fuzzy method applied to healthcare organizational structures.

This chapter is organized as follows. First an introduction to the main issue is provided, and then the second section provides a reminder of some principles and concepts of corporate governance to IT governance. In Section 18.3, we evoke the enterprise architect and organizational audit. After a brief context definition in Section 18.4, we present the case study with some first results and discussion. Finally, we synthesize in the conclusion some future works and perspectives.

18.2   CORPORATE GOVERNANCE TO IT GOVERNANCE

The objective of this part is to show the relationship between corporate governance and information systems (IS) governance. It seems appropriate to start with a presentation of corporate governance and its measures. Such a presentation is essential insofar as it will allow us to justify the relationship that exists between the objectives of corporate governance and the different components of IS governance.

Several definitions have been given for the term "corporate governance", according to (Ramírez and Tejada 2018). Corporate governance encompasses the oversight mechanisms, including the processes, structures, and information used to direct and monitor the management of a society. It includes ways to hold board members and senior executives accountable for their actions and for establishing and implementing oversight functions and programs.

As organizations evolved, it became apparent that one of the elements of corporate governance had its origins in the control of information, with: access to information being a strategic resource (Bounfour et al. 2006). This is how the corporate information officers (CIOs) wanted to take ownership of the general approach by applying it to information systems (IT governance). There are several reasons for this:

- First, IT governance was a wonderful opportunity to specify operating rules and control methods, highlighting the contribution of IS to value creation: the workhorse of CIOs.
- It also responded to Solow's paradox (Capirossi 2002) by making it possible to shift the IS problem; from a cost center vision to that of value creation and support ofthe company's vital processes.
- By shifting IS on strategic actions, the IT department hoped to "go up a notch" and thus get out of the spiral in which it had confined itself to the "cost center".
- The implementation of IS governance rules also had the advantage of defining the respective roles and duties of the stakeholders, organized around good practices to meet the company's ambitions. These rules and controls had a pedagogical role for its actors: by demystifying the organization of the IT department and demonstrating its potential "performance". They also made it possible to work together at last.

Among the missions of corporate governance (Del Baldo 2012), we can mention:

- Elaborate the strategy.
- Make the company's "policy" through the adoption of a management type and the development of decision-making processes.
- Manage risks.
- Ensure compliance with regulations, whether legal, accounting and tax, normative, or specific to the company and with the principles and statutes of established governance.
- Steer the performances.
- Do the reporting and the audit.

To achieve these missions, fundamental rules of corporate governance must be respected, namely:

- The transparency.
- The realization of a perfectly effective decision-making process, providing each actor with the powers and information to act at his or her level.

- The implementation of a sufficiently complete performance evaluation system to understand performance synthetically, in its entirety and in its details.
- Finally, specific management to the governance itself to ensure its compliance and sustainable functioning, complemented with an efficiency audit ensuring the creation of value over time.

In relation to the governance of information systems, it consists of defining and managing a set of processes supporting the objectives of corporate governance. Traditionally, its purpose is to:

- Support value creation objectives, by promoting the innovation dynamic.
- Align information system processes with the business strategy and ensure that they achieve optimal performance.
- Manage this performance by fully controlling the costs and gains relating to the IS function.
- Manage the risks inherent in information systems and establish a security policy that harmoniously complements risk management at the company level.
- Anticipate the evolution of solutions and skills according to the company's development prospects but also the evolution of tools and emerging technologies.
- Promote stakeholder communication to ensure that the information systems function is really one with all the company's functions.

Governance therefore involves formalizing the rules of the game: from strategy to operations, from the company to the IT department, from the business departments to the IT department. This is a double alignment, that of:

- IS on the company.
- The IT department on the best practices listed.

18.3   ENTERPRISE ARCHITECT AND ORGANIZATIONAL AUDIT

Each organization marks out a path with a certain trajectory by taking some decisions that impact its business processes. Then, that also impacts its information system by transforming digitally those processes (business and IT processes).

Structured frameworks known as enterprise architectures capture and manage the complexity of modern organizations. Modern organizations today are based on the connectivity of the enterprise system model and the corresponding information system model.

These integrated management systems are very complex systems consisting of elements such as goals, data, people, processes, and technology. These systems require coordination and integration to manage the interdependencies that exist between all these elements.

18.3.1 ENTERPRISE ARCHITECTURE

Enterprise architecture frameworks present a conceptual blueprint necessary for building a business model that is integrated and managed by the information system of interest. They require credentials of the organization from different points of view. The main viewpoints are data, functions, networks, organizational structures, schedules, and strategy. Each of its business viewpoints will be progressively implemented in components of a future information system. This process is supported by the second dimension of the concept map containing some levels., Executives constantly support various managerial tasks such as business process improvement, workflow management, software engineering, and, enterprise system management in general.

Building enterprise architectures involves modeling techniques to capture the ontology of the organization, the organization adequate methodology to establish a foundation for enterprise system management, and a concept life cycle of the IS, as well as the integrated modeling tool to build and maintain these architectures.

18.3.2 GOVERNANCE OF AN INFORMATION SYSTEMS ORGANIZATIONAL AUDIT

The specificity of each organization translates the way of running the company. Therefore, to maintain and strengthen their positioning, organizations are part of a continuous improvement process of their maturity. Consequently, to maintain and strengthen their positioning, organizations are part of a continuous improvement process of their maturity. In this context, they must carry out regular evaluations and internal audits of their system and their organizational structure. This assessment makes it possible to discover the real organizational structure of the company and to compare it with that established by the top management.

The organizational audit makes it possible to identify potential dysfunctions within the management and administrative bodies and information systems. The Information system process governance provides enterprise governance through performance and compliance. To respond to compliance, we took an interest in the organization of the company and its evaluation in an audit process to measure the degree of noncompliance with the strategy and objective set out by top management.

The search for better performance for the company led us to study the organizational structure of the company. We focused exclusively on the organizational restructuring process. Gradually, our field of reflection widened to integrate the process areas of the CMMI-Dev good practices in the proposed organizational redesign process.

18.4   CASE STUDY: CONTEXT DEFINITION

In general, digital transformation solves traditional problems with technology such as:

- Customers focus over product focus on technology decision making.
- Internal customer experience when employees have consumer-grade technology.
- Organizational structure that supports a seamless customer experience.
- Logistics and supply chain that improve efficiencies.

- The change management and cultural transformation.
- Data security, privacy, and data ethics where standards are the focus.
- Transformational when leadership drives technology decisions.
- Evolution of products, services, and processes around delivery.
- Technology decisions involve the entire c-suite.
- Digitization of the business
- Integration of all data systems.
- Personalization that guides the customer.

In this chapter we focus on digital transformation of the healthcare area. The trends of digital transformation in healthcare that have come into existence with the introduction of the greater technology adoption by digital health start-ups are:

- Telemedicine that uses technologies for patient-doctor communication, patient monitoring, and e-prescription.
- By 2025, telehealth doctor visits replace the traditional annual physical.
- Big data for several applications, such as preventing human errors, rate prediction, strategic planning, and patient admission.
- Also using the Internet of Things (IoT) for remote patient monitoring, risk management, and health condition prediction.
- Virtual reality based treatment for better patient experience and, pain reduction techniques.
- Also, the use of artificial intelligence like pathology image analytics or drug and vaccine research.
- Finally, the use of collaborative robots for waiting reduction time or medication alerts.

Figure 18.1 reveals the interaction between the patient and Internal transformation using datadriven intelligence and design from digital experience.

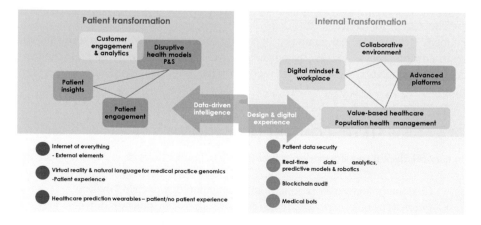

FIGURE 18.1   Health digital transformation.

18.5   CASE STUDY OF AN INTUITIONISTIC FUZZY METHOD APPLIED TO A HEALTHCARE ORGANIZATIONAL STRUCTURE

There has been an affected disruption to the traditional care model in which a patient visits their doctor's office or a hospital to seek medical treatment. The healthcare workforce is being augmented by technology to help meet the needs of a changing patient population. More patients are seeking care under the Affordable Care Act; at the same time, there is a shortage of physicians and medical workers. A growing number of healthcare facilities have expanded their use of telemedicine to deliver services to patients in hospitals as well as in remote locations.

The case study refers to a large university hospital center, located in an urbanized area in Morocco. A preliminary analysis is performed to detect main patterns from the data, used in the business process simulation.

Following interviews with managers, doctors, and nurses, and through an accurate quantitative analysis, we were able to build the As-Is process model of the emergency department, which is illustrated in Figure 18.2.

Hospitals require precision in the execution of job responsibilities and multiple layers of accountability to function. To accomplish this, hospitals use a vertical organizational structure with many layers of management, with a kind of organizational structure as a matrix.

Understanding the hospital organizational structure ensures that hospital employees know their own responsibilities, the responsibilities of those around them, to whom they report, and whom to talk to about particular responsibilities or fields of knowledge.

Hospitals are represented by different organizations, depending on their status (public, private, nonprofit-making, or private profit-making). Finally, pharmaceutical manufacturers and producers of medical devices each have their own structure. Figure 18.3 represents the Moroccan hospital center university's organizational chart. It shows a particular shape of the actual organizational structure.

In terms of processes, digitization had already started before the occurrence of COVID-19. It now offers solutions for maintaining business relationships in times of crisis and may continue to grow thereafter.

From both organizational structures associated with a selection of a hospital process, we define the following collaboration matrix (organizational unit × activity: Table 18.1).

This matrix will be an input for the proposed structural engine. It helps to extract specific processes that should be optimized or changed. The aim is to get a better organization that is aligned with the top management strategy or future ones.

Before that, we need to analyze this collaboration relationship to guess the new shape of the organizational structure

18.5.1   DIGITALIZED PROCESSES

Digital health solutions are autofit to support patient access to treatment and can considerably reduce the pressure on healthcare infrastructure both during and after

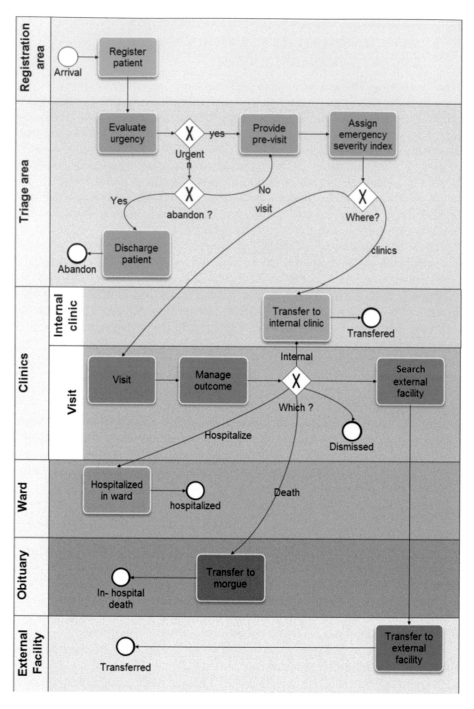

FIGURE 18.2 Hospital processes of emergency unit of Moroccan's HCU.

FIGURE 18.3 Moroccan hospital center university's organizational chart.

TABLE 18.1
Collaboration Relationship between Activities and Organizational Units of the Selected Processes

	U_1	U_2	U_3	U_4	U_5	U_6	U_7	U_8	U_9	U_{10}
A_1	1	1	0	0	0	1	0	0	0	0
A_2	0	0	1	1	1	1	0	1	1	0
A_3	0	1	1	0	0	0	1	1	1	0
A_4	0	0	0	1	0	1	0	1	0	0
A_5	0	0	1	0	1	1	0	0	0	1
A_6	0	0	1	0	0	0	1	1	1	1

Health-Tech-Cluster		Health-Tech-Capability
Patient-centric	Education	• Health Information Platform • Consumer Education
	Triage	• Medical concierge • Chatbots • Tracks & trace apps
	Telemedicine	• Tele-consultation
	Distribution	• Consumer Marketplaces
Diagnostic-centric	Chronic Disease Management	• Digital therapeutics • Disease management
	Point of-care Testing Diagnostics	• On-demand lab test • Medical Diagnostics
	Screening	• Medical imaging • Teleradiology
R&D-centric	Research	• Drug and vaccine discovery
	RCT	• Research clinical trials

FIGURE 18.4   Emerging digital solutions.

the pandemic. Here some digital solutions emerged during COVID-19. Figure 18.4 shows a set of emerging digital solutions in the field.

We can notice that every part of the organization is linked to those digitalized processes from patient-centric, diagnostic, to research and development (R&D) with both health-tech-cluster and health tech capability.

18.5.2   ORGANIZATIONAL UNITS RELATED TO DIGITALIZED PROCESSES

Those digitalized processes lead to new organization units. Here we extract a sample of the new shape of the relationship between activities and the new organizational units (Table 18.2).

TABLE 18.2

Collaboration Relationship between Activities and Organizational Units of the Digitized Processes

$$\begin{pmatrix}
 & U_1' & U_2' & U_3' & U_4' \\
A_1 & 1 & 0 & 0 & 0 \\
A_2 & 0 & 1 & 1 & 1 \\
A_3 & 1 & 1 & 0 & 1 \\
A_4 & 0 & 1 & 0 & 1 \\
A_5 & 0 & 1 & 1 & 0 \\
A_6 & 0 & 1 & 0 & 1
\end{pmatrix}$$

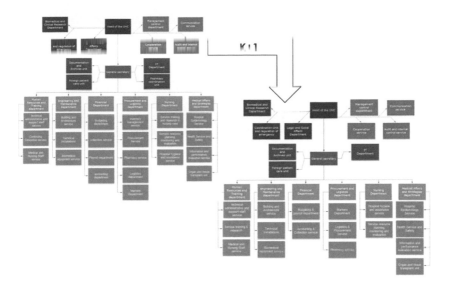

FIGURE 18.5   The iteration K+1 of the new organizational shape.

With those new organizational units, we can establish the new shape of the organizational structure of the university hospital center. The next step is to make a real overhaul then re-assess the emerged organizational structure conducted by digital transformation to confirm the intuition proposed before.

18.5.3   THE ORGANIZATIONAL STRUCTURE CONDUCTED BY DIGITAL TRANSFORMATION

Figure 18.5 presents the iteration K+1 of the new organizational shape.

The next step is to discover possibilities of the new organizational structure after we explore the intuitionistic fuzzy collaboration relationship matrices defined by the relation between the organizational unit (group of performers) and a set of activities (the activities defined by the extracted processes to assess and then to develop) (Table 18.3).

TABLE 18.3

Intuitionistic Fuzzy Input Matrix for Discovering Organizational Structure

$$\Lambda^{(k+1)} = \begin{pmatrix} & U_1 & U_2 & U_3 & U_4 & U_5 & U_6 & U_7 & U_8 & U_9 & U_{10} & U_1' & U_2' & U_3' & U_4' \\ A_1 & 1 & 1 & 0 & 0 & 0 & 1 & 0 & 0 & 0 & 0 & 1 & 0 & 0 & 0 \\ A_2 & 0 & 0 & 1 & 1 & 1 & 1 & 0 & 1 & 1 & 0 & 0 & 1 & 1 & 1 \\ A_3 & 0 & 1 & 1 & 0 & 0 & 0 & 1 & 1 & 1 & 0 & 1 & 1 & 0 & 1 \\ A_4 & 0 & 0 & 0 & 1 & 0 & 1 & 0 & 1 & 0 & 0 & 0 & 1 & 0 & 1 \\ A_5 & 0 & 0 & 1 & 0 & 1 & 1 & 0 & 0 & 0 & 1 & 0 & 1 & 1 & 0 \\ A_6 & 0 & 0 & 1 & 0 & 0 & 0 & 1 & 1 & 1 & 1 & 0 & 1 & 0 & 1 \end{pmatrix}$$

Those matrices will be the input of the proposed intuitionistic fuzzy analysis approach, Once we result the k+1 iteration previously defined.

We can compute some metrics, such as degree centrality (Besri and Boulmakoul 2021). Besides, the important one is to compute the degree of hesitation to be or not to be part of the new organizational structure.

18.5.4 Intuitionistic Fuzzy Method Applied to a Healthcare Organizational Structure

Following, we present necessary reminders (Zadeh 1965, Atanassov 1999, Zuo et al. 2019) to understand the new approach. Then we develop the process of calculating collaboration intuitionistic matrices. The Intuitionist fuzzy hesitation degree (Zadeh 1965) will be evaluated on those collaboration matrices to generate communities corresponding to the emerging organizational structures.

- μ : degree of membership

- $$\mu^{(k+1)}(\omega_i,\omega_j) = \frac{\left|\Gamma_\alpha^{(k)}(\omega_i) \cap \Gamma_\alpha^{(k)}(\omega_j)\right|}{\left|\Gamma_\alpha^{(k)}(\omega_i) \cup \Gamma_\alpha^{(k)}(\omega_j)\right|}$$

$$= \frac{\sum_{\omega \in A^{(k)}} \Pi_{\Lambda^k \cap \Lambda^{(k+1)}}(\omega_i,\omega) \times \Pi_{\Lambda^k \cap \Lambda^{(k+1)}}(\omega_j,\omega)}{\sum_{\omega \in A^{(k)}} \max\left[\Pi_{\Lambda^k \cap \Lambda^{(k+1)}}(\omega_i,\omega), \Pi_{\Lambda^k \cap \Lambda^{(k+1)}}(\omega_j,\omega)\right]}$$

- $\Gamma_\alpha^k(\omega_i) = \left\{ and \begin{smallmatrix} \omega \in A/\Lambda^{(k)}(\omega_i,\omega) \neq 0 \\ \Lambda^{(k+1)}(\omega_i,\omega) \neq 0 \end{smallmatrix} \right.$

- ν: degree of non-membership

- $$\nu^{(k+1)}(\omega_i,\omega_j) = \frac{\left|P_\beta^{(k+1)}(\omega_i) \cap P_\beta^{(k+1)}(\omega_j)\right|}{\left|P_\beta^{(k)+1}(\omega_i) \cup P_\beta^{(k+1)}(\omega_j)\right|}$$

- $P_\beta^{(k+1)}(\omega_i) = \Gamma^{(k+1)}(\omega_i)^{\cdot\cdot}.(\Lambda^{(k+1)} - (\Lambda^{(k+1)} \cap \Lambda^{(k)}))$ with $\cdot\cdot$ denotes restrictions

- We have $0 \le \mu^{(k+1)}(\omega_i,\omega_j) + \nu^{(k+1)}(\omega_i,\omega_j) \le 1$

- Hesitation: $\Pi_{ij} = 1 - [\mu^{(k+1)}(\omega_i,\omega_j) + \nu^{(k+1)}(\omega_i,\omega_j)]$

18.5.5 Results and Discussion

The case study covered the time between 2019 and 2021. The following discussion looks at how the digital leadership process developed using the four platforms discussed in Figure 18.1 and provides some general recommendations or strategies for organizations to support digital transformations today.

The case study allows us to get further with this approach as it takes natural behavior into making and executing decisions. Every low hesitation degree conducts to preserve the core Hospial Central University (HCU) processes (Table 18.4). Then the other ones are on the focus to be changed, transformed, or withdrawn.

TABLE 18.4
Fuzzy Intuitionist Metrics Resulting from the Collaboration Relationship Analysis

Organizational Unit\Metric	Membership	Non-membership	Hesitation
U1	0,5	0,25	0,25
U2	0,5	0,25	0,25
U3	0,5	0,25	0,25
U4	0,5	0,25	0,25
U5	0,5	0,25	0,25
U6	0,5	0,25	0,25
U7	0	1	0
U8	0,33	0,33	0,34
U9	0,33	0,33	0,34
U10	0,33	0,33	0,34
U'1	0,8	0,12	0,08
U'2	0,8	0,12	0,08
U'3	0,8	0,12	0,08
U'4	0,8	0,12	0,08

The aim to study was a new approach to drive the transformation of the organization regarding to digital transformation. Indeed, we propose a fuzzy intuitionist approach to discover organizational structures that support digital transformation. The theory of intuitionist fuzzy sets handles uncertain situations, by extension of the classical fuzzy theory, with the integration of the concept of hesitation. The chapter presents such a process to conduct this method to discover this new emergent organizational structure that supports the digital transformation of the company.

18.5.6 Some Propositions for a New Organizational Structure

There has been an affected disruption to the traditional care model in which a patient visits their doctor's office or a hospital to seek medical treatment. The healthcare workforce is being augmented by technology to help meet the needs of a changing patient population. More patients are seeking care under the Affordable Care Act; at the same time, there is a shortage of physicians and medical workers. A growing number of healthcare facilities have expanded their use of telemedicine to deliver services to patients in hospitals as well as in remote locations.

The future objective of this study would then be to propose the model intended for management teams or those in charge of organizational responsibilities in health establishments or participating in the public hospital service. This model should allow, by integrating the specificities of the business sector, the construction of organizational change projects. This model should also promote the implementation of the approach that continuously mobilizes a dual dynamic: substantive and procedural, but also considering different collective and individual, structural, emotional, capitalizing, and innovative dimensions.

18.6 CONCLUSION

The rise of the digital transformation is now accelerated by COVID-19. And the organizational trajectory represents the path taken by an organization challenged by the social, the cultural, and the scientific changes of the modern era. Today, the search for a trajectory is undeniably a choice, and the rooted knowledge of the actors can have an influence on its advantageous evolution.

Today, IT governance is a core activity embraced or expected by most organizations to control the behavior of IT assets. However, this discipline faces a growing gap between the views, priorities, and practices of academics and practitioners. In this area of research, we present a consolidated view of the functionalities enabling IT governance to be implemented within an organization. In previous works, we had evaluated these capabilities in the practice of companies in the field of container port logistics, as well as in the urban transport area Besri and Boulmakoul 22016, Besri et al. 2017, Boulmakoul et al. 2017a).

In this chapter we present a critical field of the healthcare domain. Key gaps in IT governance capability adoption are discussed and research insights are provided to align theory and practice. We propose a new approach to drive the transformation of the organization regarding digital transformation. Indeed, we proposed a fuzzy intuitionist approach to discover organizational structures that support digital transformation. In this research work, we focus on the processes and on the organizational units resulting from an analytical approach to show that the development of digital strategies reveals new processes and new organizational structures resulting from emerging strategies. The purpose is to understand, through some case studies, how digital transformation on some organizational structures operates. In this chapter, we deal with a case study of an intuitionistic fuzzy method applied to healthcare organizational structures.

REFERENCES

Atanassov, K.T. (1999). Intuitionistic fuzzy sets. In Intuitionistic Fuzzy Sets. Springer, Berlin, Germany, Pages 1–137.

Ben-David, S., U. von Luxburg, and D. P'al (2006). A sober look at clustering stability. Learning Theory - Lecture Notes in Computer Science. Volume 4005/2006, Pages 5–19.

Besri, Z., and A. Boulmakoul (2016). Discovery and diagnosis of the organization for the governance of information logistics and transport business systems. IEEE Xplore. 3 November 2016. DOI: 10.1109/GOL.2016.7731696. INSPEC Accession Number: 16430461.

Besri, Z., and A. Boulmakoul (2017a). Contrôle de la conformité organisationnelle basée sur la mesure des distances de partitions de l'ensemble des complexes simpliciaux. Conférence Magrébine sur les Avancées des Systèmes Décisionnels. Tabarka, Tunisie. 27–29 Aprili 2017. ISBN 978-9938-12-460-6. pages 157–170.

Besri, Z., and A. Boulmakoul (2017b). Framework for organizational structure re-design by assessing logistics' business processes in harbor container terminals. Elsevier Transportation Research Procedia. Volume 22C, Pages 164–173.

Besri, Z., and A. Boulmakoul (2017c). La connectivité combinatoire et l'analyse du graphique spectral pour l'analyse criminelle. Innovation and New Trends in Information Systems 6th edition EMSI Casablanca 24 et 25 November 2017. ISBN: 978-9954-34-378-4.

Besri, Z., and A. Boulmakoul (2021). An intuitionist fuzzy method for discovering organizational structures that support digital transformation. In: Kahraman C, Cevik Onar S, Oztaysi B, Sari I, Cebi S, Tolga A. (eds), Intelligent and Fuzzy Techniques. Smart and Innovative Solutions. INFUS 2020. Advances in Intelligent Systems and Computing, Vol. 1197. Springer, Cham.

Beth Chrissis, M., M. Konrad, and S. Shrum (2013). CMMI for Development, 3rd ed.: Pearson.

Boulmakoul, A., and Z. Besri (2013a). Performing enterprise organizational structure redesign through structural analysis and simplicial complexes framework. The Open Operational Research Journal. 2013, Volume 7, Pages 11–24. DOI: 10.2174/1874245613070100011

Boulmakoul, A., and Z. Besri (2013b). Scoping enterprise organizational structure through topology foundation and social network analysis. Innovation and New Trends in Information Systems 3rd edition. Ryad mogador Tangier, 29–30 November 2013. ISBN 978-9-98-113000-1, pages 3–17.

Boulmakoul, A., and Z. Besri (2015). Patent No. WO 2015/174811 A1.

Boulmakoul, B., Z. Besri, L. Karim, A. Boulmakoul, and A. Lbath (2017a). Combinatorial connectivity and spectral graph analytics for urban public transportation system. Transportation Research Procedia. Volume 27C, 2017, Pages 1154–1162.

Boulmakoul, B., Z. Besri, L. Karim, A. Boulmakoul, and A. Lbath (2017b). Galois's algebraic structure and bipartite graph spatio-structural analytics for urban public transportation system assessment. Procedia Computer Science. Volume 109C, 2017, Pages 172–179.

D'yachkov, A., V. Rykov, D. Torney, and S. Yekhanin (2006). On application of the partition distance concept to a comparative analysis of psychological or sociological tests. Stochastic Analysis and Applications. Volume 24, Pages 61–78.

Deiser, R. (2018). Digital Transformation Challenges in Large and Complex Organizations. CFFO Press, Claremont, CA.

Del Baldo, M. (2012). Corporate social responsibility and corporate governance in Italian SMEs: The experience of some "spirited businesses". Journal of Management and Governance. Volume 16, Pages 1–36. https://doi.org/10.1007/s10997-009-9127-4

Gusfield, D. (2002). Partition-distance: A problem and a class of perfect graphs arising in clustering. Information Processing Letters. Volume 82, Pages 159–164.

Holt, T. (2018). CEO Siemens Power Generation Services. Roland Deiser. Digital Transformation Challenges in Large and Complex Organizations. Deiser.

Konovalov, D.A., B. Litow, and N. Bajema. (2005). Partition-distance via the assignment problem. Bioinformatics. Volume 21(20), Pages 3912–3917.

Krishnan and al, Digitization isn't stimulating productivity growth—yet. June 4, 2018 | Article. McKinsey Quarterly.

Laloux, F. (2016). Reinventing Organizations. Nelson Parker. ISBN 978-2-9601335-6-1.

McMahon, P.E. (2010). Integrating CMMI and Agile Development. Pearson Education.

Pinto da Costa, J., and P. Rao (2004). Central partition for a partition-distance and strong pattern graph. REVSTAT – Statistical Journal. Volume 2(2), Pages 127–143.

Ramírez, Y., and Á. Tejada (2018). Corporate governance of universities: Improving transparency and accountability. International Journal of Disclosure and Governance. Volume 15, Pages 29–39. https://doi.org/10.1057/s41310-018-0034-2

Scott Morton, M.S., and T.J. Allen (1994). Information Technology and the Corporation of the 1990s: Research Studies. Oxford University Press.

Sommerfeld, B., and R. Moise-cheung (2016). The digitally fit organization. InSiDE Magazine, issue 7, Part 01. From digital perspective.

Valeur et Performance des SI: une nouvelle approche du capital inmatériel" (Bounfour et Epinette, 2006) Ref : A.bounfour and G. Epinette, Dunod 01 Informatique, ISBN 2 10 05 0022 8 , Octobre 2006, page 244.

Yu, C., B.C. Ooi, K.-L. Tan, and H. V. Jagadish (2001). Indexing the distance: An efficient method to KNN processing. In Proceedings of the 27th International Conference on Very Large Data Bases, VLDB '01, pages 421–430, 2001.

Zadeh, L.A. (1965). Fuzzy sets. Information and Control. Volume 8, Pages 338–353.

Zuo, C., A. Pal, and A. Dey (2019). New concepts of picture fuzzy graphs with application. Mathematics. Volume 7 (470), doi:10.3390/math7050470.

19 Dynamic Detection of Fuzzy Subcongested Urban Traffic Networks

Fatima-ezzahra Badaoui, Azedine Boulmakoul, and Rachid Oulad Haj Thami

CONTENTS

DOI: 10.1201/9781003255635-19

19.1 INTRODUCTION

Traffic management is an essential component of modern mobility, as it enables the best possible use of the existing transport network. It monitors and controls different types of traffic to avoid congestion and improve traffic conditions. As a consequence, road users can reach their destination as quickly and safely as possible and reduce the negative impacts of mobility on the environment, such as air pollution. The use of complex network theory in the analysis of the characteristics and dynamics of urban transportation networks has attracted and continues to attract great interest (Boulmakoul et al. 2021; Badaoui et al. 2021; Lu and Lu 2022; Karim et al. 2020). Urban traffic systems networks are composed of a set of roads intersections interconnected by roads segments, represented as nodes and edges in the terminology of complex networks. The contribution of each junction (node) in the analysis of the traffic state differs depending on different concepts. In this study we focus on the contribution of nodes in the rapid spread of information, which is the spread of traffic congestion. These nodes are playing the role of bridges connecting other nodes together, which are also defined as nodes spreaders. Traffic congestion can be measured using different characteristics such as speed, flow, or travel time or the combination of these metrics. In this chapter we used the travel time as a measure of congestion, proposing the fuzzification of this measure to take into account the imprecision in the calculations. Therefore, we proposed a fuzzy weighted betweenness centrality for the detection of node spreaders, i.e., roads junctions responsible for the spread of traffic congestion. Based on these results, we'll be creating communities of subcongested areas. In this chapter, we begin by defining traffic characteristics in Section 19.2. Section 19.3 introduces the basics of fuzzy logic theory, while Section 19.4 discusses complex network theory. In Section 19.5, we discuss the proposed method and then conclude.

19.2 TRAFFIC CHARACTERISTICS

Intelligent transportation systems are established to manage, monitor, and control traffic networks to ensure better use of the existing road infrastructure and to improve traffic comfort. The management of transportation systems is based on the calculation and analysis of traffic measurements, which characterize and describe the traffic situation. These measurements are classified into three levels, the microscopic level, where the measurements collected represent vehicle-level data, such as the speed of each vehicle. At the macroscopic level, the collected measurements reflect the traffic condition from a global point of view, which is calculated as an aggregation of the microscopic data. And there is the mesoscopic level that lies between the two. In this section, we will present the following macroscopic metrics: density, flow, and travel time.

19.2.1 Density

The density of a road is defined as the number of vehicles given the length of a lane in a road segment, expressed as follows:

$$\rho = \frac{n}{L} \qquad (19.1)$$

Where n is the number of vehicles and L is the length of lane. It should be noted that the length of vehicles is not taken into consideration in the calculation of this measure, which influences the results of the calculation. Another alternative for the calculation of density using the velocity is defined as follows:

$$\rho = \frac{\delta}{v} \qquad (19.2)$$

Where δ is the traffic flow, and v is the velocity. This measure is fundamental in the analysis of transportation road networks. It represents the state of the traffic, i.e., if a road segment has a low value of density, the traffic is practically fluid (at its free-flow state) and the vehicles are traversing the road segment at nearly the maximum allowed speed. But if the density value is high or reaches its maximum, the stream of vehicles and their speeds decrease to zero, causing the formation of traffic congestion.

19.2.2 Traffic Flow

The study of the interaction between road users and infrastructure, which investigates the movements of drivers and vehicles between two points, is known as traffic flow. This metric is defined as the number of vehicles ΔN passing through a segment at a given time ΔT.

$$\delta = \frac{\Delta N}{\Delta T} \qquad (19.3)$$

One of the oldest traffic flow theories is the Grienshields model proposed in (Greenshields et al. 1934). This theory is based on the assumption that there is a linear relationship between density and velocity. It is represented with the following equation:

$$v(\rho) = v_f \left(1 - \frac{\rho}{\rho_c} \right) \qquad (19.4)$$

Where v_f is the velocity at the free flow (the free speed) and ρ_c is the maximum density, also called congestion density. So the flow function can be defined as follows:

$$\delta = \rho \times v \qquad (19.5)$$

$$\Rightarrow \emptyset(\rho) = \rho \times v_f \times \left(1 - \frac{\rho}{\rho_c} \right) \qquad (19.6)$$

$$\Rightarrow \emptyset(\rho) = \rho \times v_f - \frac{\rho 2 \times v_f}{\rho_c} \qquad (19.7)$$

Grienshields (Greenshields et al. 1934) proposed a representation of the relationships between traffic flow, density, and velocity measures in the form of fundamental diagrams of "speed-density", "speed-flow", and "flow-density". Figure 19.1 represents the flow-density fundamental diagram, which will be the focus of our study.

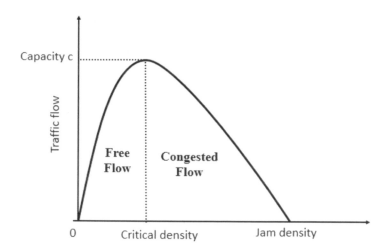

FIGURE 19.1   Fundamental diagram: flow-density.

The two measurement frameworks for flow and density are different: Flow has a temporal measurement at a specific point in space. While density, on the other hand, refers to a spatial measurement at a specific point in space (Hall 1996). The flow density diagram is used to depict the traffic situation on a road. This phenomenon occurs when the traffic demand exceeds the supply, which can be caused by several reasons such as severe weather conditions or roads closure due to accidents or public works. Therefore, the important number of jammed vehicles on the road segment leads to slower velocity and lower traffic flow.

19.2.3 Travel Time

Travel time is a very important metric. It shows how long it takes a traveler to reach his or her destination and provides crucial information to help save time by choosing the best routes, avoiding overtime, and reducing transportation costs. The primary concern of traffic managers is to avoid traffic jams, which are the main reason for most travel delays, whether it is a personal or business trip, as in the case of logistics (ensuring reliable delivery time). In literature, there are several travel time measurements proposed (Li and Chen 2014; Afrin and Yodo 2020; Kondyli et al. 2018; U.S. Bureau of Public Roads 1964). In this study, we will use link performance function, denoted "Bureau of Public Roads (BPR)" (U.S. Bureau of Public Roads, 1964). Which approximates the travel time within a link by means of the relationship between the time of travel and the flow. It is assumed that the travel time can be modeled by the method proposed by the U.S. Bureau of Public Roads (U.S. Bureau of Public Roads, 1964). The method is expressed as follows:

$$TT_a = TT_0 \left[1 + \alpha \left(\frac{\delta_a}{C_a} \right)^\beta \right],$$

(19.8)

Where TT_a is the actual travel time of link a, TT_0 is representing the travel time at free flow, δ is the traffic flow, C_a represents the practical capacity of link a and α and β are function parameters, generally at 0.15 and $\beta = 4.00$.

19.3 FUZZY LOGIC

In the mid-1960s–1970s, Lotfi Zadeh (Zadeh 1965) introduced to the field of science and technology a new theoretical concept of fuzzy logic based on fuzzy sets, which is capable of dealing with imprecision and uncertainty that are often a major concern in the study of various decision problems in real life. Many extensive applications of fuzzy logic can be found in various fields, such as transportation, artificial intelligence, pattern recognition, etc.

19.3.1 FUZZY SETS

Fuzzy sets are defined as a generalization of classical crisp sets characterized by a membership function that represents the grade of belongingness to the set. The elements of fuzzy sets are represented by a membership function with the degree to which the element belongs to the fuzzy set, in the closed interval [0,1]. Let X be a universal set and A a fuzzy set in X characterized by a membership function μ_A which represents the grade of belongingness of x in A. The fuzzy set is expressed by the set of pairs (x, $\mu_A(x)$).

$$\mu_A : X \rightarrow [0,1]$$
$$x \rightarrow (x, \mu_A(x))$$

(19.9)

19.3.2 FUZZY NUMBERS

Among the various types of fuzzy sets, those which are defined on the universal set R of real numbers are of particular importance. They can, under certain conditions, be considered fuzzy numbers, which reflects the human perception of an uncertain numerical quantification (Hanss 2005). A fuzzy set is called a fuzzy number if it satisfies the following properties:

a. It is a convex fuzzy set.
b. It is a normalized fuzzy set.
c. There is exactly one $\bar{x} \in \mathbb{R}$ with $\mu_{\tilde{A}}(\bar{x}) = 1$, that is $core(\tilde{A}) = \bar{x}$.
d. It is piecewise continuous.

Where $\bar{x} = core(\tilde{A})$ which shows the maximum degree of membership $\mu_{\tilde{A}}(\bar{x}) = 1$. In accordance with the notation of the value that appears most frequently in the data samples, it is referred to as the modal value of the fuzzy number. The modal value is also known as the peak value, central value, or mean value (Hanss 2005).

A fuzzy number is called symmetric if its membership function $\mu_{\tilde{A}}$ satisfies the condition:

$$\mu_{\tilde{A}}(\bar{x}+x) = \mu_{\tilde{A}}(\bar{x}-x), \ \forall x \in \mathbb{R}, \tag{19.10}$$

The last two properties are preferably used for symmetric fuzzy numbers.

19.3.3 Types of Fuzzy Numbers

In fuzzy logic theory, an important number of fuzzy sets were proposed. Among which some membership functions $\mu_{\tilde{A}}(x)$ have more significant importance, notably with regard to use as fuzzy numbers in applied fuzzy arithmetic. Due to the ease of the membership functions and the best depiction of actual problems, we will use the most significant and widely used fuzzy numbers, represented in this section.

19.3.3.1 Triangular Fuzzy Number

Triangular fuzzy numbers (TFNs) are also called, linear fuzzy numbers due to their simple linear type membership function. As an abbreviation, we can introduce the notation: $\tilde{A} = tfn(l,k,r)$.

We can define a TFN with the membership function:

$$\mu_{\tilde{A}}(x) = \begin{cases} 0 & for \ x < l \\ \dfrac{x-l}{k-l} & for \ l \leq x \leq k \\ \dfrac{r-x}{k-l} & for \ k \leq x \leq r \\ 0 & for \ x > r \end{cases}, \ \forall x \in \mathbb{R} \tag{19.11}$$

The parameter k represents the fuzzy number's modal value, and the parameters l and r represent the left-hand and right-hand worst-case deviations from the modal value, respectively, represented in Figure 19.2.

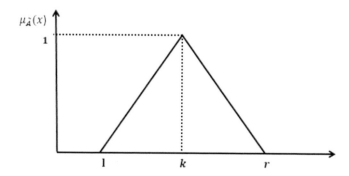

FIGURE 19.2 Triangular fuzzy number.

19.3.4 ARITHMETIC OPERATIONS FOR TRIANGULAR FUZZY NUMBERS

In this section, we represent the arithmetic operations for the TFNs used in our study resulting in another TFN. Let $A = (l_A, k_A, r_A)$ and $B = (l_B, k_B, r_B)$ be two TFNs.

19.3.4.1 Addition
The addition of two TFNs is defined as follows:

$$A + B = (l_A + l_B, k_A + k_B, r_A + r_B) \tag{19.12}$$

19.3.4.2 Multiplication
Let α be a scalar, the scalar multiplication of a TFN is given by:

$$\alpha A = \begin{cases} (\alpha l_A, \ \alpha k_A, \ \alpha r_A) & \text{if } \alpha \geq 0 \\ (-\alpha r_A, \ \alpha k_A, \ -\alpha l_A) & \text{if } \alpha < 0 \end{cases} \tag{19.13}$$

19.3.4.3 Canonical Representation of Operations on Fuzzy Numbers
The canonical representation of operations on TFNs is based on the graded mean integration representation method (Chou 2003), and is defined as follows:

Given a TFN $\tilde{A} = (l_{\tilde{A}}, k_{\tilde{A}}, r_{\tilde{A}})$, the graded mean integration representation of TFN $\tilde{A}$ is defined as:

$$P(\tilde{A}) = \frac{1}{6}(l_{\tilde{A}} + 4 \times k_{\tilde{A}} + r_{\tilde{A}}) \tag{19.14}$$

Let $\tilde{A} = (l_{\tilde{A}}, k_{\tilde{A}}, r_{\tilde{A}})$, $\tilde{B} = (l_B, k_B, r_B)$ be two TFNs. By applying the previous equation, the graded mean integration representation of TFNs $\tilde{A}$ and $\tilde{B}$ can be obtained, respectively, as follows:

$$P(\tilde{A}) = \frac{1}{6}(l_{\tilde{A}} + 4 * k_{\tilde{A}} + r_{\tilde{A}}), \ P(\tilde{B}) = \frac{1}{6}(l_{\tilde{B}} + 4 * k_{\tilde{B}} + r_{\tilde{B}})$$

The representation of the addition operation $\oplus$ on TFNs $\tilde{A}$ and $\tilde{B}$ can be defined as:

$$P(\tilde{A} \oplus \tilde{B}) = P(\tilde{A}) + P(\tilde{B}) = \frac{1}{6}(l_{\tilde{A}} + 4 * k_{\tilde{A}} + r_{\tilde{A}}) + \frac{1}{6}(l_{\tilde{B}} + 4 * k_{\tilde{B}} + r_{\tilde{B}}) \tag{19.15}$$

19.3.5 FUZZY GRAPHS

A graph illustrates a specific relationship between the elements of a set V in terms of the magnitude of the relationship between any two elements of V. As long as the appropriate weights are known, we can solve this problem by using a weighted graph. However, the relations in real-world problems are not to be known, where the

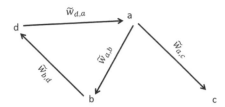

FIGURE 19.3 Fuzzy graph.

relationships are fuzzy in nature. Most often, the weights and relations are "fuzzy" in the most natural sense. Therefore, a fuzzy relation is the right solution to deal with the situation in a better way (Mathew et al. 2018). Fuzzy graphs were introduced for the first time by Kaufman (Kauffmann 1973). Rosenfeld has introduced subgraphs, paths, connectedness, cliques, bridges, forests, and trees, among other fuzzy analogs of graph theory (Rosenfeld 1975). The concepts of connectivity in fuzzy graphs, including vertex and edge connectivity, were independently proposed by (Yeh and Bang 1975), as they were the first to use fuzzy graphs in data clustering.

A fuzzy graph $\tilde{G}$ has different definitions. In general it is defined as $\tilde{G} = (\tilde{V}, \tilde{E})$, where $\tilde{V}$ is a nonempty set of fuzzy vertices and $\tilde{E}$ is a fuzzy set of edges between vertices, or it can be defined with a crisp set of vertices and fuzzy edges. In our study, we are defining our fuzzy graph based on fuzzy edges weights, which are defined as $\tilde{G} = (V, E, \tilde{W})$, where V is the set of vertices, E is the set of edges connecting the vertices, and $\tilde{W}$ are the fuzzy weights of the edges, defined with a fuzzy membership function $\mu_{\tilde{A}}(x)$, see Figure 19.3.

19.4   COMPLEX NETWORKS

Network analysis is an effective tool for describing complex phenomena such as transportation, epidemiology, social problems, and a variety of other challenges. Since their collective behavior cannot be predicted from their individual components, these systems are called complex systems. On the other hand, to be able to control and eventually anticipate these complex systems, a clear understanding of the mathematical description is crucial. Despite the varieties of studied complex phenomena, they have all a similar representation as a graph composed of a set of vertices representing the actors of the network and a set of edges representing the type of relationships and interactions between the actors. One advantage of modeling a system as a graph is that problems become simpler and more tractable.

19.4.1   CENTRALITY MEASURES

In graph theory, the importance of nodes is characterized by means of the concept of centrality measures, which play an important role in understanding and characterizing real-world systems. Several measures of centrality, including degree centrality, proximity and interdependence centralities, and eigenvector centrality, among others, have been proposed in the literature.

Let G be a graph with n × n vertices, and A the binary adjacency matrix if the graph in unweighted, which consists of elements a_{ij}, where

$$a_{ij} = \begin{cases} 1 & \text{the vertices } i \text{ and } j \text{ are connected} \\ 0 & \text{otherwise} \end{cases}$$

Or if the graph is weighted, the adjacency matrix is defined with $a_{ij} = w_{ij}/w_{ij} \in \mathbb{R}$ representing the weights.

19.4.1.1   Local Centrality Measures

The degree centrality measure studies the network from a local viewpoint, i.e., it considers only the direct neighbors of the studied node. Thus, it is classified as a local centrality measure. Its computation is as follows:

$$C_D(i) = \sum_{j=1}^{n} a_{ij} \tag{19.16}$$

This centrality measure reflects the importance of the nodes in relation to the spatial geography of the network.

19.4.1.2   Global Centrality Measures

These measures analyze the network from a global point of view, taking into consideration the whole topology of the network instead of only the direct connection. Global centrality measures are closeness and betweenness centralities and eigenvector centrality. The centrality measures of closeness and betweenness are based on the calculation of shortest paths (geodesics). The former metric measures the rate at which information spreads from a node to all other nodes, whereas the latter depicts the node's influence on information transmission in the network. Another type of centrality is based on eigenvectors, which rank nodes' importance by not only computing their degree but also taking into account the importance of their neighbors. Several eigenvector centrality variants have been proposed, including Katz score (Katz 1953), alpha-centrality (Bonacich and Lloyd 2001), and other variants based on the random walk, the most famous of which is PageRank (Page et al. 1999). The degree centrality, closeness, betweenness, and eigenvector are the most important and significant centrality measures in urban transportation networks. These measures have received a great deal of attention in the analysis of traffic conditions, or more specifically, the study of traffic congestion. In this chapter, we will be studying the betweenness centrality measure in the study of traffic congestion and the detection of most congested road intersections.

19.4.1.2.1   Betweenness Centrality

Centrality metrics are essential in understanding network structure, since they capture the relative importance of individual nodes in the overall network. The identification of a network's most central nodes is a fundamental problem in network analysis. Betweenness centrality (BC) measures the importance of a node according to the number of shortest paths that passes through it. The higher the BC of a node, the more the node participates in the spread of the information through shortest paths. For instance, this includes identifying important intersections in road networks or influencers in social networks. Formally, the BC of a node v is defined as:

$$C_B(v) = \frac{1}{n(n+1)} \sum_{s \neq v \neq t} \frac{\sigma_{st}(v)}{\sigma_{st}} \tag{19.17}$$

Where n is the number of nodes, σ_{st} is the number of shortest paths between two nodes s and t, and $\sigma_{st}(v)$ is the number of these paths that go through node v.

Because of the high cost of computing BC (typically $O(n^3)$) due to the computation of all shortest paths for each node, Brandes (Brandes 2001) has proposed a faster computation of betweenness, through the notion of single-source shortest path (SSSP) with the complexity of $O(nm)$ for unweighted networks and $O(nm+n^2\log(n))$ for weighted networks, where n and m are the number of nodes and edges, respectively.

Brandes's algorithm starts with the computation of all shortest path for each source vertex $s \in V$ once using breadth-first search (BFS) for unweighted and Dijkstra for weighted graphs. During the exploration of shortest paths, the set of predecessors is created as well as the number of shortest paths between each vertex, and its predecessors $P_s(w)$ are computed as σ_{sv}. Afterward, the dependencies of every source vertex are computed as follows:

$$\delta_{st}(v) = \frac{\sigma_{sv}}{\sigma_{sw}} (1 + \delta_{st}(w)) \tag{19.18}$$

And finally, the BC of a vertex is defined as the sum of all its dependencies, expressed as follows:

$$\delta_s(v) = \sum_{w:v \in P_s(w)} \frac{\sigma_{sv}}{\sigma_{sw}} (1 + \delta_s(w)) \tag{19.19}$$

The authors in (Henry et al. 2019) considered both spatial and temporal characteristics in the analysis of correlations between dynamic edge BC and traffic flow, proposing different variants of edge BC. The availability of real-time sensors and small devices distributed across geographical areas allows for the discovery of spatiotemporal patterns in urban traffic, in which valuable information is transmitted. In recent years, several dynamic BC measures for evolving networks were proposed. They can be divided into exact algorithms and approximation algorithms of the BC. These dynamic proposed algorithms are either semi-dynamic algorithms that recompute an approximation of betweenness in connected graphs after batches of edge insertions (Bergamini 2015a) proposing incremental approximation algorithms for both weighted and unweighted networks, or fully dynamic approximation algorithms (Bergamini et al. 2015b) applied for edges addition and weight decrease. In (Chernoskutov et al. 2015), the authors proposed an approximation to the computation of the BC for dynamically changing graphs, allowing a reduction in the cost of the computation of the BC after each update of the network, which is the basis of the graph enlargement. This approach is divided into two steps: the first step is to create a condensed graph based on the k vertices with the highest centrality values, which leads to a smaller representation of the graph, then the second step is to apply the updates on the condensed graph by adding new edges and finally deducing the approximate intercommunity centrality using the calculated intercommunity centrality before and after the changes.

Urban traffic networks are in most cases represented as weighted networks that are highly dynamic, with continuous updating of data as a function of weights. Therefore, the proposed dynamic betweenness is not suitable for reducing updates in the networks, since the weight of the entire network changes over time.

19.4.2 COMMUNITY DETECTION

Community detection, proposed by Girvan and Newman (Girvan and Newman 2002), is one of the most popular topics in the field of complex network analysis. A community is defined as a set of densely connected nodes and edges forming sub-networks, with weak connections to the rest of the network. The algorithm proposed by (Girvan and Newman 2002) is based on removing the edges with the highest edge-betweenness, which are the peripheries of the community, to construct highly connected actors of the community. Several algorithms were proposed to tackle the problem of community detection, such as InfoMap, Walktrap, and Louvain community detection that is based on modularity, which tries to maximize the difference between the actual number of edges in a community and the expected number of edges in the community. However within some communities, certain nodes play a more important role, called leaders, whose detection facilitates the forming of communities via the discovery of his followers. Most leaders' detection methods are based on the computation of centrality measures, which reflects the importance of the node in the network. In this context, wide research studies have been proposed. In (Ahajjam et al. 2018), the authors have proposed two approaches for the leaders community detection algorithms based on the detection of leaders using centrality measures for undirected and unweighted networks. In order to avoid community overlapping, they have proposed to delete each detected community from the network and then continue the process of community detection on the remaining nodes in the network.

19.5   PROPOSED STUDY

In this study, we are proposing an algorithm for the detection of subcongested networks in an urban transportation system based on the proposed fuzzy weighted inversed BC and the nearest neighbors.

19.5.1   URBAN TRAFFIC NETWORK

Urban traffic networks are known to be highly dynamic, with the continuous flow of vehicles passing through different directions. Dynamic networks are often represented as a series of static graphs in different timestamps, named snapshots. In this study, the urban traffic network is represented as a dynamic directed fuzzy weighted network, with road intersections as nodes and road segments as edges, having as a weight the estimated travel time. Let $G_{t_i}(V_{t_i}, E_{t_i}, \tilde{W}_{t_i})$ be our dynamic transportation network, and V_{t_i}, E_{t_i} are the set of vertices and edges at time period t_i, respectively, where $\tilde{W}_{t_i}$ is the set of the edges' weights. The graph $[G_{t_i}]_{(i=1.n)}$ is defined over different snapshots together representing time series dynamic graphs.

19.5.2 Travel Time Fuzzification

Due to the uncertainties in data collection from logistic development tools, it is worth paying attention to deal with the computation of travel time in a fuzzy manner. We propose the fuzzification process via the aggregation of travel time of a time period ΔT defined in the interval $[t - \Delta t, t]$ to be converted into symmetric TFNs as follows:

$$TFN_a^{\Delta T}(TT) = TFN_a^{\Delta T}(k-l,k,k+r) = TFN(\tau_a^{\Delta T} - \sigma_a^{\Delta T}, \tau_a^{\Delta T}, \tau_a^{\Delta T} + \sigma_a^{\Delta T}), \quad (19.20)$$

Where the kernel is represented by the mean travel time denoted $\tau_a^{\Delta T}$, the distance from the mean to the left and right hand spreads is the standard deviation denoted as $(\tau_a^{\Delta T} - \sigma_a^{\Delta T}, \tau_a^{\Delta T} + \sigma_a^{\Delta T})$, respectively:

$$\tau_a^{\Delta T}(t) = \frac{1}{n} \sum_{i=0}^{n-1} TT_a(t - i\Delta T), \quad \sigma_a^{\Delta T}(t) = \sqrt{\frac{1}{n} \sum_{i=0}^{n-1} [TT_a(t - i\Delta T) - \tau_a^{\Delta T}(t)]^2} \quad (19.21)$$

19.5.3 Leader Node Detection

BC has been applied in several studies of traffic networks, cited in Section 19.3. It is a very important measure in the study of nodes spreaders. In our study we propose a fuzzy weighted BC for the detection of nodes spreaders based on a fuzzy Dijkstra algorithm to deal with our fuzzy weights. Authors (Deng et al. 2012) have proposed a fuzzy Dijkstra algorithm for the exploration of shortest paths in a fuzzy environment using the graded mean integration representation of fuzzy numbers represented in Section 19.2. We propose a fuzzy modification of Brandes's BC by incorporating a fuzzy Dijkstra to detect network leaders. The main feature of this measure is that it detects node propagators, yet it is based on the calculation of shortest paths, which implies that in the case of transport networks, it will detect nodes located on the shortest paths with the lowest weights, i.e., the shortest uncongested paths. But the goal of our study is the detection of the junction lying on the shortest path with the heaviest weights representing the congestion. Therefore, we are proposing to inverse the value of the computed fuzzy BC. The process of leaders' detection is defined as follows:

1. Computation of fuzzy weighted BC for every node in the network. The network leaders in our study are responsible for the fast propagation of traffic congestion.

2. The computation of the average of the computed inversed fuzzy BC, which is defined as

$$Avg_{FBC} = \frac{1}{n}\sum_{i=1}^{n} C_{FBC}(v_i)^{-1} \qquad (19.22)$$

3. After computing the fuzzy weighted BC for all nodes, we selected the most important ones as leaders, the ones with BC $\geq Avg_{FBC}$.

Due to the high dynamics of transportation networks, and especially in the updates in the traffic measures such as the estimated travel time of road segments, we have opted for fuzzy Brandes BC, which is recomputed at every snapshot.

19.5.4 COMMUNITY DETECTION

The occurrence and propagation of traffic jams only affects a part of the network, so there are congested and uncongested areas. Therefore, in our study, in order to reduce the computations, we will focus only on the most important road intersections detected on the basis of their BC, which represents their contribution to the rapid propagation of traffic congestion. The detection of communities is done as follows:

1. Ranking of selected nodes according to their inversed BC value.
2. Starting from the node with the highest centrality, we compute the neighbors of degree = 1 (directly connected neighbors) to form a local community.
3. Delete the leader and the community members from the list of selected nodes.
4. Repeat steps 2-3 until no nodes are left.

As a result, we get subcongested networks of congested junctions without overlapping, with the degree of congestion of the community determined according to the degree of congestion of the community leader.

19.5.5 STUDIED AREA

The chosen area for the study is represented in Figure 19.4, with generated data of vehicle flows using tracI, an online simulator of SUMO.

Our network is composed of 676 nodes representing junctions and 1531 edges representing the road segments, represented in Figure 19.4. The red dots in Figure 19.4b are representing road intersections. Recurrent traffic congestion happens during morning rush hours or at evening rush hours, causing congestion in different zones. Due to the lack of the availability of real-world data, the data are generated from tracI SUMO in real time.

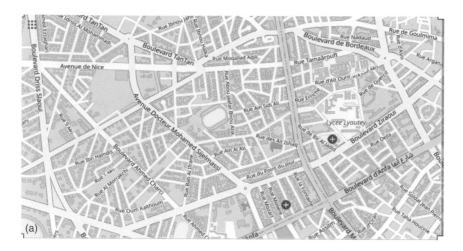

FIGURE 19.4   Studied zone of Casablanca (OpenStreetMap 2017). a) Studied region of Casablanca, b) Studied area of Casablanca from SUMO.

19.6   CONCLUSION

The study of urban traffic systems is crucial for the detection of traffic congestion to help urban planners in the management and control of the traffic. In this study we have proposed a community detection algorithm of subcongested networks based on focusing only on the junctions with the higher contribution of the propagation of traffic congestion. The selection of these junctions was based on the computation of fuzzy BC, which takes as a weight fuzzy estimated travel time. Our solution provides a resilient solution for the detection of congested areas during different periods of time. As a future work, the implementation of the proposed method using parallel computation of fuzzy BC to form communities representing congestion will be undertaken.

REFERENCES

Afrin, Tanzina, and Nita Yodo (2020). A survey of road traffic congestion measures towards a sustainable and resilient transportation system. Sustainability 12, no. 11: 4660.

Ahajjam, Sara, Mohamed El Haddad, and Hassan Badir (2018). A new scalable leader-community detection approach for community detection in social networks. Social Networks 54: 41–49.

Badaoui, Fatima-Ezzahra, Azedine Boulmakoul, and Rachid Oulad Haj Thami (2021). Fuzzy dynamic centrality for urban traffic resilience. In 2021 International Conference on Data Analytics for Business and Industry (ICDABI), pp. 12–16. IEEE, Bahrain.

Bergamini, Elisabetta, and Henning Meyerhenke (2015). Fully-dynamic approximation of betweenness centrality. In Algorithms-ESA 2015, pp. 155–166. Berlin, Heidelberg: Springer.

Bergamini, Elisabetta, Henning Meyerhenke, and Christian L. Staudt (2015). Approximating betweenness centrality in large evolving networks. In 2015 Proceedings of the Seventeenth Workshop on Algorithm Engineering and Experiments (ALENEX), pp. 133–146. Society for Industrial and Applied Mathematic, San Diego, CA, USA.

Bonacich, Phillip, and Paulette Lloyd (2001). Eigenvector-like measures of centrality for asymmetric relations. Social Networks 23, no. 3: 191–201.

Boulmakoul, Azedine, Fatima-ezzahra Badaoui, Lamia Karim, Ahmed Lbath, and R. O. Thami (2021). Fuzzy spatiotemporal centrality for urban resilience. In International Conference on Intelligent and Fuzzy Systems, pp. 796–803. Springer, Cham.

Brandes, Ulrik (2001). A faster algorithm for betweenness centrality. Journal of Mathematical Sociology 25, no. 2: 163–177.

Chernoskutov, Mikhail, Yves Ineichen, and Costas Bekas (2015). Heuristic algorithm for approximation betweenness centrality using graph coarsening. Procedia Computer Science 66: 83–92.

Chou, Ch-Ch (2003). The canonical representation of multiplication operation on triangular fuzzy numbers. Computers & Mathematics with Applications 45, no. 10–11: 1601–1610.

Deng, Yong, Yuxin Chen, Yajuan Zhang, and Sankaran Mahadevan (2012). Fuzzy Dijkstra algorithm for shortest path problem under uncertain environment. Applied Soft Computing 12, no. 3: 1231–1237.

Girvan, Michelle, and Mark E. J. Newman (2002). Community structure in social and biological networks. Proceedings of the National Academy of Sciences 99, no. 12: 7821–7826.

Greenshields, Bruce Douglas (1934). The photographic method of studying traffic behavior. Proceedings of the 13th annual meeting of the highway research board, vol. 13, pp. 382–399, Washington, D.C., USA.

Hall, Fred L (1996). Traffic stream characteristics. Traffic Flow Theory. U.S. Federal Highway Administration, vol. 36, Washington, D.C., USA.

Hanss, Michael (2005). Applied fuzzy arithmetic. Springer-Verlag, Berlin Heidelberg.

Henry, Elise, Loïc Bonnetain, Angelo Furno, Nour-Eddin El Faouzi, and Eugenio Zimeo (2019). Spatio-temporal correlations of betweenness centrality and traffic metrics. In 2019 6th International Conference on Models and Technologies for Intelligent Transportation Systems (MT-ITS), pp. 1–10. IEEE, Krakow, Poland.

Karim, Lamia, Azedine Boulmakoul, Ghyzlane Cherradi, and Ahmed Lbath (2020). Fuzzy centrality analysis for smart city trajectories. In International Conference on Intelligent and Fuzzy Systems, pp. 933–940. Springer, Cham.

Katz, Leo (1953). A new status index derived from sociometric analysis. Psychometrika 18, no. 1: 39–43.

Kauffmann, A. (1973). Introduction to the theory of fuzzy sets. Vol. 1, Orlando, Florida: Academic Press.

Kondyli, Alexandra, Bryan St. George, and Lily Elefteriadou (2018). Comparison of travel time measurement methods along freeway and arterial facilities. Transportation Letters 10, no. 4: 215–228.

Li, Chi-Sen, and Mu-Chen Chen (2014). A data mining based approach for travel time prediction in freeway with non-recurrent congestion. Neurocomputing 133: 74–83.

Lu, Jian, and Dengtian Lu (2022). Modelling traffic congestion propagation based on data analysis method. In Sixth International Conference on Electromechanical Control Technology and Transportation (ICECTT 2021), vol. 12081, pp. 959–964. SPIE, Chongqing, China.

Mathew, Sunil, John N. Mordeson, and Davender S. Malik (2018). Fuzzy graph theory. Vol. 363. Berlin, Germany: Springer International Publishing.

OpenStreetMap contributors. (2017). Planet dump retrieved from https://planet.osm.org.

Page, Lawrence, Sergey Brin, Rajeev Motwani, and Terry Winograd (1999). The PageRank citation ranking: Bringing order to the web. Stanford InfoLab, Stanford, USA.

Raymond T. Yeh, and Sung Y. Bang (1975). Fuzzy graphs, fuzzy relations and their applications to cluster analysis. In Fuzzy Sets and Their Applications to Cognitive and Decision Processes. pp. 125–149, New York, NY.

Rosenfeld, Azriel (1975). Fuzzy graphs. In Fuzzy sets and their applications to cognitive and decision processes, pp. 77–95. Academic Press, New York USA.

U.S. Bureau of Public Roads (1964). Traffic assignment manual for application with a large, high speed computer. Vol. 2. US Department of Commerce, Bureau of Public Roads, Office of Planning, Urban Planning Division, USA.

Zadeh, Lotfi A (1965). Fuzzy sets. Information and Control 8 (3): 338–353.

20 Multiagent Modeling for Pedestrian Risk Assessment

Maroua Razzouqi and Azedine Boulmakoul

CONTENTS

20.1 INTRODUCTION

Traffic accidents aren't just a local or national problem that affects one or a few countries. It is a global problem from which all countries of the world suffer, but with a difference in the incidence of accidents and their negative repercussions on society. Road accidents have been around since the invention of cars, but their treatment was not taken seriously until 1960 when developed countries adopted scientific methods to solve this problem (Peden et al. 2004). This has, indeed, reduced the accident rate in these countries. But despite this, the number of road accidents remains high and the losses are enormous.

Road accidents differ in the frequency of their occurrence and in the two parties that collide. They might include two cars colliding, a car colliding with a fixed object on the road, with an animal, or with a pedestrian, which is the most common in urban areas. In this chapter, we are interested in this last case of an accident which includes a car and a pedestrian in the urban environment. The World Health Organization's studies and statistics have shown that pedestrians and cyclists account for about half

DOI: 10.1201/9781003255635-20

317

of all road crash victims worldwide, with a fatality rate for pedestrians of 23% of all road fatalities (World Health Organization 2015). Therefore, they must be given more attention and extra care to reduce the number of deaths.

In order to assess the pedestrian's behavior, pedestrian dynamics theories typically assume a decision structure with three levels of behavior: *strategic, tactical,* and *operational*. The pedestrian decides on the activities he or she intends to do in his environment and, if applicable, the scheduling between these activities at the strategic level. Because these decisions are generally made before entering the pedestrian area, simulation models of pedestrian behavior consider them exogenous. Strategically determined activities can be carried out in a variety of ways and locations. The tactical level determines the precise manner in which these activities are carried out as well as the routes used to reach these locations. The actual movement of pedestrians is described at the operational level.

Models that simulate pedestrian behavior are classified in a variety of ways. One of the most common is the scale-based classification, which classifies the models into *microscopic* models, *mesoscopic* models, and *macroscopic* models, as shown in Figure 20.1.

The macroscopic crowd simulation models were the first to appear. The goal is to reduce the simulation scenario to a graph network and describe pedestrians as cumulative and fluid densities. To put it another way, they simulate a large number of pedestrians without taking into account their behavior and interactions. Although this method allows for quick simulations, it has a low spatial resolution. Microscopic models are used to identify interactions between pedestrians as well as interactions between pedestrians and their surroundings. They enable a more in-depth examination of pedestrian design and interaction. Pedestrian behavior models based on microscopic analysis can be approached through simulation or analysis. The mesoscopic models are a hybrid of the two models. Pedestrians are modeled as a group based on some shared characteristics such as the same departure time, the same route, or simply walking at the same speed.

Microscopic simulation can be approached in a variety of ways. We can look to the cellular-based models pioneered by Gipps and Marksjö (1985). Each pedestrian

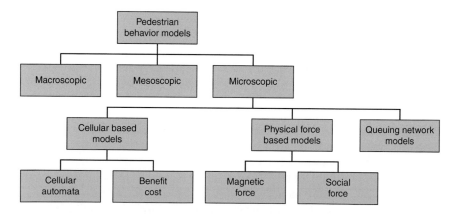

FIGURE 20.1   The scale-based classification of pedestrian behavior models.

in a cell is represented by a particle. The surfaces of pedestrian traffic are represented by a square grid. Each cell can only have one pedestrian, to which a score describing the pedestrian's proximity is assigned. This score represents the repelling effect of nearby pedestrians and must be balanced against the pedestrian's gain as he moves toward his destination. When two pedestrians pass through a field, the score in each cell equals the sum of the scores produced by each pedestrian (Gipps and Marksjö 1985). Then there are physical force models. The magnetic force was the first physical force model, and it is based on Coulomb's law and the assumption that pedestrians and other obstacles are positively charged particles, while their destinations are negatively charged particles that attract them. Each pedestrian's movement was simulated by the attraction between positively and negatively charged particles (Okazaki & Matsushita 1993). The social force model is another physical force model that is commonly used in the simulation of pedestrian behavior. Henderson's fluid crowd modeling method served as the foundation for Helbing and Molnar's social force model.

The multiagent approach is still useful for modeling the complexity of this system in which information, control, processing, or all of them are distributed rather than centralized.

Godara et al. created a multiagent model that accounts for both pedestrian and vehicular traffic, as well as the interactions that occur as a result of their actions. To simulate vehicular traffic, they used a cellular automata model. Simple behavioral rules are used in conjunction with an agent-based approach to simulate pedestrian traffic (Godara et al. 2007). Liu et al. introduced a microscopic pedestrian behavior model that takes into account different interactions on pedestrian dynamics at crosswalks. A modified social force model was developed that took into account evasion behavior with counter-flow pedestrians, following behavior with leader pedestrians, and collision avoidance behavior with vehicles (Liu et al. 2017). Shaaban and Abdelwarith investigated various pedestrian characteristics such as gender, age, clothing type, carrying bags, using mobile phones, and crossing in a group. They concluded that the risk factor was lower in the case of females and pedestrians crossing in a group and that there were no differences in other attributes in this study (Shaaban & Abdelwarith 2020). Gupta et al. also presented an agent-based simulation model for pedestrian negotiations in autonomous vehicles. They were able to demonstrate the critical role of pedestrian-vehicle communication in improving traffic scenario efficiency (Gupta et al. 2018). They also took an in-depth look at vehicle-pedestrian negotiations considering social rules for both risk-taking and risk-averse pedestrians' cases, extending the previous model to multiple pedestrian-vehicle interactions to reduce the waiting time for vehicles at intersections (Gupta et al. 2019). In addition, Yang et al. proposed a framework for dealing with vehicle-pedestrian interactions (VPIs) in uncontrolled crossing scenarios. They did this by incorporating a multistate social force pedestrian motion model into the framework (Yang et al. 2020).

As indicated in the title, in this chapter, we present a multiagent model that will allow us to assess the risk incurred by pedestrians using measures called risk indicators. In order to get into the thick of things, we must first dissect this title into two parts: a first part to understand the multiagent modeling and another section to define these risk indicators. Thus, our chapter is presented as follows: In the second section, we provide an introduction to distributed artificial intelligence and multiagent modeling. Then, in a third section, we present the risk indicators, and more precisely the fuzzy risk indicators. This is followed by a presentation of the simulation based on the model that we have developed, as well as the results obtained in Section 20.4. Finally, Section 20.5 summarizes the work and offers concluding remarks.

20.2 MULTIAGENT MODELING

Multiagent systems are one of the categories of distributed artificial intelligence (DAI), which is an extension of artificial intelligence (AI).

20.2.1 An Introduction to Distributed Artificial Intelligence

What does the term DAI add to the definition of AI? Considering a problem where we expect simultaneous actions of several autonomous components, AI validates this need for concurrency and parallelism by an immediate modification of the state variables of the system. This affects the coherence of the system and autonomy of its components, especially if it requires large data. DAI comes to solve this problem by distributing the problem to autonomous processing nodes (agents).

The distinction is that whereas AI models the intelligent behavior of a single agent, DAI is concerned with the intelligent behavior that is the result of several agents' cooperative activity.

This transition to collective behavior has given rise to new properties and new behaviors. Indeed, DAI seeks to address the shortcomings of the traditional AI approach by proposing the distribution of expertise among a group of agents who must be able to work and act in a shared environment and resolve potential conflicts. As a result, new AI concepts such as cooperation, action coordination, negotiation, and emergence have come to the fore.

DAI can then be defined as the branch of AI that is interested in modeling intelligent behaviors through cooperation between a group of agents (Bouquet et al. 2015).

In general, research in DAI can be classified into three main categories:

- *Parallel Artificial Intelligence* (PAI): It concerns the development of parallel languages and algorithms for DAI. PAI aims at improving the performance of AI systems without focusing on the nature of the reasoning or the intelligent behavior of a group of agents. However, the development of concurrent languages and parallel architectures can indeed have an important impact on DAI systems.
- *Distributed Problem Solving* (DPS): It is focused on how to divide a particular problem into a set of distributed and cooperating entities. It is also dealing with the way to share the knowledge of the problem and to obtain the solution.
- *Multiagent Systems* (MAS): These are made up of autonomous creatures known as agents and are the subject of this chapter. Agents, like computational units in DPS, collaborate to solve problems, but their innate ability to learn and make autonomous decisions gives them more freedom.

We will present MAS in detail in the following sections, but first, let's take some time to properly define an agent.

20.2.2 WHAT'S AN AGENT?

We can't define MAS without defining an agent. An agent is defined as a physical or abstract entity that can act on itself and its environment; has a partial representation of its environment; can communicate with other agents; and whose behavior is the result of its observations, knowledge, and interactions with other agents (Wooldridge 2009). In other words, it can figure out for itself what it needs to do to satisfy its design objectives rather than being told explicitly what to do at any given time.

The goal of each agent is to complete its assigned task while also adhering to some additional constraints. To achieve this goal, the agent first senses parameters in the environment to learn about the latter. An agent may also benefit from its neighbors' knowledge. This knowledge, along with the agent's previous actions and the goal, is fed into an inference engine, which determines the appropriate action to take.

The following characteristics allow agents to be broadly applicable and solve complex tasks:

- Sociability: Agents can share their knowledge and request information from other agents to improve their performance and achieve their goals.
- Autonomy: Each agent can execute the decision-making process and take appropriate action independently.
- Proactivity: Each agent predicts possible future actions based on its history, sensed parameters, and information from other agents. These predictions enable agents to take effective actions that achieve their objectives. This capability implies that the same agent can act in a variety of ways when placed in different environments.

20.2.3 WHAT ARE MULTIAGENT SYSTEMS?

While an agent working alone can take actions (based on autonomy), the true benefit of agents can only be realized when they collaborate with other agents (Dorri et al. 2018). MAS are made up of multiple agents that work together to solve a complex task. To interact successfully, these agents must be able to cooperate, coordinate, and negotiate with one another in the same way that we do with other people in our daily lives.

The idea behind learning is that the agent's perceptions should be used not only to choose actions but also to improve the agent's ability to act in the future. Learning is critical for an agent because it allows it to evolve, adapt, and improve.

20.2.4 WHY MULTIAGENT MODELING?

Because it is possible to incorporate high-quality behavior, agent-based simulation of pedestrian behavior that produces each individual as an independent and (more or less) intelligent actor in a simulated environment is an appealing way to generate pedestrian strength. Modeling the agent's rate of movement is also appealing because it can be done efficiently in terms of memory and calculation time. Another significant advantage is that an agent-based simulation separates pedestrian behavior from a specific spatial layout. This is due to the concept of a pedestrian being a self-contained, autonomous entity that has been introduced as an agent in its environment. As a result, environmental changes can be made with a degree in planning or train schedules without affecting pedestrian practice. Such adaptation is less common.

20.2.5 How Do We Model Our System?

After we've addressed the "what" and "why," it's time to address the "how".

The interaction of a vehicle and a pedestrian can be interpreted as a process of two agents mutually recognizing and affecting each other. Both agents can be conceptually divided into three perception layers: perception, interaction, and motion. All of those layers are linked to one another. To put it another way, the vehicle's perception layer employs a predictor to forecast the pedestrian's future movement. And the pedestrian perception layer employs an estimator for the time difference between the vehicle's real-time position and the pedestrian's crossing position. In the interaction layer, the pedestrian changes speed and direction, and the vehicle modifies its longitudinal speed control to avoid any unpleasant movement. Finally, the motion layer employs dynamics to carry out the desired actions.

The general architecture of the MAS is depicted in Figure 20.2. To collect information from the outside world, each agent has its own sensors (the five senses in the case of a pedestrian or driver, while for a vehicle it can be a camera, a stereo, a RADAR, a LIDAR, or any other type of sensor). The data collected from the environment (including other agents) via sensors are then analyzed in the inference mechanism, with exchanges with the knowledge base, and are stored to keep track of them. These data are then converted into action orders in the actuators. Using the shared information, each agent can interact with the other agents and update its knowledge base.

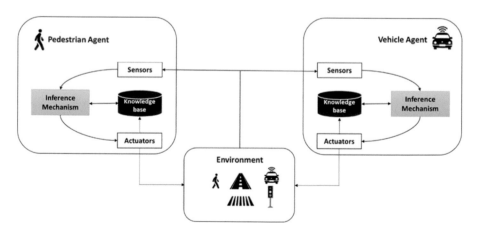

FIGURE 20.2 The system's architecture.

20.3   RISK INDICATORS

Risk indicators are used to study more frequent critical traffic events, making such incidents easier to analyze. Each indicator is evaluated based on its ability to consider collision risk, which is further subdivided into an event's initial conditions, the magnitude of any evasive action, and the injury risk in any traffic event.

The mutual exposure measurement of accident risk can be applied to a road network to identify the most dangerous areas as well as those with low or moderate risk. This will allow civil engineering safety operators to change the architecture and layout of these sections of the road network—for example, the construction of crossing bridges or even signs on streets or sidewalks to assist children in crossing. However, these safety operators will also need to investigate the modification of road traffic rules in order to adapt them to these purposes.

20.3.1   THE THEORY BEHIND RISK INDICATORS

Every traffic event that involves two or more road users in close proximity poses a risk. These occurrences are distinguished by their frequency and severity. The two latter measures are linked in such a way that we can calculate the frequency of extremely severe (including fatalities) but infrequent events based on the known frequency of less severe but more frequent incidents (Hydén 1987) (Svensson & Hydén 2006).

The probability that a traffic event will result in an accident can be calculated as a function of the initial conditions and the avoidance actions defined by those conditions. If this probability is greater than zero, the event is considered critical (Davis et al. 2011).

Johansson et al. assumed that in order to accurately estimate the risk of collision, a risk indicator should ideally reflect both aspects mentioned earlier (Johnsson et al. 2018). This means that the critical events are identified by an estimation of the severity of the initial conditions or proximity to the crash and then the assessment of the effectiveness of any potential avoidance action taken by the road users involved based on the initial conditions. Finally, these factors are combined to produce a final estimate of the crash risk in a specific traffic event.

Research results show that various indicators and their combinations can reflect different aspects of any traffic event. However, there appears to be no existing indicator that covers all bases.

20.3.2   RISK INDICATORS

Many risk indicators have been developed. In this section, we will present some of the most commonly used risk indicators in order to understand the wide range of aspects that are reflected.

In 1971, Hayward and Horst introduced the Time to Collision (TTC) indicator (Hayward 1971) (Horst 1991). As its name suggests, it represents the amount of time until a collision occurs. It can be calculated for the vehicle and for the pedestrian, respectively, as:

$$TTC_v(t) = \frac{D_{v_y}(t)}{V_V(t)} \text{ and } TTC_p(t) = \frac{D_{v_x}(t) - D_p(t)}{V_p(t)} \tag{20.1}$$

where $D_{v_y}(t)$ is the longitudinal distance between the vehicle and the crossing at the instant t, $V_V(t)$ is the vehicle speed at the instant t, $D_{v_x}(t)$ is the lateral vehicle distance at instant t, $D_p(t)$ is the pedestrian position on crossing at instant t, and $V_p(t)$ is the pedestrian's speed at instant t.

The two most commonly used indicators based on TTC are TTC_{min}, which is the minimum TTC value, and the Time to Accident (TA), which is the time elapsed between the start of the evasive action and the collision (Allen et al. 1978). They are used to assess the severity of the traffic conflict.

Another indicator used to describe the driver's behavior as he approaches the zebra crossing is the Time to Zebra (TTZ), calculated as:

$$TTZ(t) = \frac{D_z(t)}{V(t)} \tag{20.2}$$

where $D_z(t)$ is the distance to the zebra crossing and $V(t)$ is the speed.

TTZ_{arr} is the time remaining for the car to arrive at the zebra crossing when the pedestrian arrives at the curb. And the TTZ_{brake} value indicates how much time is left when braking begins which will allow indicating the driver's readiness to stop before the zebra crossing.

The Post Encroachment Time (PET) represents the time elapsed between when the pedestrian leaves the conflict area and when the conflicting vehicle arrives (Gipps & Marksjö 1985).

The Deceleration to Safety Time (DST) is the deceleration required for the second road user to arrive at the conflicting location after the first road user has left it (Hupfer 1997). The DST of a vehicle I between the points j and k for a required safety time distance of x seconds can be calculated as follows:

$$DST_x = \frac{2.(S_{jk} - V_{ij}.t_{DST_x})}{t_{DST_x}^2} \tag{20.3}$$

where S_{jk} is the distance between point j and k, V_{ij} is the speed of vehicle I at point j, and t_{DST_x} is the time to travel from point j to point k for vehicle I.

20.3.3 FUZZY RISK INDICATORS

Numerous human activities involve some degree of inaccuracy associated with numerical quantities. This is due to a lack of information or specific qualities about the actor's state and/or environment that would allow one to act correctly in accordance with intellectual, moral, or even aesthetic ideals. This imprecision characterizes one's level of knowledge of a given situation and is caused by a lack or absence of information. Fuzzy set theory is a natural modeling tool for this type of situation. Because precise numerical quantities are represented by real numbers, it stands to reason that imprecise or ambiguous quantities can be represented by "fuzzy real numbers". In 1964, Zadeh first developed the concept of a fuzzy set, describing the mathematics of fuzzy set theory and, by extension, fuzzy modeling. Traditional knowledge representation systems lack the ability to express fuzzy notions. As a result, logic-based and classical probability-based techniques do not provide an adequate foundation for dealing with knowledge representation. Because such knowledge is lexically imprecise and emphatic by nature, a proper framework for dealing with knowledge representation is required. Zadeh's theory attempts to solve this issue by employing opportunity distribution, or fuzzy intervals in other words. The values "True" and "False" were proposed to be operated on the interval of real numbers [0,1].

The term "exposure" is used in epidemiology to describe a situation in which an agent, positioned in a specific location, is exposed to a potentially harmful circumstance or chemical. In our situation, pedestrians are exposed to an accident in a section of the road where there is a flow of vehicles during the crossing time. Mandar and Boulmakoul defined the first formulation of pedestrian exposure as follows:

$$Exp_{p/V} = \rho_V \cdot v_{max} \cdot t_p - \frac{\rho_V^2 \cdot v_{max} \cdot t_p}{\rho_{max}}, \tag{20.4}$$

where ρ_V is the vehicle's density, ρ_{max} is the vehicle's maximum density, v_{max} is the maximum value of the vehicle's speed, and t_p is the necessary time for the pedestrian to cross the road (Mandar & Boulmakoul 2014).

Now, considering the safety stop time for a vehicle and the pedestrian crossing time as triangular fuzzy numbers due to the fact that their knowledge is not deterministic and remains imprecise (Mandar et al. 2017a), the risk exposure becomes:

$$\widetilde{Exp}' = \left(T_r + \frac{v_{max}}{2\gamma} \left(1 - \frac{\rho}{\rho_{max}} \right) \right) \cdot q - \tilde{T}_p \cdot \rho \cdot v = (\tilde{\alpha} - \beta v) \cdot q \tag{20.5}$$

with $\tilde{\alpha} = T_r - \tilde{T}_p$ and $\beta = \frac{1}{2\gamma}$, where T_r is the reaction time of the driver, v is the vehicle's speed, v_{max} is the maximum vehicle's speed, γ is the vehicle's acceleration, ρ is the vehicle's density, ρ_{max} is the maximum vehicle's density, q is the vehicle's flow, and $\tilde{T}_p$ is the pedestrian's crossing time represented as a triangular fuzzy number.

Mandar et al. explored the behavioral psychology of pedestrians and drivers using intuitive methods based on intuitionistic fuzzy set theory to calculate new indicators for pedestrian exposure.

The inclusion of behavioral aspects linked to the perception of space and the decision of the two antagonists is completely absent when calculating the risk in VPIs. The intuitionist approach allows the antagonists to connect the two realities they perceive. The method takes into account the driver's and pedestrian's objective and subjective information. The two conflicting perspectives were combined into intuitionistic fuzzy numbers, and a relative ranking method was developed to derive risk exposure indicators (Mandar et al. 2017b) (Mandar et al. 2018).

20.4 SIMULATION

The primary goal of this research is to assess pedestrian behavior in order to reduce the risk of a collision with a vehicle. These findings will benefit both autonomous and nonautonomous vehicles. In the case of self-driving cars, this simulator will be used to improve decision making. Otherwise, this model will serve as a useful tool in better understanding the risks that these pedestrians face, so that the authorities involved can take the necessary precautions to reduce these risks.

Figure 20.3 summarizes the application's global architecture in the case of an autonomous car. The data gathered by the various sensors must be synchronized and processed (segmentation, object detection, mapping, etc.). The next goal is to be able to understand our surroundings and how we move through them. Any agent or element on the road must first be identified. Then, in order to predict its next move, we must first comprehend its movement. However, before we can classify and localize the environmental elements relevant to the driving task, we must first determine the autonomous vehicle's current position in space. The perception module is in charge of both of these tasks. Based on all of the information provided by the perception and the results provided by our modules, the planning module makes all decisions about what to do and where to drive. As a result, it should provide a planned path for the vehicle to follow to its destination that is safe, efficient, and comfortable. Finally, the execution module determines the best steering angle, gas pedal position, brake pedal position, and speed settings to follow the planned path precisely.

Our simulation was carried out using Unity, the leading open-source interactive 2D, 3D, and virtual reality content development platform. Unlike other existing simulators, Unity allowed us to create, integrate, and configure the simulation's

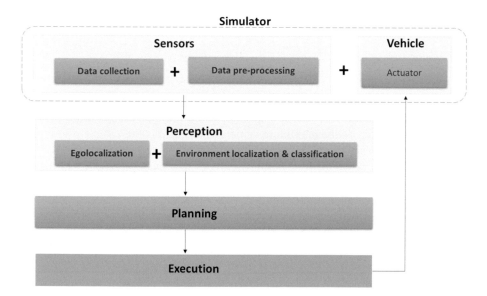

FIGURE 20.3 System architecture and components.

FIGURE 20.4   Two different scenarios after the pedestrian crosses the road.

various components. Using our multiagent model and fuzzy risk indicators allowed us to simulate various possible scenarios. Figure 20.4 shows a scenario of the pedestrian crossing the road when the vehicle stopped in the first screenshot of the simulation, while in the second screenshot, none of the involved road users has stopped, which involved a crash!

As shown in Figure 20.5, we used three different cameras: a driver camera, a pedestrian camera, and a main camera. The driver camera is seen from the driver's vision, and it's the necessary camera in the case of the self-driving cars' study. The pedestrian camera is seen from the pedestrian's point of view, and the main camera shows the whole plan, which is important to better assess the behavior of the pedestrian.

In this work, we were able to model the different agents (pedestrian, driver, and environment) and their interactions while taking into account the vehicle and pedestrian movements. In terms of future works, we intend to incorporate more variables into the scenarios, such as road features, environmental conditions, and the

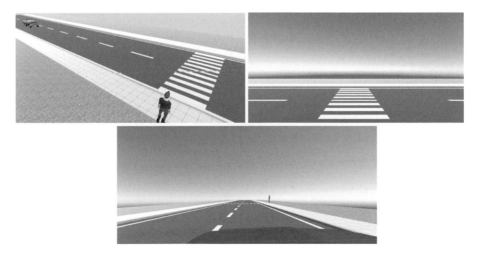

FIGURE 20.5   The three different cameras: main camera, pedestrian camera, and driver camera.

pedestrian's posture, which provides a better understanding of his or her behavior. Indeed, pedestrian preimpact posture has a significant impact on the risks and severity of injury outcomes. To that end, we will need to extract more time histories of the pedestrian's kinetic and kinematic features (Li et al. 2021).

20.5 CONCLUSION

This work is part of a National Center for Scientific and Technical Research project dealing with the dangers of VPI. The goal of this research is to identify and analyze critical conditions as well as appropriate solutions for pedestrian safety in urban areas. To that end, we intend to model the risk of pedestrian accidents by employing a logical approach that captures the behavioral semantics of pedestrians and drivers sharing the road space. This chapter describes our multiagent microsimulation model for pedestrian behavior. We were successful in modeling the various agents (pedestrian, driver, and environment) and their interactions. This allowed us to investigate different scenarios. We intend to incorporate more variables into the scenarios, such as road features, environmental conditions, and the pedestrian's posture, to gain a better understanding of his behavior.

ACKNOWLEDGMENTS

This work was partially funded by Ministry of Equipment, Transport, Logistics and Water, Kingdom of Morocco, the National Road Safety Agency (NARSA), and National Center for Scientific and Technical Research (CNRST). Road Safety Research Program# An intelligent reactive abductive system and intuitionist fuzzy logical reasoning for dangerousness of driver-pedestrians interactions analysis: Development of new pedestrians' exposure to risk of road accident measures.

REFERENCES

Allen, B., T. Shin, and P. Cooper. 1978. Analysis of traffic conflicts and collisions. No. HS-025 846.

Bouquet, F., S. Chipeaux, C. Lang, N. Marilleau, J. Nicod, and P. Taillandier. 2015. Introduction to the agent approach. In Agent-based Spatial Simulation with NetLogo, pp. 1–28. Elsevier, Oxford, UK. doi:10.1016/B978-1-78548-055-3.50001-0.

Davis, G. A., J. Hourdos, H. Xiong, and I. Chatterjee. 2011. Outline for a causal model of traffic conflicts and crashes. Accident Analysis & Prevention 43, no. 6 (2011): 1907–1919. doi: 10.1016/j.aap.2011.05.001.

Dorri, Ali, S. S. Kanhere, and R. Jurdak. 2018. Multi-agent systems: A survey. IEEE Access 6: 28573–28593.

Gipps, G., and B. Marksjö. 1985. A micro-simulation model for pedestrian flows. Mathematics and Computers in Simulation 27, no. 2–3: 95–105.

Godara, A., S. Lassarre, and A. Banos. 2007. Simulating pedestrian-vehicle interaction in an urban network using cellular automata and multi-agent models. In Traffic and Granular Flow'05, pp. 411–418. Springer, Berlin, Heidelberg.

Gupta, S., M. Vasardani, and S. Winter. 2018. Negotiation between vehicles and pedestrians for the right of way at intersections. IEEE Transactions on Intelligent Transportation Systems 20, no. 3: 888–899. doi:10.1109/tits.2018.2836957.

Gupta, S., M. Vasardani, B. Lohani, and S. Winter. 2019. Pedestrian's risk-based negotiation model for self-driving vehicles to get the right of way. Accident Analysis & Prevention 124: 163–173, doi 10.1016/j.aap.2011.01.003

Hayward, J. 1971. Near misses as a measure of safety at urban intersections. Pennsylvania Transportation and Traffic Safety Center.

Horst, R. V. D. 1991. Time-to-collision as a cue for decision-making in braking. Vision in Vehicles–III.

Hupfer, C. 1997. Deceleration to safety time (DST): A useful figure to evaluate traffic safety. In ICTCT Conference Proceedings of Seminar, vol. 3, pp. 5–7.

Hydén, C. 1987. The development of a method for traffic safety evaluation: The Swedish Traffic Conflicts Technique. Bulletin Lund Institute of Technology, Department 70.

Johnsson, C., A. Laureshyn, and T. De Ceunynck. 2018. In search of surrogate safety indicators for vulnerable road users: a review of surrogate safety indicators. Transport Reviews 38, no. 6: 765–785. doi: 10.1080/01441647.2018.1442888

Li, Q., S. Shang, X. Pei, Q. Wang, Q. Zhou, and B. Nie. 2021. Kinetic and kinematic features of pedestrian avoidance behavior in motor vehicle conflicts. Frontiers in Bioengineering and Biotechnology 9. doi:10.3389/fbioe.2021.783003. ISSN=2296-4185.

Liu, M., W. Zeng, P. Chen, and X. Wu. 2017. A microscopic simulation model for pedestrian-pedestrian and pedestrian-vehicle interactions at crosswalks. PLoS One 12, no. 7: e0180992. doi: 10.1371/journal.pone.0180992.

Mandar, M., A. Boulmakoul, and A. Lbath. 2017. Pedestrian fuzzy risk exposure indicator. Transportation Research Procedia 22, no. 2017a: 124–133.

Mandar, M., A. Boulmakoul, and L. Karim. 2018. Indicateurs de risque d'accidents piétons: Vers une décision floue intuitionniste.

Mandar, M., and A. Boulmakoul. 2014. Virtual pedestrians' risk modeling. International Journal of Civil Engineering and Technology 5, no. 10: 32–42.

Mandar, M., L. Karim, A. Boulmakoul, and A. Lbath. 2017. Triangular intuitionistic fuzzy number theory for driver-pedestrians' interactions and risk exposure modeling. Procedia Computer Science 109, no. 2017b: 148–155.

Okazaki, S., and S. Matsushita. 1993. A study of simulation model for pedestrian movement with evacuation and queuing. In International Conference on Engineering for Crowd Safety, 271–280.

Peden, M., R. Scurfield, D. Sleet et al. 2004. World Report on Road Traffic Injury Prevention. World Health Organization, Geneva, Switzerland.

Shaaban, K., and K. Abdelwarith. 2020. Pedestrian attribute analysis using agent-based modeling. Applied Sciences 10, no. 14: 4882. doi: 10.3390/app10144882.

Svensson, Å., and C. Hydén. 2006. Estimating the severity of safety related behaviour. Accident Analysis & Prevention 38, no. 2: 379–385. doi: 10.1016/j.aap.2005.10.009

Wooldridge, M. 2009. An Introduction to Multiagent Systems. John Wiley & Sons, Oxford, UK.

World Health Organization. 2015. Global Status Report on Road Safety 2015. World Health Organization, Geneva. https://apps.who.int/iris/handle/10665/189242

Yang, D., K. Redmill, and U. Özgüner. 2020. A multi-state social force based framework for vehicle-pedestrian interaction in uncontrolled pedestrian crossing scenarios. In 2020 IEEE Intelligent Vehicles Symposium (IV), pp. 1807–1812. Las Vegas, NV, USA.

Index